Materials' ageing and degradation in light water reactors

Related titles:

Nuclear corrosion science and engineering (ISBN 978-1-84569-765-5)

Understanding and mitigating ageing in nuclear power plants: Materials and operational aspects of plant life management (PLiM) (ISBN 978-1-84569-511-8)

Nuclear fuel cycle science and engineering (ISBN 978-0-85709-073-7)

Details of these books and a complete list of titles from Woodhead Publishing can be obtained by:

- visiting our web site at www.woodheadpublishing.com
- contacting Customer Services (e-mail: sales@woodheadpublishing.com; fax: +44 (0) 1223 832819; tel.: +44 (0) 1223 499140 ext. 130; address: Woodhead Publishing Limited, 80, High Street, Sawston, Cambridge CB22 3HJ, UK)
- in North America, contacting our US office (e-mail: usmarketing@woodheadpublishing.com; tel.: (215) 928 9112; address: Woodhead Publishing, 1518 Walnut Street, Suite 1100, Philadelphia, PA 19102-3406, USA)

If you would like e-versions of our content, please visit our online platform: www.woodheadpublishingonline.com. Please recommend it to your librarian so that everyone in your institution can benefit from the wealth of content on the site.

We are always happy to receive suggestions for new books from potential editors. To enquire about contributing to our energy series, please send your name, contact address and details of the topic/s you are interested in to sarah.hughes@woodheadpublishing.com. We look forward to hearing from you.

The Woodhead team responsible for publishing this book:

Commissioning Editor: Sarah Hughes
Publications Co-ordinator: Steven Matthews
Project Editor: Rachel Cox
Editorial and Production Manager: Mary Campbell
Production Editor: Richard Fairclough
Project Manager: Bhavani Ganesh Kumar, Newgen Knowledge Works Pvt Ltd
Freelance Copy-editor: Deborah Webb
Proofreader: Suma George, Newgen Knowledge Works Pvt Ltd
Cover Designer: Terry Callanan

Woodhead Publishing Series in Energy: Number 44

Materials' ageing and degradation in light water reactors

Mechanisms and management

Edited by
K. L. Murty

WOODHEAD PUBLISHING

Oxford Cambridge Philadelphia New Delhi

Published by Woodhead Publishing Limited,
80 High Street, Sawston, Cambridge CB22 3HJ, UK
www.woodheadpublishing.com
www.woodheadpublishingonline.com

Woodhead Publishing, 1518 Walnut Street, Suite 1100, Philadelphia,
PA 19102-3406, USA

Woodhead Publishing India Private Limited, G-2, Vardaan House, 7/28 Ansari Road,
Daryaganj, New Delhi - 110002, India
www.woodheadpublishingindia.com

First published 2013, Woodhead Publishing Limited

British Library Cataloguing in Publication Data
A catalogue record for this book is available from the British Library.

Library of Congress Control Number: 2012954753

ISBN 978-0-85709-239-7 (print)
ISBN 978-0-85709-745-3 (online)
ISSN 2044-9364 Woodhead Publishing Series in Energy (print)
ISSN 2044-9372 Woodhead Publishing Series in Energy (online)

The publisher's policy is to use permanent paper from mills that operate a sustainable forestry policy, and which has been manufactured from pulp which is processed using acid-free and elemental chlorine-free practices. Furthermore, the publisher ensures that the text paper and cover board used have met acceptable environmental accreditation standards.

Typeset by Newgen Knowledge Works Pvt Ltd, India

Contents

Contributor contact details

(* = main contact)

Editor, Chapters 1 and 3

K. L. Murty
North Carolina State University
Raleigh
NC 27695-7909
USA

Email: murty@ncsu.edu

Chapter 2

Thierry Couvant
EDF R&D
Materials and Component department
France

Email: Thierry.couvant@edf.fr

Chapter 3

Dr Srikant Gollapudi
77 Massachusetts Ave
Massachusetts Institute of Technology
Cambridge
MA 20139
USA

Dr Kishore Ramaswamy
Head, Material Evaluation Section
Post Irradiation Examination Division
Bhabha Atomic Research Centre
Mumbai 400 085
India

Email: kishorer@barc.gov.in

Dr M. D. Mathew
Head, Mechanical Metallurgy Division
Indira Gandhi Centre for Atomic Research
Kalpakkam, TN 603 102
India

Email: mathew@igcar.gov.in

I. Charit
University of Idaho
USA

Chapters 4 and 5

R. B. Adamson*
Zircology Plus
36848 Montecito Dr
Fremont
CA 94536
USA

Email: rbadamson@gmail.com

P. Rudling
ANT International
Mölnlycke
Sweden

Email: peter.rudling@antinternational.com

Chapter 6

Dr H. M. Hashemian
Analysis and Measurement Services Corporation
AMS Technology Center
9119 Cross Park Drive
Knoxville
TN 37923
USA

Email: hash@ams-corp.com

Chapter 7

Dr Yong H. Jeong and Dr Seong S. Hwang*
Nuclear Materials Division
Korea Atomic Energy Research Institute
989-111 Daedeokdaero
Yuseong
Daejeon 305-353
Korea

Email: yhjeong@kaeri.re.kr; sshwang@kaeri.re.kr

Chapter 8

Dr Tamás János Katona
MVM Paks Nuclear Power Plant Ltd
P.O. Box 71
Paks 7031
Hungary

Email: katonat@npp.hu

Chapter 9

Sumit Ray*
Westinghouse Electric Company LLC
Bluff Road
Drawer R
Columbia
South Carolina 29250
USA

Email: rays@westinghouse.com

Edward Lahoda
Westinghouse Electric Company LLC
1000 Cranberry Woods Drive
Cranberry Township
Pennsylvania 16066
USA

Email: lahodaej@westinghouse.com

Woodhead Publishing Series in Energy

1 **Generating power at high efficiency: Combined cycle technology for sustainable energy production**
Eric Jeffs
2 **Advanced separation techniques for nuclear fuel reprocessing and radioactive waste treatment**
Edited by Kenneth L. Nash and Gregg J. Lumetta
3 **Bioalcohol production: Biochemical conversion of lignocellulosic biomass**
Edited by K. W. Waldron
4 **Understanding and mitigating ageing in nuclear power plants: Materials and operational aspects of plant life management (PLiM)**
Edited by Philip G. Tipping
5 **Advanced power plant materials, design and technology**
Edited by Dermot Roddy
6 **Stand-alone and hybrid wind energy systems: Technology, energy storage and applications**
Edited by J. K. Kaldellis
7 **Biodiesel science and technology: From soil to oil**
Jan C. J. Bart, Natale Palmeri and Stefano Cavallaro
8 **Developments and innovation in carbon dioxide (CO_2) capture and storage technology Volume 1: Carbon dioxide (CO_2) capture, transport and industrial applications**
Edited by M. Mercedes Maroto-Valer
9 **Geological repository systems for safe disposal of spent nuclear fuels and radioactive waste**
Edited by Joonhong Ahn and Michael J. Apted
10 **Wind energy systems: Optimising design and construction for safe and reliable operation**
Edited by John D. Sørensen and Jens N. Sørensen
11 **Solid oxide fuel cell technology: Principles, performance and operations**
Kevin Huang and John Bannister Goodenough

12 **Handbook of advanced radioactive waste conditioning technologies**
Edited by Michael I. Ojovan

13 **Nuclear safety systems**
Edited by Dan Gabriel Cacuci

14 **Materials for energy efficiency and thermal comfort in buildings**
Edited by Matthew R. Hall

15 **Handbook of biofuels production: Processes and technologies**
Edited by Rafael Luque, Juan Campelo and James Clark

16 **Developments and innovation in carbon dioxide (CO_2) capture and storage technology Volume 2: Carbon dioxide (CO_2) storage and utilisation**
Edited by M. Mercedes Maroto-Valer

17 **Oxy-fuel combustion for power generation and carbon dioxide (CO_2) capture**
Edited by Ligang Zheng

18 **Small and micro combined heat and power (CHP) systems: Advanced design, performance, materials and applications**
Edited by Robert Beith

19 **Advances in clean hydrocarbon fuel processing: Science and technology**
Edited by M. Rashid Khan

20 **Modern gas turbine systems: High efficiency, low emission, fuel flexible power generation**
Edited by Peter Jansohn

21 **Concentrating solar power technology: Principles, developments and applications**
Edited by Keith Lovegrove and Wes Stein

22 **Nuclear corrosion science and engineering**
Edited by Damien Féron

23 **Power plant life management and performance improvement**
Edited by John E. Oakey

24 **Electrical drives for direct-drive renewable energy systems**
Edited by Markus Mueller and Henk Polinder

25 **Advanced membrane science and technology for sustainable energy and environmental applications**
Edited by Angelo Basile and Suzana Pereira Nunes

26 **Irradiation embrittlement of reactor pressure vessels (RPVs) in nuclear power plants**
Edited by Naoki Soneda

27 **High temperature superconductors (HTS) for energy applications**
Edited by Ziad Melhem

28 **Infrastructure and methodologies for the justification of nuclear power programmes**
Edited by Agustín Alonso
29 **Waste to energy (WtE) conversion technology**
Edited by Marco Castaldi
30 **Polymer electrolyte membrane and direct methanol fuel cell technology Volume 1: Fundamentals and performance of low temperature fuel cells**
Edited by Christoph Hartnig and Christina Roth
31 **Polymer electrolyte membrane and direct methanol fuel cell technology Volume 2: *In situ* characterization techniques for low temperature fuel cells**
Edited by Christoph Hartnig and Christina Roth
32 **Combined cycle systems for near-zero emission power generation**
Edited by Ashok D. Rao
33 **Modern earth buildings: Materials, engineering, construction and applications**
Edited by Matthew R. Hall, Rick Lindsay and Meror Krayenhoff
34 **Metropolitan sustainability: Understanding and improving the urban environment**
Edited by Frank Zeman
35 **Functional materials for sustainable energy applications**
Edited by John A. Kilner, Stephen J. Skinner, Stuart J. C. Irvine and Peter P. Edwards
36 **Nuclear decommissioning: Planning, execution and international experience**
Edited by Michele Laraia
37 **Nuclear fuel cycle science and engineering**
Edited by Ian Crossland
38 **Electricity transmission, distribution and storage systems**
Edited by Ziad Melhem
39 **Advances in biodiesel production: Processes and technologies**
Edited by Rafael Luque and Juan A. Melero
40 **Biomass combustion science, technology and engineering**
Edited by Lasse Rosendahl
41 **Ultra-supercritical coal power plant: Materials, technologies and optimisation**
Edited by Dongke Zhang
42 **Radionuclide behaviour in the natural environment: Science, impacts and lessons for the nuclear industry**
Edited by Christophe Poinssot and Horst Geckeis

43 Calcium and chemical looping technology for power generation and carbon dioxide (CO_2) capture: Solid oxygen- and CO_2-carriers
P. Fennell and E. J. Anthony

44 Materials' ageing and degradation in light water reactors: Mechanisms and management
Edited by K. L. Murty

45 Structural alloys for power plants: Operational challenges and high-temperature materials
Edited by Amir Shirzadi, Rob Wallach and Susan Jackson

46 Biolubricants: Science and technology
Jan C. J. Bart, Emanuele Gucciardi and Stefano Cavallaro

47 Wind turbine blade design and materials: Improving reliability, cost and performance
Edited by Povl Brøndsted and Rogier Nijssen

48 Radioactive waste management and contaminated site clean-up: Processes, technologies and international experience
Edited by William E. Lee, Michael I. Ojovan, Carol M. Jantzen

49 Probabilistic safety assessment for optimum nuclear power plant life management (PLiM): Theory and application of reliability analysis methods for major power plant components
Gennadij V. Arkadov, Alexander F. Getman and Andrei N. Rodionov

50 Coal utilization in industry
Edited by D. G. Osborne

51 Coal power plant materials and life assessment: Developments and applications
Edited by Ahmed Shibli

52 The biogas handbook: Science, production and applications
Edited by Arthur Wellinger and David Baxter

53 Advances in biorefineries: Biomass and waste supply chain exploitation
Edited by K. W. Waldron

54 Geoscience of carbon dioxide (CO_2) storage
Edited by Jon Gluyas and Simon Mathias

55 Handbook of membrane reactors Volume 1: Fundamental materials science, design and optimisation
Edited by Angelo Basile

56 Handbook of membrane reactors Volume 2: Reactor types and industrial applications
Edited by Angelo Basile

57 **Alternative fuels and advanced vehicle technologies: Towards zero carbon transportation**
Edited by Richard Folkson

58 **Handbook of microalgal bioprocess engineering**
Christopher Lan and Bei Wang

59 **Fluidized-bed technologies for near-zero emission combustion and gasification**
Edited by Fabrizio Scala

60 **Managing nuclear projects: A comprehensive management resource**
Edited by Jas Devgun

61 **Handbook of process integration: Energy, water, waste and emissions management in processing and power industries**
Edited by Jiří Klemeš

62 **Membranes for clean and renewable power applications**
Edited by Annarosa Gugliuzza and Angelo Basile

Foreword

The ageing of materials in the light water reactor fleet around the world is a major factor in ensuring not only the safe and economical operation of these power plants, but also preserving and extending their substantive contribution to carbon-free electricity production. 'Ageing' refers to the change in character or properties of components or systems with time. Ageing can occur in benign environments, such as those experienced by concrete under ambient conditions. It can also occur in harsh or extreme environments such as components in the core of a reactor where they are exposed to high temperature, high stresses, an aggressive chemical environment and a high level of radiation. These additional stressors can initiate, accelerate and generally shape the ageing process over that in inert environments.

Classes of components for which an understanding of ageing is important are generally divided into four categories: core components including the reactor pressure vessel, other plant components, electric cables, concrete and piping. This book focuses on core components and electric cables and the processes by which ageing occurs. Couvant and Murty address two of the key ageing modes for core components – corrosion and creep deformation – in chapters that illuminate the major processes and their consequences. Adamson and Rudling focus on the ageing of zirconium alloy fuel bundles that are critical components in the containment of the fuel and extraction of energy for electricity production. Next, Hashemian reviews the ageing of electric cables, hundreds of miles of which are installed in each nuclear plant.

But the last part of the book on assessment strategies for managing the ageing of materials in reactors is what makes it unique. While ageing of materials is unavoidable, and in fact ubiquitous, the key is to understand the ageing process and how the various components of the environment can affect or accelerate that process. Only by understanding the ageing process can mitigation or amelioration be considered. Jeong and Hwang, Katona, and Ray and Lahoda present perspectives on the evaluation of plant ageing including development of ageing management programs, proactive materials management, mitigation and repair methods, international cooperative activities and finally, integration of these programs into a system that ensures the safe, long-term operation of the power plant.

Gary S. Was
Walter J. Weber, Jr. Professor of Sustainable Energy, Environmental and Earth Systems Engineering
University of Michigan

Preface

Nuclear power plants provide around 20% of the total electrical supply in the United States and roughly around the same level across the world, helping to reduce harmful greenhouse gases (GHG). Many commercial nuclear reactors operating worldwide are of the Generation-II category and the majority of this generation is light water type; Generation-III type reactors are at an advanced stage of commercialization and deployment. The main reasons for downtime of the light water reactors (LWRs) currently operating are materials-related issues primarily due to material ageing and degradation. Degradation of materials is caused by the very aggressive environments to which the LWR structures are exposed including high neutron fluences, high temperatures along with aggressive environmental factors such as water and steam. A major objective of this book is to bring forth issues confronting the nuclear industry in terms of materials ageing and degrading with particular emphasis on mechanisms and management. This book is a compilation of chapters written by experts in the field. The book is divided into three different parts: Part I on 'Fundamental ageing issues and degradation mechanisms', Part II on 'Materials ageing and degradation in particular light water reactor (LWR) components' and Part III on 'Materials management strategies for light water reactors (LWRs)'. Each of these three parts contains three chapters.

In Chapter 1, Murty and Ramaswamy present an overview of various materials issues with discussions on fundamental aspects along with pertinent references to various materials of different LWR structures. The chapter covers briefly all the seven components (fuel, structure, moderator/reflector, control, coolant, shields and safety systems) comprising an LWR with references that deal with more details. Corrosion and stress corrosion cracking (SCC) are the most commonly limiting factors and damaging phenomena that are covered in Chapter 2 by Couvant. This is an important chapter that summarizes the corrosion phenomena encountered in LWRs. Murty *et al.* discuss in detail the time-dependent permanent deformation known as creep in Chapter 3. Any structure that is exposed to high temperatures and loads experiences creep deformation, and both the creep mechanisms and creep-life prediction methodologies are important aspects covered here, referencing their applications to LWR structural materials such as Zr-based alloys, stainless steels and Ni-based superalloys.

Part II on Materials ageing and degradation of specific light water reactor components comprises three chapters commencing with Chapter 4 by Adamson and Rudling on the zirconium-based alloys that are commonly used as thin-walled tubing to clad radioactive UO_2 fuel. This chapter starts with the basic crystallography of Zr leading to many degradation phenomena often noted in operating reactors of both PWR and BWR type such as PCI (pellet-cladding interaction), oxidation and hydriding, crud formation, radiation growth and creep, grid-to-rod-fretting (GTRF), fuel rod and assembly bow. Rudling and Adamson continue the issues of Zircaloy cladding in Chapter 5 with emphasis on performance and inspection of fuel bundle components. Issues of possible degradation and ageing of various electrical cables are dealt by Hashemian in Chapter 6; it is to be noted that the various aspects covered in this chapter are usually found only in specialized treatises.

Part III covers Materials management strategies wherein Chapter 7 by Jeong and Hwang deals with PWR management in Korea while Katona describes similar aspects for Russian VVERs in Chapter 8. In the final chapter, Ray and Lahoda cover materials problems facing operating LWR vendors following which the needs of nuclear technology and industry are pointed out.

The uniqueness of the book lies in the fact that, while fundamental materials aspects/phenomena are dealt with initially, other content is not easily found in the technical journals on nuclear materials, especially the management strategies of LWR vendors covered in Part III. The various materials science aspects described in these articles for predicting the life of nuclear structures echo the comment made by Placid Rodriguez during his Presidential address delivered at the Golden Jubilee Celebration of the Indian Institute of Metals in 1966: *To be able to predict the life of an engineering component accurately, …* [one needs to] *take into account the synergistic effects of and interactions between a variety of damaging processes like creep, fatigue, dynamic strain ageing, environmental effect and microstructural degradation.* The importance and significance of knowledge and background in nuclear materials are nicely summed up by Norman Hilberry, the former director of Argonne National Laboratory, who made the following statement way back in the 1950s: *We physicists can dream up and work out all the details of power reactors based on dozens of combinations of the essentials, but it's only a paper reactor until the metallurgist tells us whether it can be built and from what. Then only, one can figure whether there is any hope that they can produce power.*

Acknowledgements are due to the efforts and continued persistence of Messrs. Steven Mathews, Sarah Hughes and Rachel Cox of Woodhead

Publishing in arranging for various authors to contribute their chapters in an appropriate time frame and in making this publication a reality.

K. Linga (KL) Murty
Professor and Director of Graduate Programs
Department of Nuclear Engineering
North Carolina State University

Part I

Fundamental ageing issues and degradation mechanisms

1
Overview of ageing and degradation issues in light water reactors (LWRs)

K. L. MURTY, North Carolina State University, USA and
K. RAMASWAMY, Bhabha Atomic Research Center, India

DOI: 10.1533/9780857097453.1.3

Abstract: A typical light water reactor (LWR) has components like the clad, the internals, the reactor pressure vessel (RPV), the heat exchanger tubes, etc., made from different materials. Some of these components experience pressure and temperature effects while others experience an additional contribution from high neutron flux. These components undergo degradation to various extents based on the severity of service conditions and their inherent material properties. This chapter presents an overview of the various deformation modes that materials are known to undergo under reactor operating conditions, and the known theoretical or empirical relations between the crucial material and environmental parameters are outlined. Materials degradation phenomena briefly described in the chapter include radiation damage, plastic deformation, fracture and fatigue, following which radiation effects on these phenomena, as well as corrosion are enumerated. Degradation mechanisms of concern to specific nuclear reactor structures are detailed in the last section with emphasis on fuel, cladding and internals.

Key words: nuclear reactor, damage, degradation, hydride embrittlement, life prediction, mechanical property, creep, fatigue, fracture, irradiation creep, corrosion, irradiation assisted stress corrosion.

1.1 Introduction

This chapter provides a review of materials ageing and degradation encountered in light water reactors (LWRs). Ageing of any engineering structure – through exposure to pressure, temperature and environment – can manifest as changes in the material properties which may be classified into three major categories: (1) changes in dimensions or shape of the structure, (2) changes in material weight due to oxidation, corrosion and erosion and (3) changes in physical or mechanical properties without any noticeable change in dimensions. The in-service component(s) may undergo more than one of the above changes simultaneously, and when these changes affect plant safety, production efficiency or economy they are viewed as degradation. In a thermal energy based power plant (nuclear or fossil fired), various energy transfer stages with complex heavy engineering are involved before the final stage generation of electric energy is achieved. At each stage of energy transfer, the machinery involved undergoes ageing and the material properties undergo degradation with continuous use. The severity of degradation may vary from simple and minor to serious and complex. For the core components of a nuclear reactor in a nuclear power plant (NPP), there is an additional influence of the severe radiation environment that accelerates the ageing. The types of nuclear reactors vary in their design features according to the type of fuel and coolant used. The choice of materials for their construction differs according to the reactor design as well as to previous experience in operating nuclear reactors. The components in the reactor core must tolerate exposure to the coolant media (high-temperature water, liquid metals, gas or liquid salts), stresses and vibrations as well as an intense field of high energy neutrons. Ageing of materials under this extreme environment can lead to reduced performance and, in the worst cases, sudden failure of the components.

A common consideration given in a power plant design at any installation (nuclear/thermal) is the safety requirement. The concern for safety increases as the material properties get degraded from their initial values with prolonged exposure to service conditions. Thus, intermittent surveillance campaigns are mandatory in an operating installation for the evaluation of the health of the components – even if the initial design adhered to strict safety norms. For this, we must be able to identify the critical components that can possibly undergo ageing degradation and decide the frequency of the inspection campaigns. The outcome of such campaigns can forewarn of any impending failure and suggest replacement of components such that the designed life of the plant can be reached – and if reached, the campaign can advise if the life of the component can be extended beyond the design life. In the worst case, the campaign outcome may suggest shutdown of the plant if safe continued operation of the component cannot be ensured. The cost of such campaigns and subsequent component replacements should be recoverable by putting the plant back into operation. The following statement with regard to nuclear installations is pertinent:[1]

> With the present 60-year licenses beginning to expire between the years of 2029 and 2039 for the first group of NPP that came online between 1969 and 1979 utilities are likely to initiate planning of base-load replacement power by 2014 or earlier. If the option to extend current plant lifetimes is not available, strategic planning and investment required to maintain the current LWR fleet may not happen in a sustainable manner. The research window for supporting the utility's decisions to invest in lifetime extension and to support NRC decisions to extend the license must start now and is likely to extend through the following 20-year period (i.e. 2010 to 2029), with higher intensity for the first 10 years. The LWR's R&D Program represents the beginning of timely

collaborative research needed to retain the existing nuclear power infrastructure of the United States.

Our understanding of the behaviour of the service material in that environment is based on years of operating experience of a reactor. Sustained research and development is required to develop newer materials. Further, it is from the examination of the ageing/aged materials we learn the role of new environmental parameters that were unthought-of during the design stage, and allow us to modify our safety codes in future designs. Specific ageing and degradation mechanism depends on the component in question and the various conditions such as temperature, load and environment to which the materials are exposed. A typical NPP can be considered to consist of seven different components: (i) fuel, (ii) structural components, (iii) moderator/reflector, (iv) control, (v) coolant, (vi) shields and (vii) safety systems. Each of these components has specific requirements and selection criteria based on which suitable and economic materials are chosen.[2]

Fission based nuclear reactors can be classified as thermal and fast, based on the energy of the neutrons and the thermal reactors can further be categorized as boiling water reactor (BWR) and pressurized water reactor (PWR). The latter type can further be classified as light water cooled and heavy water cooled. We will confine ourselves here to the LWRs that use the steam-cycle conversion system wherein the steam produced by nuclear fission drives a conventional turbine generator to produce electricity. A steam generator is used in PWRs to produce steam while the direct cycle BWRs generate steam in the reactor core thereby not requiring a separate steam generator; Fig. 1.1a and 1.1b are schematics of PWR and BWR, respectively, with important structural components indicated.[3]

In a typical PWR which uses ceramic fuel, the fuel is separated from the coolant by a physical barrier that prevents their direct contact. The barrier, called the clad, has adequate thermal conductivity to transfer the fission-heat to the coolant and has a low thermal neutron absorption characteristic to allow fission neutrons to sustain a chain reaction. The cladded fuel is immersed in a pressurized pool of coolant flowing at an average temperature of ~300°C under a pressure of around 16 MPa thus preventing the water from boiling. Two separate water systems, the primary and secondary, are contained in the steam generator which is a heat exchanger consisting of a large number (~3000) of nickel-based super alloy tubes in a large steel shell. Depending on the vendor, PWRs may have two, three or four loops with respective coolant circuits, each with its own steam generator. In BWRs, on the other hand, water is circulated through the reactor core producing saturated steam that runs the turbine generator. Nuclear reaction in BWRs is controlled using steel-clad boron carbide control rods that are inserted from the bottom of the core while control rod cluster assemblies

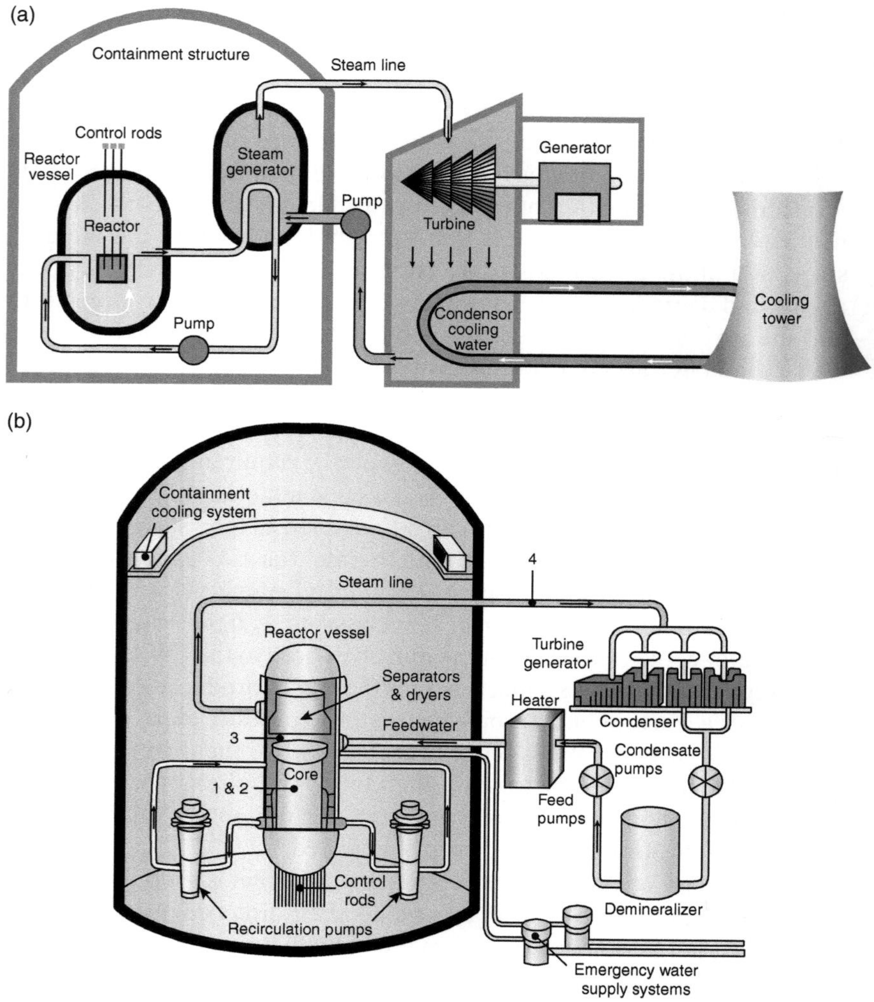

1.1 Schematics of (a) PWR (www.aboutnuclear.org) and (b) BWR (www.nrc.gov/reactors/bwrs.html).

containing either B_4C or AgInCd are inserted from the top in PWRs. While the PWRs contain about the same weight of fuel and cladding as BWRs, the number of assemblies is one third of that in the BWRs since the number of rods per assembly is greater in the former (17 × 17 in PWRs vs 8 × 8 in BWRs – these numbers vary slightly depending on the vendor and the generation type). Table 1.1 summarizes the important characteristics of these reactors that include Russian VVER and RBMK.

As outlined earlier and presented in Fig. 1.1a and 1.1b, different materials are selected for the manufacture of various components and the selection

Table 1.1 Reactor types and characteristics

Parameter	PWR	VVER	BWR	RBMK
Coolant	Pressurized water	Pressurized water	Boiling water	Boiling water
Average power rating (kW/L)	80–125	83/108	40–57	5
Fast neutron flux average (n/cm^2·s)	6–9 × 10^{13}	5 × 10^{13}/7 × 10^{13}	4–7 × 10^{13}	1–2 × 10^{13}
Temperature (°C)	320–350	335–352	285–305	290

Table 1.2 Components, requirements and possible candidate materials

Component	Requirements	Possible materials
Moderators and reflectors	Low neutron absorption Large energy loss by neutron per collision High neutron scattering	Water – H_2O, D_2O Beryllium – BeO Graphite – C
Control materials	High neutron absorption Adequate strength Low mass (for rapid movement) Corrosion resistance Stability under heat and radiation	Boron – B Cadmium – Cd Hafnium – Hf Rare earths – Eu, Gd, Dy, etc.
Coolants	Low neutron absorption Good heat-transfer properties Low pumping power (Low T_M) Stability under heat and radiation Low induced radioactivity Non-corrosiveness	Gases – Air, H_2, He, CO_2, H_2O Water – H_2O, D_2O Liquid Metals – Na, NaK, Bi Molten Salts (-Cl, -OH, -F) Organic Liquids
Shielding materials	Capacity to slow down neutrons Absorption of gamma radiation Absorb neutrons	Light water – H_2O Concrete, Most control materials Metals – Fe, Pb, Bi, TA, W, Boral – B and Al alloy
Structural materials	Low neutron absorption Stability under heat and radiation Mechanical strength Corrosion resistance Good heat-transfer properties	Al, Be, Mg, Zr Ferritic Steels Stainless Steels Superalloys (Ni based) Refractory metals – Mo, Nb, Ti, W, etc.

criteria are based on physical, mechanical, thermal and nuclear characteristics including the chemical and nuclear stability as well as the resistance to radiation damage and induced radioactivity. Table 1.2 summarizes the various components and major requirements along with possible materials. Based on

Table 1.3 LWR components, key materials, problems and causes

Component	Key materials	Major issues	Primary causes
Fuel cladding, assembly and channel (BWRs)	Zircaloy/UO_2	Cladding perforation Dimensional changes Bowing and dilation	Pellet cladding interaction (PCI) Oxidation, corrosion, hydriding Creep Swelling Grid-to-rod fretting (GTRF) Crud formation IGSCC and IASCC Shadow corrosion (BWR channels)
Control rod	304 SS/B_4C, AgInCd	Perforation Leachout	Stress corrosion cracking (SCC)
Pressure vessel	Low alloy steel	Integrity in presence of cracks	Radiation embrittlement Corrosion fatigue
Piping	304 SS (BWRs) C steel (PWRs)	Cracking Distortion	Corrosion Stress corrosion cracking (SCC)
Condenser	Cu-Ni alloys	Tube failures	SCC Corrosion
Turbine	NiCrMoV bainitic steel 12Cr SS	Rotor bursts Disc cracking Blade cracking	Fatigue Temper embrittlement IGSCC Corrosion fatigue SCC

these various possibilities the reactor vendors select different materials for the construction of the structures in LWRs and it is very interesting to follow the history of the development of LWR fuel cladding that resulted in the adoption of Zr-based alloys such as Zircaloy-2 for BWRs and Zircaloy-4 for PWRs.[4]

Table 1.3 provides a summary of the reactor components, the key materials, the major materials-related problems and the primary causes of failure.[3] These materials face different environmental parameters whose severity varies with location. This leads to differences in their ageing mechanisms and the intensity of their degradation. It is possible that one mechanism can be a precursor to another, leading to unexpected early failure of the material. The common ageing-related degradation can be classified as due to radiation embrittlement, loss of toughness, time dependent deformation (creep), fatigue, radiation growth, thermal ageing, corrosion, oxidation, stress corrosion cracking (SCC), intergranular and irradiation assisted stress corrosion cracking (IGSCC/IASCC). The dominant mechanism for the different structures will vary with environments. These phenomena are described in the sections which follow.

1.2 Degradation mechanisms and materials ageing issues in nuclear steam supply systems (NSSS)

As shown in Tables 1.2 and 1.3, we note that a large number of materials are used for various components in nuclear power systems. It is important to state here that relatively large structures can only be fabricated using welded joints and the designers need to account for the varied properties of the different materials and their welds; often welds are known to be more sensitive to radiation and corrosive environments. As pointed out by Roberts,[3] in many cases, nuclear grade materials are fabricated to more stringent specifications than those for other technologies and are subjected to inspection and surveillance following in-reactor exposures. Stress states experienced by different components vary depending on locations, for example biaxial stresses in thin-walled cladding tubes and more complex ones in pipes, elbows and their welds; a time dependent constant load leads to creep failure in out-of-core structural components while in-core materials experience irradiation enhanced damage thereby further shortening their life. All structures, especially massive ones such as reactor pressure vessels (RPVs), invariably contain flaws and cracks that need to be taken into consideration through fracture mechanics and structural integrity analyses. It is therefore necessary to develop appropriate constitutive laws and models taking into account the individual or combined effect of: (i) instantaneous elastic and plastic deformation, (ii) time dependent recoverable deformation (anelastic strain), (iii) time dependent plastic deformation (creep), (iv) strain accumulation due to cyclic loading (fatigue), (v) corrosive environment effects, (vi) compositional effects such as dynamic strain ageing leading to premature failures and finally (vi) radiation damage and effects. The common ageing-related degradation mechanisms are described in the subsections which follow while more details are given in various chapters of the book – Part I covers major phenomena and Part II pertains to specific structural components and varied NSSSs.

1.2.1 Radiation damage

High energy neutron exposure results in accumulation of many defects like point defects (vacancies and interstitials) and dislocations, and causes redistribution in the chemistry (phase change or radiation-induced segregation (RIS)) in the materials. These modifications lead to deterioration in mechanical and corrosion properties of the exposed materials. The microscopic defects produced in materials due to irradiation are referred to as *radiation damage*. The crystal defects thus produced modify the macroscopic properties (physical, thermal and mechanical) of materials which are referred to as *radiation effects*.

The high energy neutron knocks out a stationary atom from its equilibrium position and transfers some kinetic energy (KE) to it which in turn displaces more atoms to cause a cascade effect, resulting in a number of interstitials leading to Frenkel defects.[5] This process continues until the energy of all primary and secondary knock-on atoms is insufficient (<25 eV) to displace those from the lattice sites. The atomic displacements per atom (dpa), defined as the number of times each atom is displaced from the lattice site, is estimated using various models, among which the Kinchin–Pease model is often used:

$$\text{dpa} = \sigma_{\text{el}} \frac{\Lambda \bar{E}_n}{4E_d} \Phi, \quad \text{with } \Lambda = \frac{4A}{(1+A)^2}, \qquad [1.1]$$

where $\Phi = \phi t$ is the fluence, ϕ is neutron flux (/m^{2}s), t is irradiation time, $\bar{E}_n$ is the mean neutron energy, E_d is the atomic displacement energy, A is atomic mass (in amu) and σ_{el} is the elastic neutron scattering cross section. This dpa is a better damage evaluation unit than the commonly used fluence (Φ) since it takes into account the spectral distribution of neutron energy in a given reactor.[5] This process continues until the KE falls below that needed to cause further displacements. The interstitials thus formed segregate as small disc shaped clusters.[6] They can either dissolve by vacancy emission or coalesce to large nano-voids. Clearly, the creation of excess point defects not only changes the physical dimensions due to a reduction in density but also enhances the diffusion kinetics and the phenomena controlled by atomic diffusion such as creep, while the production of line, areal and volumetric defects result in radiation hardening and embrittlement. In RPV steels, the excess vacancies produced by irradiation favour long range diffusion of copper atoms to form a copper-rich-phase, which has a BCC structure and is coherent with the steel matrix. This phase gets enriched in iron, nickel and manganese, and increases in volume.[7] These phenomena pertaining to RPV steels are further discussed in detail in a later chapter of the book. Before discussing the radiation effects a brief description of various materials properties and phenomena is presented. It should be noted that materials issues specific to various reactor components as well as detailed descriptions on various phenomena are included in specific chapters later in the book: Part I covers various fundamental materials phenomena and Part II covers reactors and components.

1.2.2 Plastic deformation

Structural materials experiencing complex stresses due to varied external forces may suffer elastic, anelastic and plastic deformations. Elastic strain is an instantaneous and completely recoverable deformation, the extent of which depends on the elastic modulus of the material and, in a simple uniaxial loading case,

$$\varepsilon_E = \frac{\sigma}{E}, \tag{1.2}$$

where σ is the stress (load per unit area), E the modulus of elasticity (also known as Young's modulus) and ε_E is the instantaneous elastic strain (change in length per unit length). Anelastic strain is time dependent, completely reversible and generally small in magnitude – albeit non-negligible in some cases – as will be discussed in detail in Chapter 3. On the contrary, plastic strain is permanent and remains even after removal of the stresses; it is generally time- and rate-dependent. A typical stress vs strain curve under uniaxial loading is shown in Fig. 1.2a[8] and the important design parameters are the yield strength, tensile strength, uniform elongation and ductility or total elongation to fracture. The deformation beyond the elastic limit obeys a power relation between the true stress (σ) and the true plastic strain (ε_p):

$$\sigma = K(\varepsilon_p)^n, \tag{1.3}$$

where K is the strength coefficient and n is the strain-hardening exponent. The area under the stress–strain curve represents the energy to deformation and fracture (referred to as resilience and toughness in the elastic and plastic regime, respectively), and this grades a material as brittle or ductile (Fig. 1.2b). The various mechanical properties of a material are also rate dependent and the flow stress is often characterized by the strain-rate sensitivity (m):

$$\sigma\big|_{T,\varepsilon} = A\dot{\varepsilon}^m. \tag{1.4}$$

The higher the n value, the higher is the uniform elongation, while a higher m value means a higher total elongation to fracture. The maximum possible value for m is unity which corresponds to *viscous* flow as seen in fluids, and this is generally noted in metals and ceramics at relatively high temperatures and at low strain-rates (or stresses).

Time dependent plastic deformation that occurs under constant load or stress (creep) becomes important above homologous temperature ($T/T_M >$ 0.4, where T_M is the melting point in absolute temperature). The reader is referred to Chapter 3 for more detail on the underlying creep mechanisms and phenomenological descriptions of the creep rupture life. A typical creep curve is illustrated in Fig. 1.3 and design allowances are limited to the total strain accumulation in the primary and secondary regimes. Thus the strain at any instant of time is given by the sum of instantaneous recoverable elastic

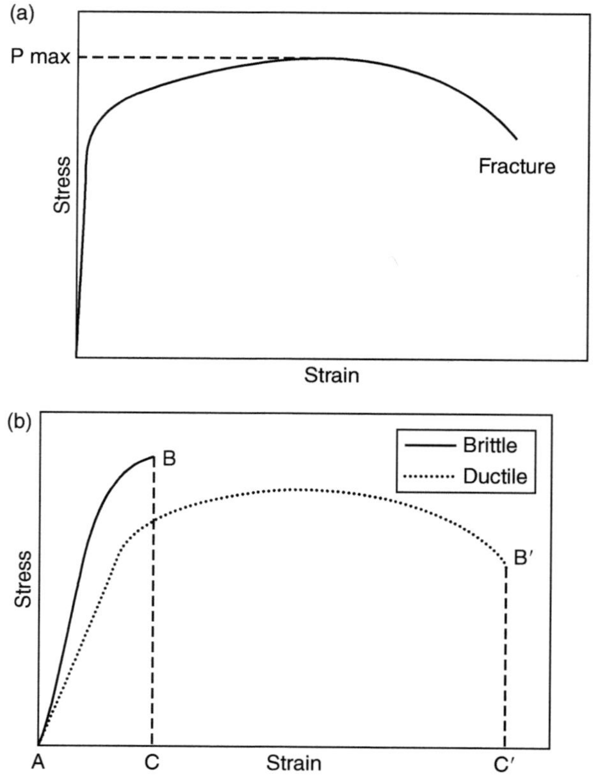

1.2 (a) Typical stress vs strain curve under unaixial loading and (b) ductile vs brittle materials.

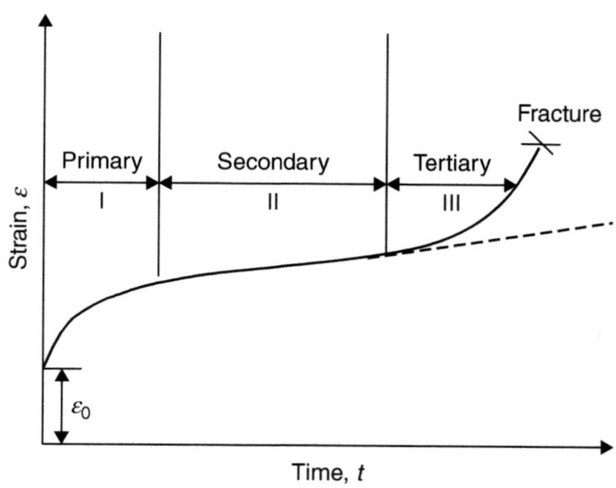

1.3 Typical creep curve.

strain, instantaneous plastic strain and time-dependent strain component from primary and secondary creep regimes:

$$\varepsilon = \varepsilon_0 + \varepsilon_t\left(1 - e^{-rt}\right) + \dot{\varepsilon}_s t, \quad [1.5]$$

where ε_0 is the instantaneous strain (the majority from elastic deformation), ε_t is the extent of primary creep strain, r is the rate at which strain decreases with time during primary creep regime and subscript 's' stands for steady-state creep rate. The steady-state creep rate is a unique function of the applied stress and temperature for a given material

$$\dot{\varepsilon}_s = A\sigma^n e^{-Q_c/RT}, \quad [1.6]$$

where Q_c is the activation energy for creep, n is the stress exponent, R is the gas constant and T is absolute temperature. The activation energy for creep can generally be matched with that for self diffusion and the above relationship can be rewritten as

$$\dot{\varepsilon}_s = A'D\sigma^n, \quad [1.7]$$

where D stands for appropriate diffusion coefficient and A' could be grain size dependent (see Chapter 3 for details). In general, lattice diffusion is temperature dependent:

$$D = \frac{\beta}{6}\alpha^2 v_D C_V e^{-Q_m/RT}, \quad [1.8a]$$

and

$$C_V = e^{-Q_V/RT}, \quad [1.8b]$$

where β is the coordination number, α is the atomic jump distance, v_D is Debye frequency, C_V is vacancy concentration and Q_m and Q_V are the activation energies for migration and formation of a vacancy, respectively. It should be noted that higher stress increases the diffusion and leads to higher creep rate with reduced rupture time.

1.2.3 Fracture toughness

All structural materials contain some types of flaw in them, the size of which can range from microscopic to mesoscopic in scale; these defects promote stress to concentrate locally around them leading to premature failure of the structure. The toughness of structures, in presence of inherent defects, is evaluated through a fracture mechanics approach. While most of the codes use a linear elastic fracture mechanics (LEFM) approach, small structures and ductile materials require elastic-plastic fracture mechanics (EPFM) formulations. The validity of LEFM compared to EPFM depends on the plastic zone size as shown in Fig. 1.4 and, in general, LEFM is not applicable when the plastic zone size is too large compared to either the crack size, the uncracked ligament or the member height.[9] In very large structures and relatively brittle materials where LEFM is valid, the stress fields are characterized by stress intensity factor, K_I, given by

$$K_I = Y\sigma\sqrt{\pi a}, \quad [1.9]$$

where a is half-crack length, σ is applied nominal stress and Y is a geometry factor which is a function of the ratio of crack length to its width (a/w). As long as K_I is lower than the plane strain critical fracture toughness K_{IC}, the structure with the crack can withstand the applied loads. In cases where LEFM is not valid (Fig. 1.4) either crack tip opening displacement (CTOD) or elastic-plastic fracture toughness (J-integral) can be conveniently adopted.

Although fracture toughness is a fundamental parameter characterizing the fracture behaviour of cracked bodies, it is often more convenient to use the ductile to brittle transition temperature (DBTT) measured using the relatively simple Charpy impact tests, to study the effect of neutron

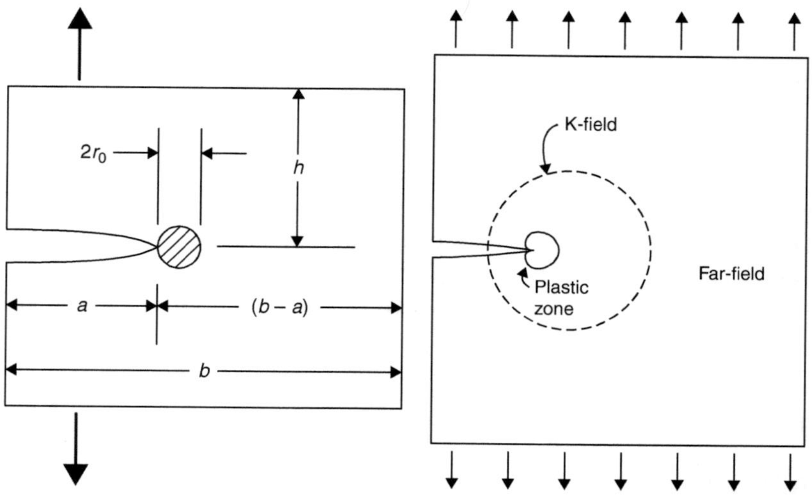

1.4 Stresses around a cracked body (a = half-crack length, r_0 = plastic zone size and $b - a - 2r_0$ = remaining ligament).

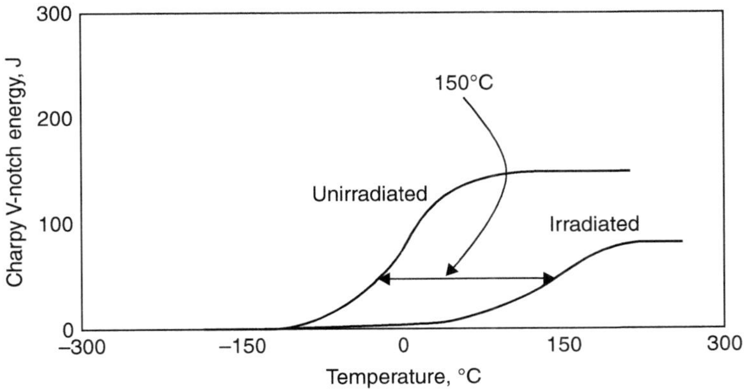

1.5 Charpy energy vs temperature – typical RPV steel and effect of neutron radiation exposure.[10]

radiation exposure (Fig. 1.5).[10] These effects are well defined in BCC metals, as against FCC metals, which exhibit a clear transition from ductile to brittle fracture behaviour as test or operating temperature decreases. It is common practice to consider a reference transition temperature corresponding to a specific Charpy impact energy of 41J (50 ft-lb) in lieu of actual nil-ductility transition (RT_{NDT}) such that brittle fracture is expected to take place below this reference temperature. As we will note later, exposure of ferritic steels to neutron irradiation leads to decreased fracture energy and increased $RT_{NDT,}$ commonly referred to as radiation embrittlement of RPV steels. Charpy impact tests are very useful and are conveniently adopted for reactor pressure vessel surveillance programmes (RVSPs). The transition temperature is a function of various factors such as the chemical composition, the temperature, the neutron flux and fluence as well as the microstructure (such as base material, heat-affected zone (HAZ) or weld metal). Validation of thermal annealing of radiation defects in RPV steels is also often established using the Charpy test method. It has been well recognized that other fracture parameters such as crack arrest fracture toughness (K_{Ia}), dynamic critical stress intensity factor (K_{Id}), etc., need to be considered in detailed analyses involving strain-rate effects that become important during a loss of coolant accident (LOCA) condition. It has also been found applicable to high-temperature crack growth, presumably because the plastic stress zone is often relatively small and linear elastic fracture mechanics are considered valid. Another fracture mechanics based parameter used to describe creep crack growth is C^*. While there are many advantages in using C^* analyses in creep of cracked bodies, these types of studies are confined more to scientific curiosities than to technological applications.

1.2.4 Fatigue

The premature fracture of materials under fluctuating load (stress/strain/temperature) is known as fatigue. Fatigue is a sudden failure exhibiting no overall ductility in the component and is known to be the cause in 90% of the total failures of structures. During each fatigue cycle the material absorbs part of the applied energy and, when the accumulated strain energy reaches the value of the surface energy of the material (in that environment), a surface forms (i.e. a crack appears). The accumulation of strain energy is facilitated by the presence of a notch or scratch and the surface energy is the minimum for the exposed crack than the embedded one. Often, the fracture surface is perpendicular to the direction of the applied stress and a compressive residual stress is beneficial in delaying the fatigue failure. Fatigue life is represented by a plot of applied stress (S) against the number of cycles to failure (N_f) known as the S–N curves. Figure 1.6 depicts the S–N curves for various metals[8] and we note that ferrous metals exhibit a distinct 'endurance' limit below which fatigue failures do not occur whereas non-ferrous metals do not seem to exhibit such a limit, albeit the slope of the S–N curve decreases at very high cycles. The stress axis can also be either the stress amplitude $(\sigma_{max} - \sigma_{min})/2$, the stress range $(\sigma_{max} - \sigma_{min})$ or mean stress $(\sigma_{max} + \sigma_{min})/2$ and it is generally seen that the fatigue life depends weakly on the R ratio ($R = \sigma_{min}/\sigma_{max}$), where σ_{max} and σ_{min} represent the maximum and minimum stresses, respectively. Depending on the number of cycles to failure the fatigue curve is classified as low cycle fatigue (LCF) and high cycle

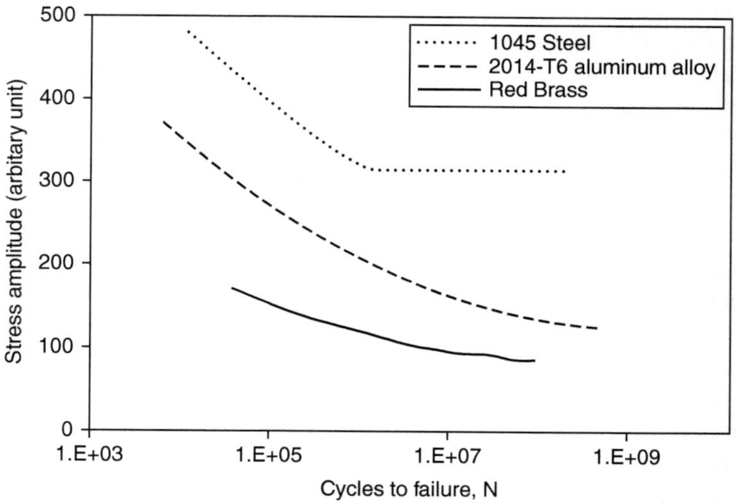

1.6 S–N curves for ferrous and nonferrous metals.

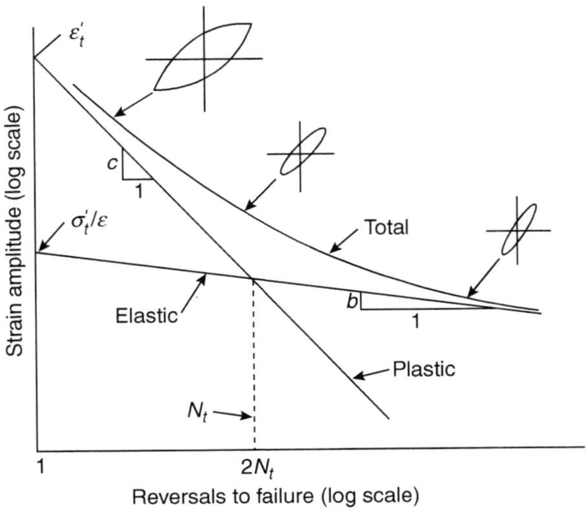

1.7 $\Delta\varepsilon$ vs N curve showing plastic and elastic strain regimes.[11]

fatigue (HCF) regions, corresponding to the plastic and elastic deformation ranges, respectively. LCF is characterized by macroscopic cyclic plastic strains and is generally limited to less than 10^4 cycles. LCF is controlled by the ductility and HCF by the strength of the material, and thus, cold-work and radiation hardening (both of which result in reduced ductility) result in decreased fatigue life in the LCF range while being beneficial in the HCF range, especially at low stresses/strains. Figure 1.7 shows a typical fatigue life plot as strain range ($\Delta\varepsilon$) against number of failure cycles (N_f) along with the corresponding stress–strain loops (broad in LCF and narrow in HCF). In the high cycle region corresponding to HCF, the Basquin equation relates the applied stress ($\Delta\sigma$) to the number of cycles:

$$N_f(\Delta\sigma)^p = C \text{ or in terms of strains } N_f(E\Delta\varepsilon)^p = C, \qquad [1.10]$$

where C and p are material constants. LCF with inelastic strains is often described by the Coffin–Manson equation

$$\Delta\varepsilon = 2A(2N_f)^c \qquad [1.11]$$

where A, a function of the ductility, and c (–0.5 to –0.7) are material constants and N_f is the number of stress/strain reversals. The Coffin–Manson equation is seen to be valid for many materials over a broad range of temperature, environment, stress history and microstructural conditions. The complete fatigue curve can be described by combining the LCF and the HCF formulations by either the universal slopes equation (Equation [1.12a]) or the characteristic slopes equation (Equation [1.12b]):

$$\frac{\Delta\varepsilon}{2} = 3.5\frac{S_u}{E}N^{-0.12} + \varepsilon_f^{0.6}N^{-0.6}, \qquad [1.12a]$$

$$\frac{\Delta\varepsilon}{2} = \frac{\sigma_f}{E}(2N)^b + \varepsilon_f(2N)^c, \qquad [1.12b]$$

where S_u is ultimate tensile strength, ε_f is true fracture strain, σ_f true fracture stress, and b and c are material constants. In terms of the characteristic slopes (Equation [1.12b]) the value of fatigue life at which the transition from low cycle (plastic) to high cycle (elastic) occurs is given by

$$2N_{\mathrm{tr}} = \left(\frac{\varepsilon_f E}{\sigma_f}\right)^{\frac{1}{b-c}}. \qquad [1.12c]$$

Fatigue crack growth rate (FCGR, da/dN) is determined by measuring the extension of a pre-crack using visual, potential drop, unloading compliance or other techniques over the elapsed number of load cycles from stress control tests conducted on either compact tension (CT) or three-point bend specimens and is related to the range of stress intensity factor (ΔK). Typical crack extension curves at two different starting stress ranges ($\Delta\sigma$) versus number of cycles are shown in Fig. 1.8a and the slopes of the curve yields da/dN. The plot on logarithmic scale of (da/dN) versus ΔK (Fig. 1.8b) clearly reveals three stages. Stage I is associated with crack blunting with very little crack growth, while crack growth in stage II can be related using Paris' law:

$$\frac{\mathrm{d}a}{\mathrm{d}N} = A(\Delta K)^p, \qquad [1.13]$$

where p is the Paris parameter/constant with values ranging from 2 to 4; this covers the majority of the crack growth event before entering the final stage (stage III) where plastic fracture occurs as crack length reaches a critical value (a_f) corresponding to the plane strain fracture toughness (K_{IC}) value:

$$a_f = \frac{K_{\mathrm{IC}}^2}{Y^2\pi^2\sigma_{\max}}. \qquad [1.14]$$

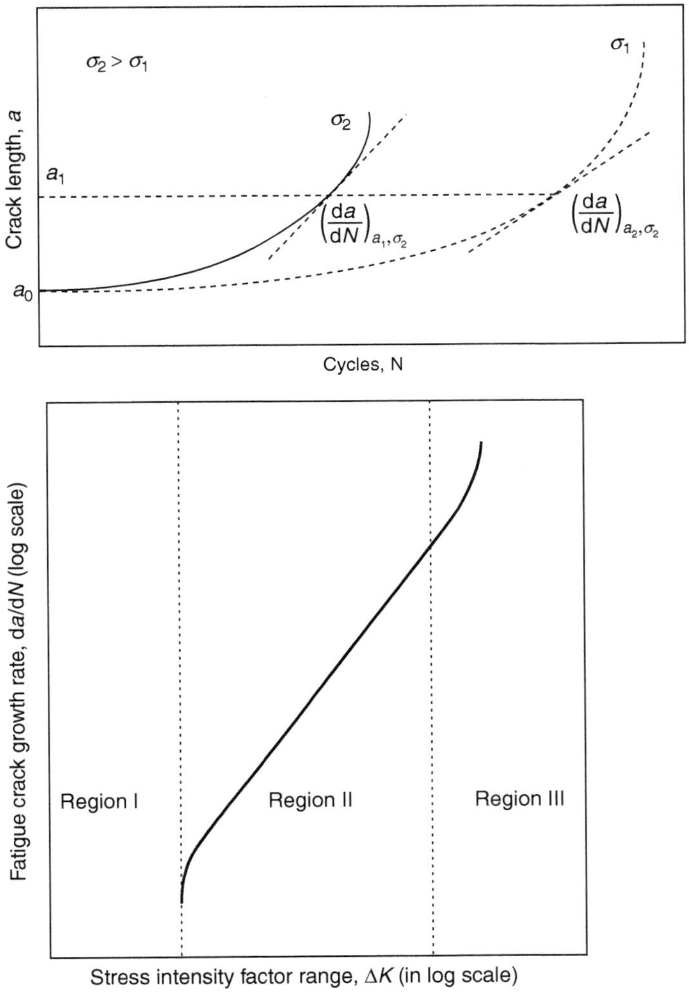

1.8 (a) Crack extension with number of cycles and (b) log–log plot of d*a*/d*N* vs Δ*K*.

In Equation [1.14], Y is the geometric factor which is a function of a/w (a is crack length and w is specimen width).

Stage I corresponds to formation of a fine crack from surface defects (such as scratches, key ways, stress concentrations) with slow initial propagation along specific crystallographic directions covering few grains before the growth enters stage II where the crack propagates at a relatively faster rate and on a plane perpendicular to the loading direction. In general, persistent slip bands (PSBs), beach marks and fatigue striations (Fig. 1.9a and 1.9b)

(a)

(b)

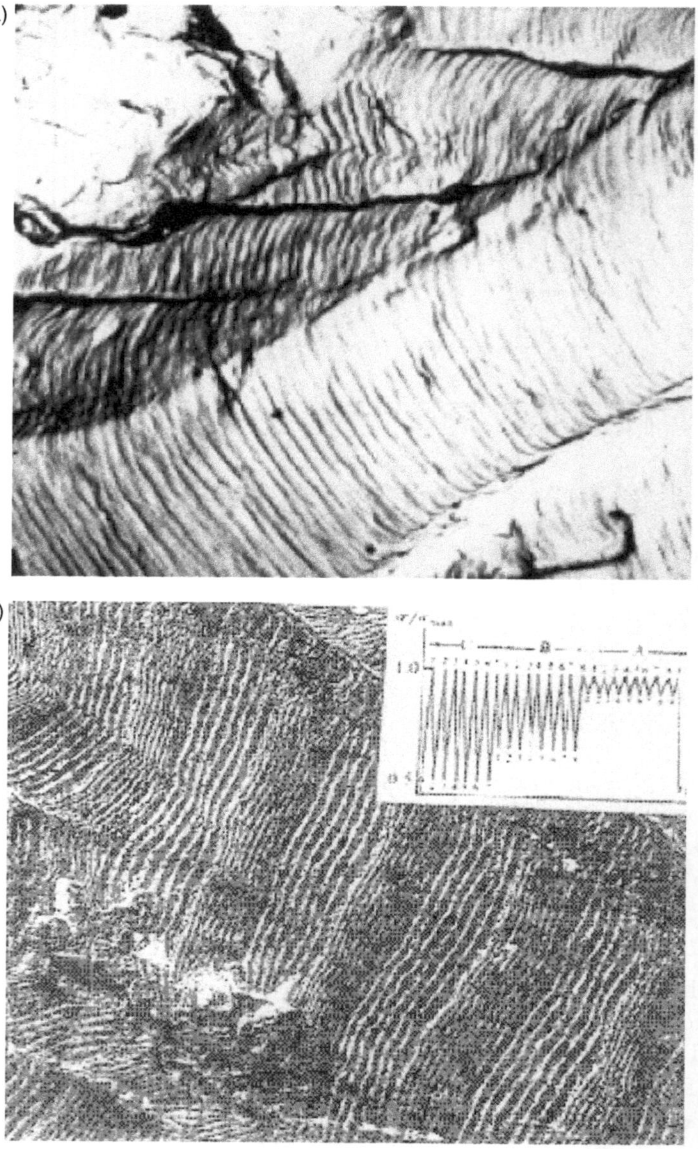

1.9 (a) Beach marks and (b) fatigue striations on the fracture surface failed under fatigue.

are characteristics of stage II crack propagation and the separation between striations depends on the stress range and frequency of loading. The total number of cycles to failure can be estimated as follows from Equations 1.15 and 1.16:

$$N_f = \frac{a_f^{-\frac{p}{2}+1} - a_o^{-\frac{p}{2}+1}}{\left(-(p/2)+1\right)A(\Delta\sigma)^p \pi^{p/2} Y^p} \quad \text{for} \quad p \neq 2 \qquad [1.15]$$

and

$$N_f = \frac{1}{A(\Delta\sigma)^2 \pi Y^2} \ln\left(\frac{a_f}{a_o}\right) \quad \text{for} \quad p = 2. \qquad [1.16]$$

Another important aspect of considering the crack growth versus ΔK is to examine the effects of superimposed environment such as corrosion and radiation. The variation of da/dN with ΔK in these cases would shift the threshold stress intensity range to lower values and the critical crack length at fracture would be indicated by K_{ISCC} instead of K_{IC}.

In strain controlled fatigue tests for life evaluation, it may be noted that the cyclic stress–strain curve leads to a hysteresis loop as depicted in Fig. 1.10a where O–A–B is the initial loading curve[11] and, on unloading, the yielding occurs at lower stress (point C as compared to A) which is known as the Bauschinger effect. The material may undergo cyclic hardening or softening; in rare cases it remains stable (Fig. 1.10b). This behaviour depends on the initial metallurgical condition of the material. According to Fig. 1.10b, as the number of cycles increases cyclic hardening leads to decreasing peak strains while the peak strains increase in the case of cyclic softening. In general, the hysteresis loop stabilizes after about 100 cycles and the stress–strain curve obtained from cyclic loading will be different from that of monotonic loading (Fig. 1.10c), but the stress–strain follows a power law relationship similar to that in monotonic loading (Equation [1.3]):

$$\Delta\sigma = K'(\Delta\varepsilon)^{n'}, \qquad [1.17]$$

where the cyclic hardening coefficient n' ranges from 0.1 to 0.2 for many metals and is given by the ratio of the parameters (b/c) (Equation [1.12b]). In some cases fatigue ratchetting occurs resulting in an increase in strain as a function of time when tested under a constant strain range (Fig. 1.11); this is often referred to as cyclic creep.[12] In a stress controlled test with non-zero mean stress, the shift in the hysteresis loop along the strain axis, as depicted in Fig. 1.11, is attributed to thermally activated dislocation movement at stresses well below the yield stress and/or due to dislocation pile up resulting in stress enhancement. Fatigue ratchetting may also occur in the presence of residual stress and in cases where microstructural inhomogeneities exist such as in welded joints.

In real situations stresses change at random frequencies and, in general, the percentage of life consumed in one cyclic loading depends on

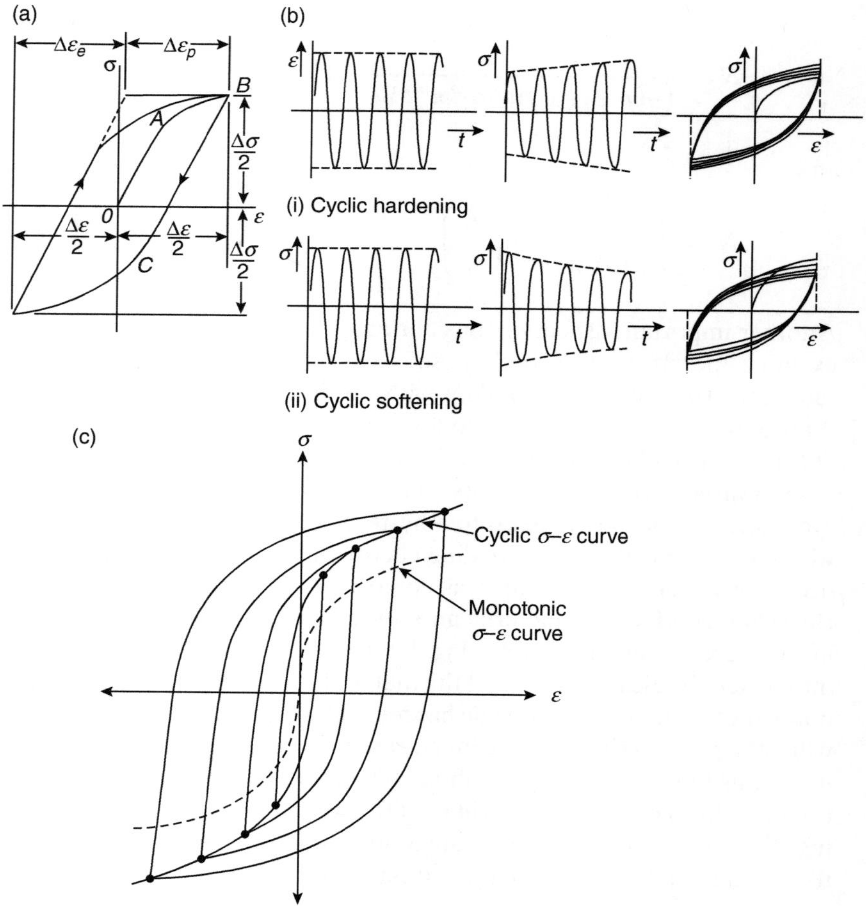

1.10 (a) Cyclic stress–strain curve illustrating hysteresis loop.[11]
(b) Hysteresis loops during cyclic hardening and cyclic softening.[12]
(c) Comparison of cyclic stress–strain curve for cyclic hardening and stress–strain curve under monotonic loading.[11]

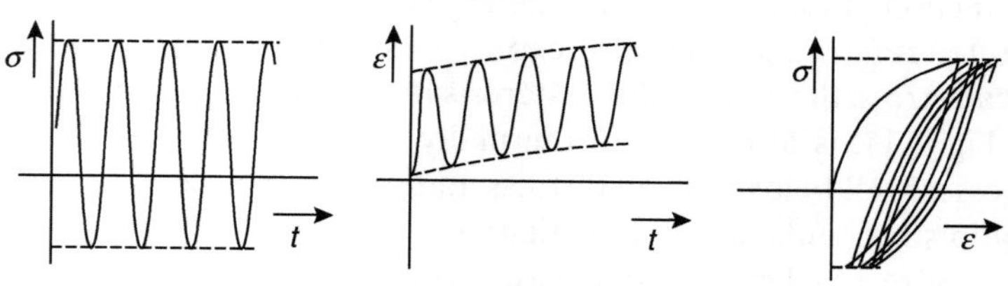

1.11 An example of ratcheting fatigue.[12]

the magnitude of stress in subsequent cycles. However, the linear cumulative damage rule, known as Miner's rule, assumes that the total life of a component can be estimated by adding up the life fraction consumed by each of the loading cycles. If N_{fi} is the number of cycles to failure at the ith cyclic loading and N_i is the number of cycles experienced by the structure then

$$\sum \frac{N_i}{N_{fi}} = 1 \quad [1.18]$$

although Miner's rule is too simplistic and fails to predict the life when notches are present. Further, it fails to predict the life when mean stress and temperature are high or cyclic frequency is low where creep deformation dominates over fatigue loading. In such situations a better approximation is given by combining Robinson's rule for creep fracture with Minor's rule;

$$\sum \frac{N_i}{N_{fi}} + \sum \frac{t_i}{t_{fi}} = 1, \quad [1.19]$$

where (t_{fi}) and fracture time (t_i) corresponding to the ith creep conditions.

It turns out that many materials exhibit deviations from this linear addition depending on whether it is cyclically hardening or softening.[13] In particular the predictions tend to be highly non-conservative for cyclically softening materials.

Fatigue strength or life of structures can be improved by reducing the mean positive stress, through appropriate design with no stress raisers and by surface finish and modifications. In particular, case hardening by carburizing and nitriding as well as shot-peening, which increase surface residual compressive stresses, result in distinct improvements in fatigue life. In comparison to pure metals, solid solution has been found to improve fatigue strength. Other factors such as interstitials inducing strain ageing could also improve fatigue life.

Environmental effects on creep-fatigue are quite complex and each case needs to be considered separately. While Equation [1.19] gives an approximate assessment, the mechanistic explanations of high-temperature fatigue effects are corrosion- or creep-related. Coffin considered the time dependent fatigue to be essentially SCC and formulated frequency-modified fatigue life–time correlations for crack initiation and propagation.[12] Manson proposed a plasticity oriented fatigue model using a strain-range partitioning method.[13] Fatigue crack growth assisted by creep cavitation at grain boundaries was considered by Majumdar and Maiya[14] to model high-temperature fatigue crack growth.

1.3 Radiation effects

As described in Section 1.2.1, exposure of materials and structures to high energy neutrons leads to the creation of microscopic defects such as vacancies, interstitials, Frenkel defects, dislocations and faulted loops, as well as voids and cavities. Figure 1.12a depicts voids and precipitates in irradiated stainless steel[15] while large Frank loops are shown in Fig. 1.12b.[16] Similar faulted Frank loops are noted in irradiated aluminium and copper as well as iron (Fig. 1.12c).[17]

Materials undergo many changes on exposure to neutron radiation: defect concentration increases, neutron transmutation occurs, chemical reactivity changes (generally gets enhanced), diffusion of the elements increases and new phases (both equilibrium and non-equilibrium) form. The extent of change in properties is, in general, proportional to radiation flux, particle energy and irradiation time, while it decreases with an increase in irradiation temperature. The creation of voids, cavities and depleted zones leads to decreased density of the material with a corresponding increase in volume known as radiation swelling. Increased defect concentration leads to increased electrical resistivity and decreased thermal conductivity while magnetic susceptibility decreases. The threshold neutron fluence or dpa that leads to extensive degradation in a material depends on the crystal structure and nature of atomic bonding – semiconductors and polymers degrade at much lower neutron fluences compared to ceramics and metals. The reader is referred to various monographs on nuclear materials and radiation effects for more details.[18] These defects result in hardening and embrittlement of the material with an increase in strength and accompanying decrease in ductility commonly referred to as *radiation hardening and radiation embrittlement*; strain hardening in the material decreases accompanied by a decreased uniform elongation and an increase in DBTT (or RT_{NDT}), which decreases the fracture toughness. The increased defect density enhances the diffusivity in the material which in turn increases the creep rates and reduces the rupture time. These various phenomena will be discussed in detail in the following sections.

1.3.1 Mechanical behaviour

It was mentioned in earlier sections that BCC materials such as iron and some steels show a distinct yield point (Fig. 1.13) where as FCC and HCP materials show a continuous transition from elastic to plastic range (Fig. 1.3). The distinct yield point is due to the locking of the dislocation sources by interstitial impurities such as C and N in low alloy steels that increases the stress resulting in a sudden increase in free or mobile dislocation density. The velocity of these dislocations decreases in order to maintain the imposed

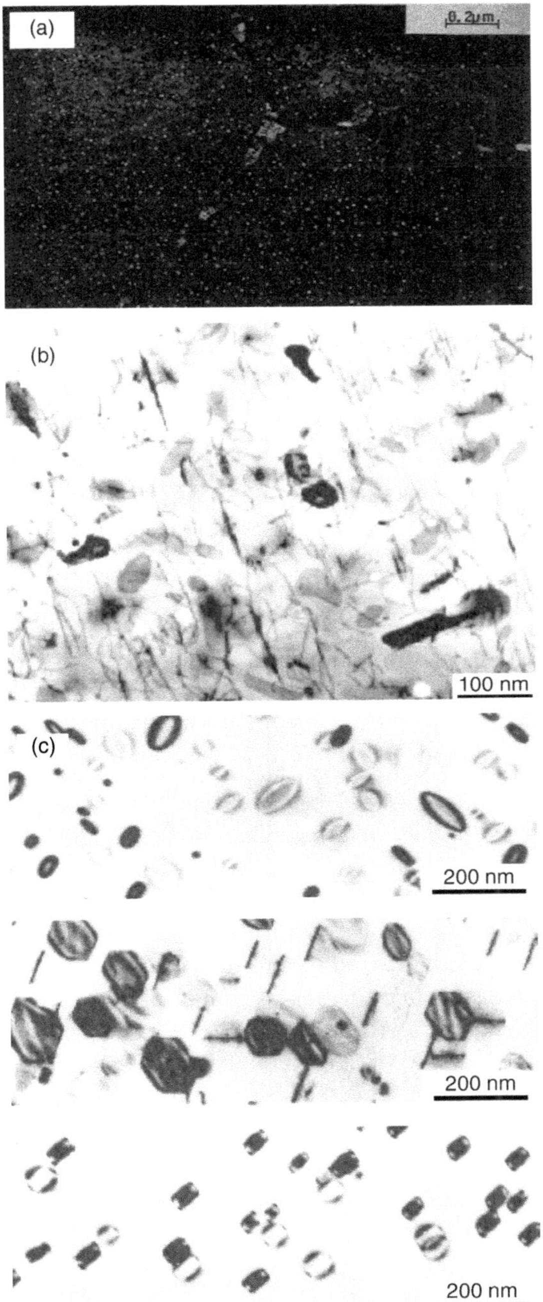

1.12 (a) Voids and precipitates in irradiated stainless steel.[15]
(b) Faulted Frank loops and dislocations in irradiated stainless steel.[16]
(c) Faulted Frank loops in irradiated aluminium, copper and iron.[17]

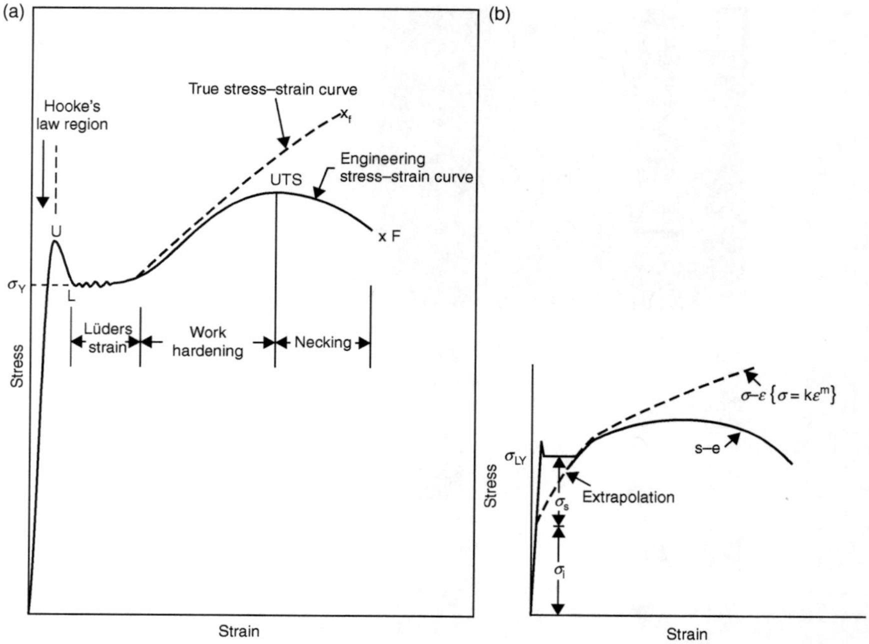

1.13 (a) Typical stress vs strain curve depicting yield point such as is observed in steels[20] and (b) extrapolation of the plastic curve to elastic line delineating the source (σ_s) and friction (σ_i) hardening terms.[22]

constant strain-rate resulting in the observed load or stress drop. The stress maximum known as the *upper yield point* is followed by deformation taking place within a relatively small region of the specimen (Luders band), with continued elongation of the specimen by the propagation of the band along the gauge section wherein the deformation is inhomogeneous. Once the entire gauge section is traversed by the band, normal strain hardening occurs with stress increasing and further deformation taking place. This distinct yield points (σ_y) in stress–strain curves can be represented as a sum of a non-zero source hardening term (σ_s) and a friction hardening term which represents the resistance experienced by the mobile dislocation (σ_i),

$$\sigma_y = \sigma_i + \sigma_s, \qquad [1.20a]$$

similar to the well-known Hall–Petch equation:

$$\sigma_y = \sigma_i + \frac{k_y}{\sqrt{d}}, \qquad [1.20b]$$

where d is the grain size and k_y is the Petch unpinning coefficient. Thus, the source hardening term (σ_s) is equivalent to the grain size dependent term which can be determined from the grain size dependence of the yield stress:

$$\sigma_s = \frac{k_y}{\sqrt{d}}. \quad [1.20c]$$

Alternatively, it can be evaluated using the Makin–Minter[19] method by extrapolating the work-hardening portion of the stress–strain curve in Fig. 1.13a to the elastic range (Fig. 1.13b). The intercept is interpreted as the friction stress (σ_i) and the difference between the yield stress and the intercept is the source hardening (σ_s). Murty[20] demonstrated the equivalence of Hall–Petch relation and Makin–Minter method from experimental results on grain size dependence of the mechanical properties of pure iron.

Effects of neutron radiation exposure in austenitic stainless steel (FCC)[21] and mild steel (BCC)[22] are shown in Fig. 1.14a and 1.14b, respectively. It can be seen that the smooth stress–strain curve in the unirradiated stainless steel (Fig. 1.14a) developed a distinct yield point subsequent to radiation exposure accompanied by a decrease in strain hardening and a decrease in the uniform and total elongations. The fact that the yield point and the Luders strain observed in unirradiated BCC mild steel (Fig. 1.14b) increases initially with increase in neutron fluence and eventually disappears after the highest value (10^{19} n/cm^2) clearly demonstrates the decrease in source hardening with increased neutron radiation dose. On the contrary, in FCC metals the yield point appeared following radiation exposure indicating an increased source hardening in the irradiated material.

Friction hardening (σ_i) arises mainly from the long range elastic interactions of moving dislocations with other (forest) dislocations as well as short range interactions with faulted dislocation loops, precipitates, etc., so that

$$\sigma_i = \sigma_{\mathrm{LR}} + \sigma_{\mathrm{SR}} = \alpha G b\sqrt{\rho} + \beta G b\sqrt{Nd}, \quad [1.21]$$

where the subscripts LR and SR represent the long-range and short-range stresses, G is shear modulus, b is the Burgers vector, ρ is dislocation density, N and d are the number density and diameter of the clusters (faulted loops, precipitates, etc.) and α and β are constants representing the strengths of long range and short range forces. In general the defect densities (ρ and N) are proportional to the fluence (or dpa) and thus

$$\sigma_{\mathrm{i}} = \alpha' G b\sqrt{\Phi}, \quad [1.22]$$

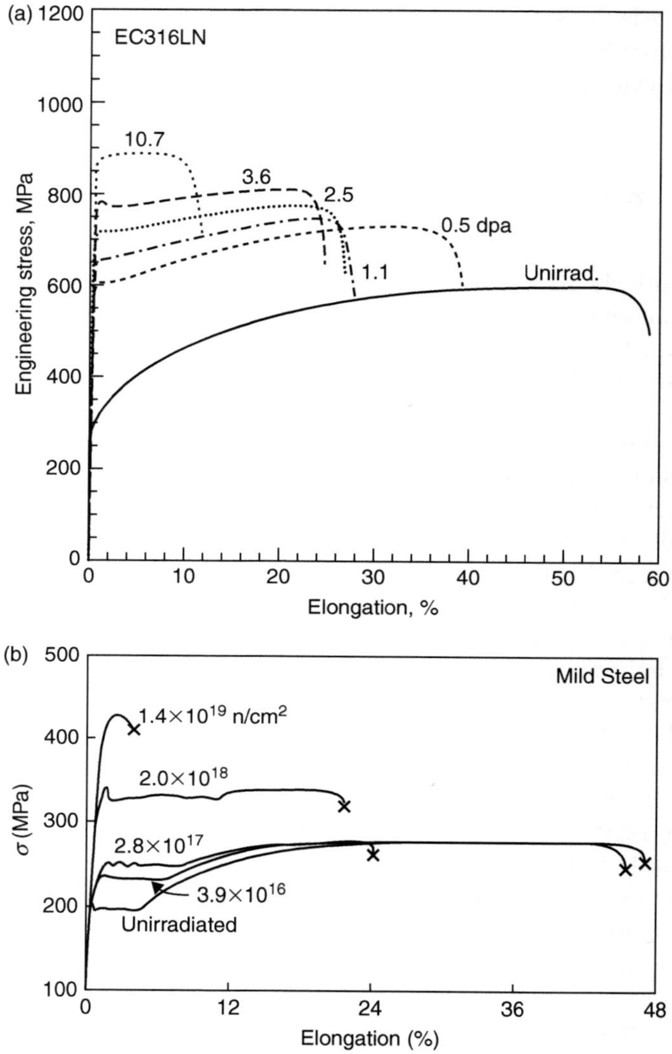

1.14 (a) Effect of neutron irradiation on stress vs strain curves for stainless steel (FCC) depicting the occurrence of yield points following radiation exposure.[21] (b) Effect of neutron irradiation on stress vs strain curves for mild steel (BCC) depicting the absence of yield points following high neutron radiation exposure.[22]

where Φ is fluence (ϕt). The stress–strain curves shown in Fig. 1.14b on mild steel at room temperature as a function of fluence reveal that the yield strength varied as cube-root of fluence (Fig. 1.15) and not the square-root. This seems to stem from the fact that friction hardening (σ_i) indeed varied

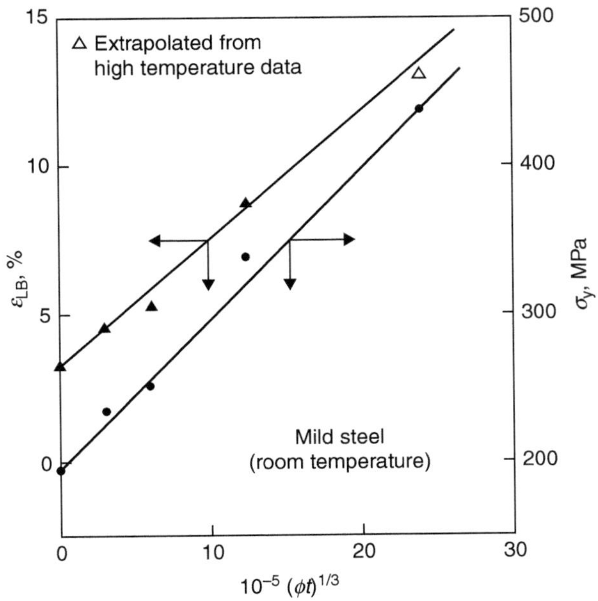

1.15 Effect of neutron irradiation on yield stress and Luders strain in mild steel.

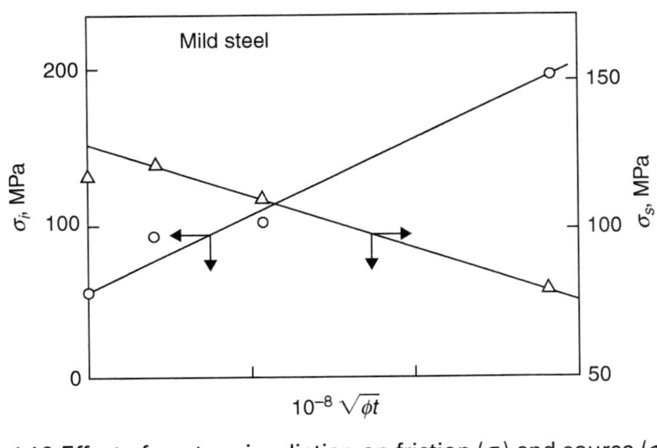

1.16 Effect of neutron irradiation on friction (σ_i) and source (σ_s) for mild steel.[23]

as square-root of fluence as noted in Fig. 1.16.[23] Since σ_s decreased with increase in fluence, the yield stress, being the sum of these two factors, should be a function of fluence raised to a power slightly less than 0.5. In FCC metals such as stainless steel, source hardening is very small before irradiation while it increases with fluence thereby resulting in small yield points at high fluences (Fig. 1.14a). It is believed that at high fluences saturation of radiation hardening occurs resulting in deviations from the square-root dependence of the hardening on the fluence.

Since the Luders strain in steels varies linearly with yield stress, it increased as cube-root of fluence (Fig. 1.15) where we note that the datum point at the highest neutron dose of 1.4×10^{19} n/cm^2 is an extrapolation from high temperatures to ambient. Thus, at room temperature the highly irradiated material exhibited severe localized deformation and failed during Luders band propagation itself before reaching the strain-hardening regime. The increase in Luders strain and the decrease in source hardening, subsequent to irradiation, imply that the work hardening should decrease as neutron dose increases. Indeed, the work-hardening exponent decreased from ~0.34 for the unirradiated mild steel to ~0.19 at a neutron fluence of 2×10^{18} n/cm^2.[23]

The fact that the source hardening in BCC metals such as steels decreases on exposure to neutron irradiation implies that the concentration of interstitial C and N in solution decreases with increased neutron fluence. Murty[24] examined the effect of incremental neutron dose on static strain ageing kinetics and demonstrated that the ageing kinetics are slowed and that fluences greater than 10^{18} n/cm^2 rendered the steel non-ageing. In a correlation between the effects of neutron irradiation and dry hydrogen treatment, Murty and Charit[25] demonstrated that the concentration of nitrogen in solution decreases with neutron fluence, reaching a value very close to zero at 10^{18} n/cm^2 (Fig. 1.17). These results imply that interstitial impurities combine with radiation-induced point defects such as vacancies and interstitials, either with individual defects or loops, to form complexes. These complexes are probably responsible for part of the increase in friction hardening and the corresponding decrease in solution hardening. McLennan and Hall[26] found from internal friction experiments that the concentration of C in solution decreased by a factor of four in steels after irradiation to about 10^{19} n/cm^2.

This is also the reason for the decrease in the intensity of dynamic strain ageing (DSA) in annealed mild steel, as depicted in Fig. 1.18a–1.18e, where the load drops in the stress–strain curves decreased with increase in radiation fluence, finally rendering the steel non-ageing after irradiation at 10^{19} n/cm^2.[27] It must be noted here that though radiation exposure results in reduced concentration of interstitial C and N in solution leading to reduced *blue brittleness*, radiation hardening and embrittlement can still occur. Thus the competing and synergistic effects of DSA and neutron irradiation could lead to increased ductility along with increased strength at appropriate temperature and strain-rates. Comparison of stress–strain for unirradiated material (~100°C) with those irradiated to different doses clearly reveals (Fig. 1.19)[28] the typical embrittlement due to DSA in the unirradiated

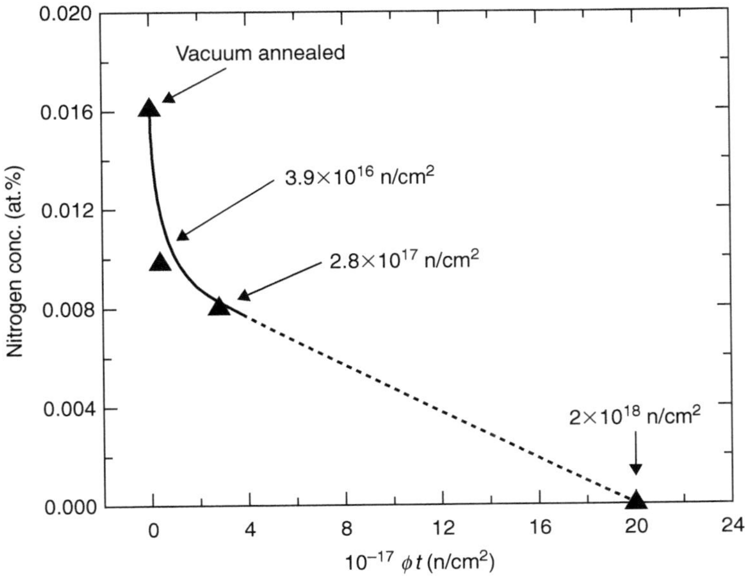

1.17 Effect of neutron irradiation on concentration of nitrogen in solution in mild steel.[25]

material, whereas increased strength and ductility are noted following irradiation to around 2×10^{18} n/cm^2; but, after irradiating at the highest fluence level of 1.4×10^{19} n/cm^2 the ductility decreased to around 2% with possible fracture during Luders band propagation. These results are in contrast to those at room temperature where no DSA or blue brittleness is noted (Fig. 1.14b). At 100°C in mild steel where jerky flow started, the ductility decreased to 11% while it increased to ~20% following neutron irradiation to 10^{18} n/cm^2. The fact that strength increased along with an increase in ductility implies that toughness (as defined by the area under the stress–strain curve) increases at temperatures where DSA is suppressed following radiation exposure. This is clearly shown in Fig. 1.20 which compares the toughness (J) for mild steel before and after neutron irradiation to 2×10^{18} n/cm^2.[28] Normal radiation embrittlement is noted at ambient temperature while an increase in toughness is observed at elevated temperatures following radiation exposure. The measured toughness is sensitive to the strain-rate of testing and a minimum toughness value is obtained when tested over a strain-rate range. This minimum in the unirradiated material occurs at higher temperatures for increased strain-rates and follows an Arrhenius relation ($\dot{\varepsilon} = Ae^{-Q/RT_c}$ where T_c is the temperature at which minimum toughness occurs) with the activation energy (Q) identifiable with that for diffusion of C and N in steel. Thus these synergistic effects of neutron irradiation

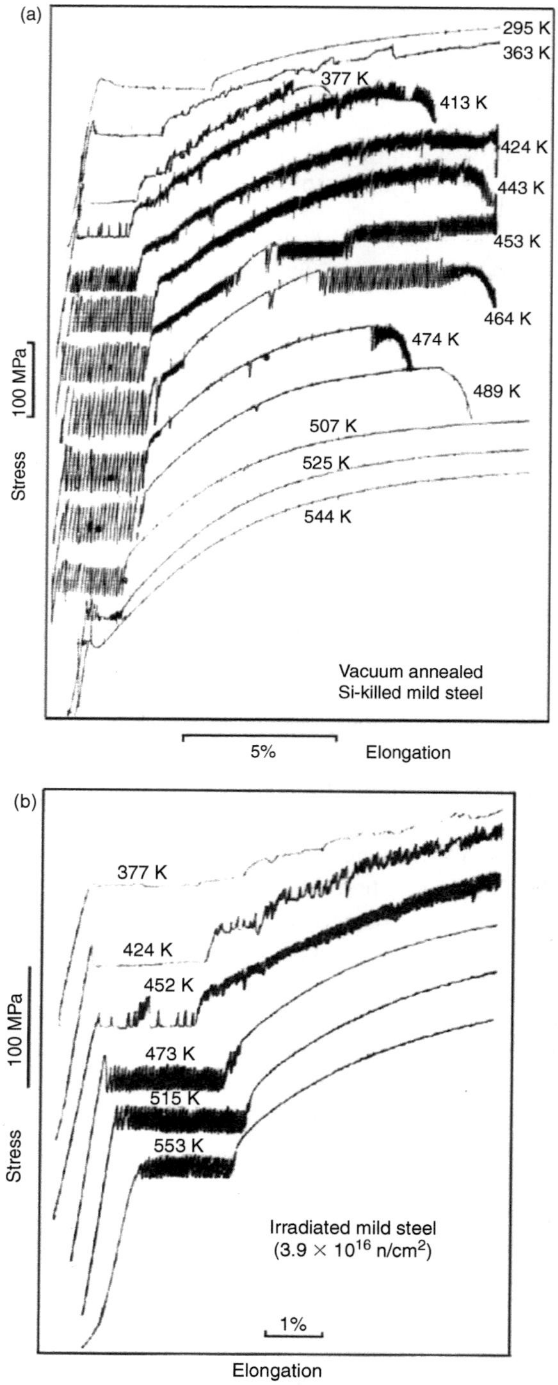

1.18 Continued on page 33. See caption on page 34.

(c)

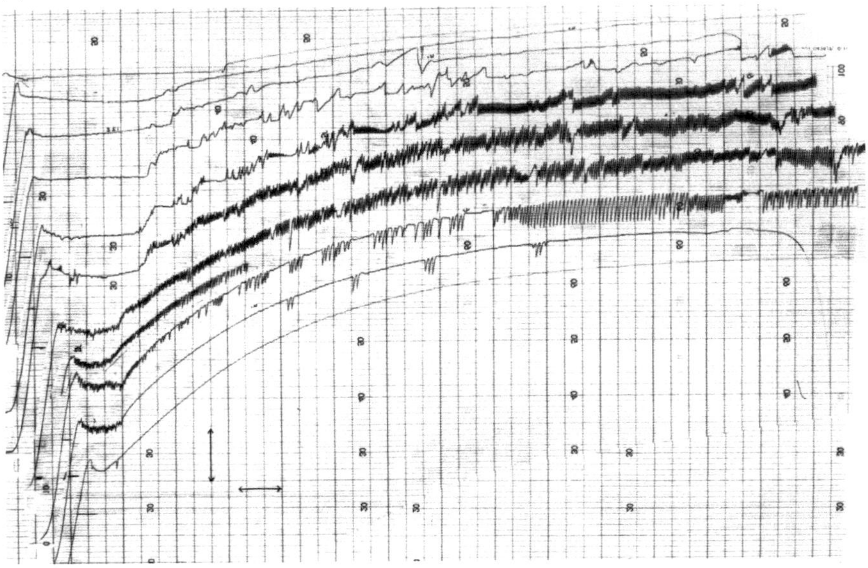

(d)

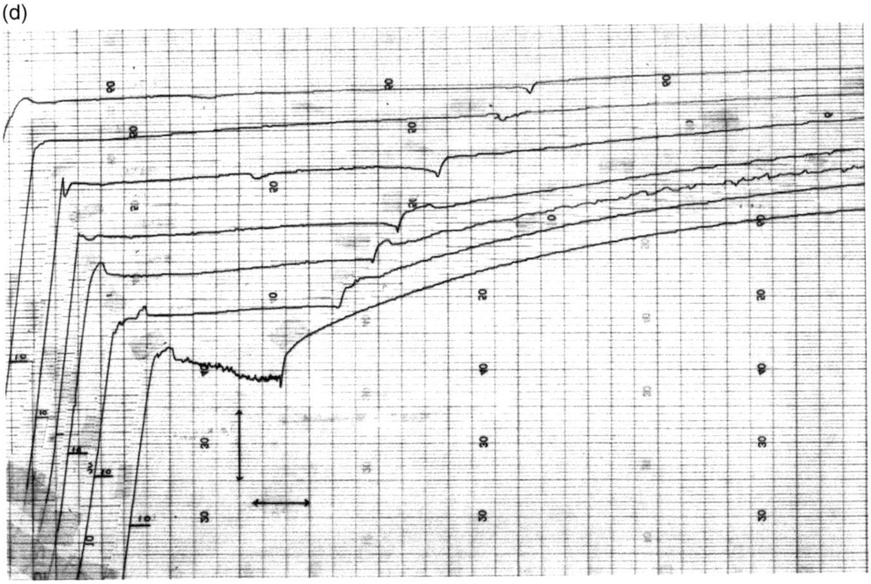

1.18 Continued

(e)

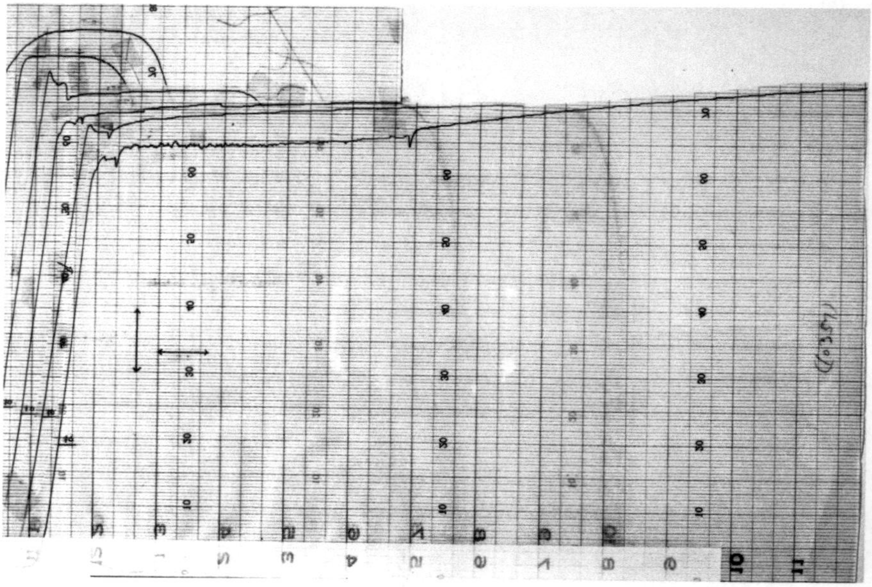

1.18 (a) Stress–strain curves for mild steel at varied temperatures before irradiation[27]; (b) stress–strain curves for mild steel at varied temperatures following irradiation (3.9×10^{20} n/m^2)[27]; (c) stress–strain curves for mild steel at varied temperatures following irradiation (2.8×10^{21} n/m^2)[27]; (d) stress–strain curves for mild steel at varied temperatures following irradiation (2.0×10^{22} n/m^2)[27]; (e) stress–strain curves for mild steel at varied temperatures following irradiation (1.4×10^{23} n/m^2).[27]

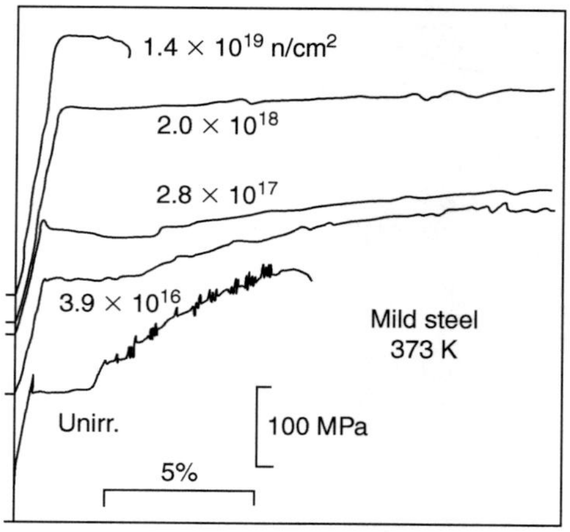

1.19 Effect of neutron irradiation on stress–strain curves for mild steel at 373 K.[28]

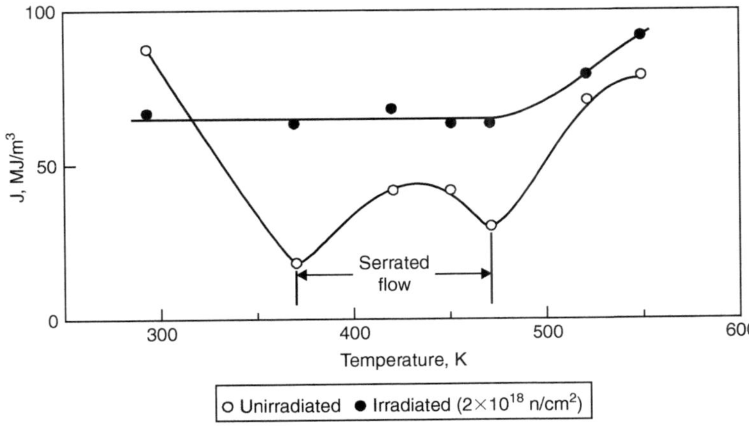

1.20 Effect of DSA and neutron on the temperature variation of energy to fracture (J).[29]

and DSA could lead to beneficial effects on strength and ductility in certain temperature and strain-rate regimes.[29] While these descriptions are limited to mild steels, such DSA and neutron radiation effects are also observed in steels used for nuclear reactor pressure boundary applications; we will discuss these in the next section under the radiation effects on radiation embrittlement of nuclear RPVs and support structures.

Kass and Murty[30] have successfully used the Hall–Petch relation and friction/source hardening concepts to explain the influence of fast and thermal neutrons, in the total neutron spectrum, on the grain size effects in pure iron and low alloy steels. They evaluated the effects of total and fast neutron spectra by irradiating samples with and without Cd-wrapping thereby eliminating low energy (<~0.5eV) neutrons in the Cd-wrapped samples. The bar chart in Fig. 1.21 shows the effect of fast and total neutron radiation on the yield stress for Armco-iron and steels (1020, A516 and A588).[30] All steels exhibited increased hardening due to the total neutron radiation exposure compared to only fast neutrons whereas in pure iron we note that as grain size decreases from 190 to 50 μm, exposure to the total neutron spectrum resulted in lower radiation hardening than only fast flux. This is explained on the basis of the Hall–Petch plot where the y-axis intercept representing friction hardening increases with neutron radiation exposure while the slope decreases following fast neutrons. Exposure to the total neutron spectrum with additional low energy neutrons would result in a further slight increase in friction hardening (or y-axis intercept) accompanied by a slight further decrease in the slope (i.e. decrease in the source hardening). This implies that the two lines (Fig. 1.22) representing the effects of fast and total (fast + low energy) fluences will cross over at a critical grain size below

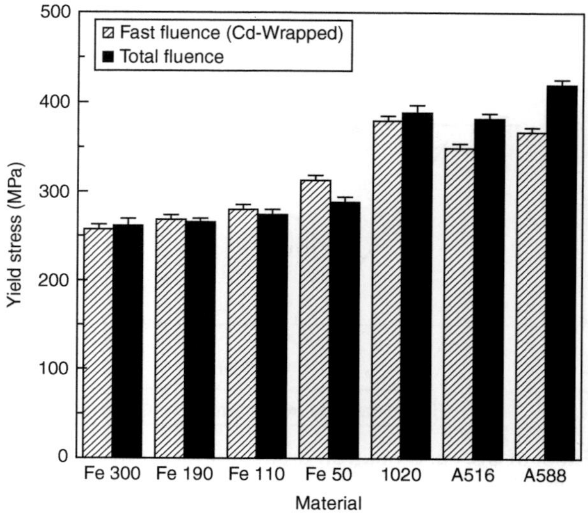

1.21 Bar chart of yield stress for Armco-iron and steels (1020, A516 and A588) following fast and total neutron radiation exposures of 2.8×10^{18} and 3.4×10^{19} n/cm^2, respectively.[30]

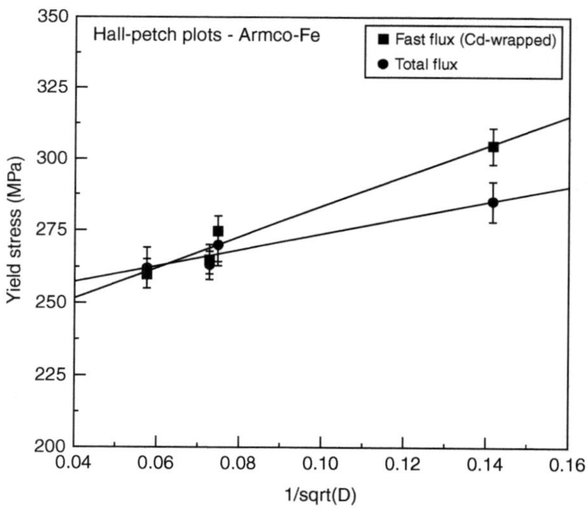

1.22 Effects of fast and total neutron fluences on Hall–Petch plots for pure iron.[30]

which (means larger values of the abscissa) fast neutron fluence would have a slightly larger effect on hardening than exposure to total flux.

Although ferritic steels such as those commonly used for pressure boundary and reactor support applications have very small grain sizes (<50 μm), radiation hardening in these steels (1020, A516 and A588) is opposite to that noted in Armco-Fe of small grain sizes. A plausible explanation for this observation lies in the fact that the source hardening and the slope of the Hall–Petch plot are very small in these steels. Once these materials are exposed to fast neutrons, the hardening will essentially be due to friction hardening with negligible contribution from source hardening thereby resulting in grain size independent yield strength (i.e. horizontal line in Hall–Petch plot). Exposure to the additional low energy neutrons along with fast neutrons is expected to result in a slight increase in friction hardening, thereby resulting in a Hall–Petch line parallel to that noted for fast fluence exposure alone. Hence, exposure to the total neutron energy spectrum in these steels will lead to an additional grain size independent hardening as observed in the bar chart in Fig. 1.21. Thus, identifying the yield stress as a sum of friction and source terms lends explanation and support to the experimental observations in both pure iron and steels.

1.3.2 Toughness loss

As described in the foregoing sections on fracture, the energy absorbed before failure is an indication of the toughness of a material. Exposure of the material to neutron fluxes reduces the toughness value and increases the transition temperature (RT_{NDT}), below which the material fails in a brittle manner (absorbing very low energy) (Fig. 1.5); this is commonly referred to as radiation embrittlement. The extent to which the RT_{NDT} is raised from the initial value is an indication of the degradation. A simple model developed using the concentrations of copper and nickel, and the value of flux and fluence could yield a fair prediction for the shift in RT_{NDT}.[31] The most severely affected region in a RPV is the belt region where fast neutron bombardment is at a maximum. Presence of copper and phosphorous are found to promote the embrittlement effect while that of nickel acting alone is unclear,[32] although in combination with copper, the effect of embrittlement due to nickel is increased. The role of other elements like Mo, Mn, As, Cr and Sn is not completely understood. The change in DBTT may be evaluated using Cottrell brittle fracture theory and is given in terms of friction and source hardening terms:

$$\Delta\mathrm{DBTT} = \frac{1+\left(2\Phi^2\sigma_y/\sigma_i\right)\left(\mathrm{d}\sigma_s/\mathrm{d}\Phi\right)}{1+\left(\sigma_y/\sigma_s\right)\left(\mathrm{d}\sigma_s/\mathrm{d}T\right)}\Delta\sigma_i,\ \Delta\sigma_i = \alpha Gb\sqrt{\phi t} = \alpha Gb\sqrt{\Phi}. \quad [1.23]$$

Thus, one needs to know the influence of neutron irradiation and test temperature on both the friction and source hardening terms to evaluate

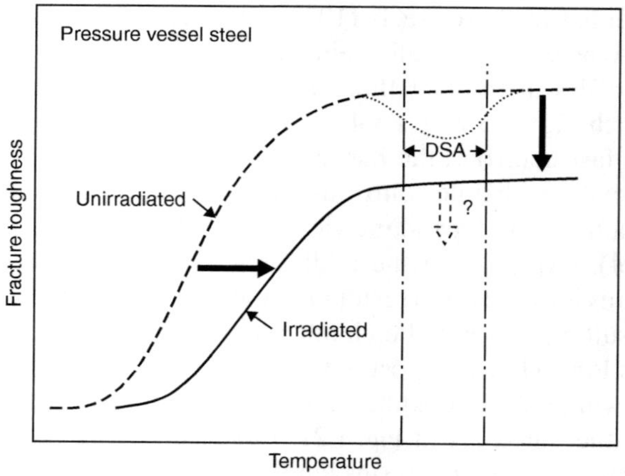

1.23 Effects of neutron irradiation on fracture toughness and possible effect of superimposed DSA.[32]

the effect of neutron radiation exposure on DBTT or RT_{NDT}. However, a lack of information on the fluence dependence of σ_s as well as other factors makes it rather difficult to apply the above formula in predicting the changes in radiation embrittlement of nuclear pressure vessels. Radiation embrittlement of ferritic steels is far more complex than using simple fracture theory. The question now arises as to the superimposed effect of DSA on fracture toughness and whether DSA adds to the increase in DBTT (Fig. 1.23). Jung and Murty[33] examined the effect of neutron irradiation on the elastic-plastic crack initiation fracture toughness (J_{IC}) using an unloading compliance method before and after radiation exposure. Figure 1.24a illustrates the effect of DSA on decreased fracture toughness in the DSA regime which happens to be in the upper shelf region where we note a dip in toughness. After irradiation the dip appeared at a slightly higher temperature (Fig. 1.24b) illustrating the fact that neutron radiation exposure results in reduced amounts of interstitial C and N atoms in solution and that radiation does not eliminate DSA but postpones its occurrence to higher temperatures. These synergistic effects of radiation-produced defects and interstitials are sensitive to the composition of the steels.[34]

1.3.3 Radiation growth and creep

As we noted earlier in Fig. 1.12 voids and precipitates are generated in materials such as stainless steel during neutron irradiation resulting in reduced density or increased volume known as radiation swelling that, in turn, leads to dimensional changes even in the absence of external stresses. Zirconium

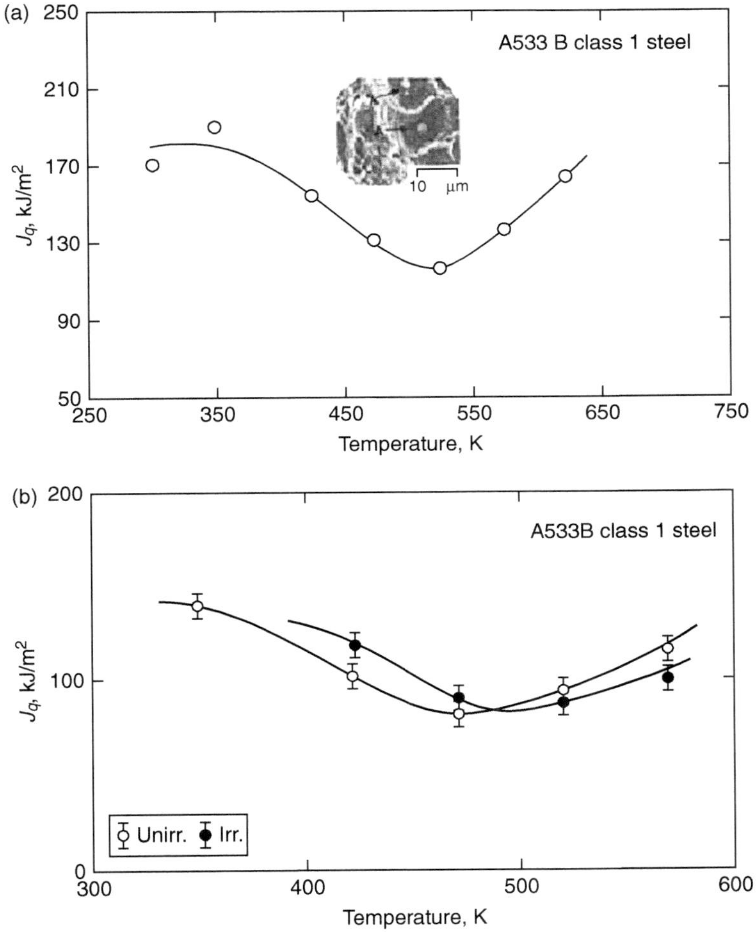

1.24 (a) Fracture toughness vs test temperature in A533B steel depicting the effect of DSA as a dip in the upper shelf energy[33]; (b) effect of neutron irradiation on temperature variation of fracture toughness for A533B steel in the upper shelf regime.[33]

alloys on the other hand resist void formation albeit undergoing stress-free radiation growth due to the inherent crystallographic texture with preferred crystallographic orientation developed during the thermo-mechanical processing of thin-walled tubing used to clad nuclear fuel (UO_2). Radiation exposure of single crystal Zr exhibits elongated a-axis with decreased c-axis thereby the single crystal becomes short and fat (Fig. 1.25a) mainly due to the formation of interstitial <a> loops on prism ($\{10\bar{1}0\}$) planes albeit the volume is unchanged. Cladding tubes typically exhibit preferred orientations

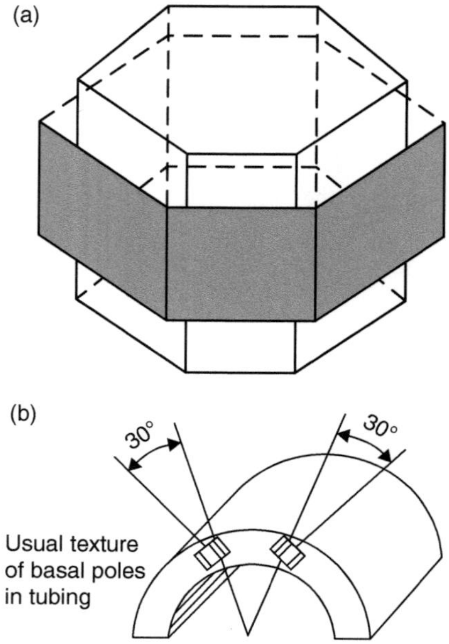

1.25 (a) Effect of neutron radiation exposure on a Zr-single crystal leading to decreased c-axis (vertical) and increased a-axis (horizontal); (b) preferred orientation (texture) in a typical Zircaloy cladding tube.

or textures such that the c-axis of the grains are mainly oriented at 30° from the radial (thickness) direction towards the hoop (transverse) direction as shown in Fig. 1.25b. This results in small, positive strains along the axial and hoop directions equal to the contractile strain along the radial or thickness direction such as that the total sum is zero:

$$\varepsilon_z + \varepsilon_\theta + \varepsilon_r = 0. \qquad [1.24]$$

Since all the radiation-induced strains are relatively small, no observable changes occur in the diameter (~9.5 mm) and thickness (~0.56 mm) while measurable changes occur along the axial direction of the ~3.9 m long cladding tubes. This lengthening of the cladding tubes due to radiation exposure is known as radiation growth with no change in volume; this is in contrast to void swelling where dimensional changes are accompanied by increased volume. Radiation growth of Zircaloy cladding leads to rod bow and, in cases where a gradient in neutron flux and/or texture along the cladding tubes exist, then the entire assembly can bow; more details may be found in the later chapters on Zr-alloys.

Deformation due to creep occurs in the presence of external stress during irradiation. At relatively low temperatures, where thermal creep is negligible, radiation may induce creep due to the increased concentration of point defects during irradiation which enhances diffusion which is thus known as *radiation-induced* creep. At high temperatures where thermal creep can take place, radiation enhances creep due to increased defect concentration and is referred to as *radiation-enhanced* creep. In an extremely simplified way one may express the vacancy concentration as due to thermal and radiation so that the creep-rate equation (Equation [1.8]) becomes

$$\dot{\varepsilon} = AD\sigma^n,\ D = \frac{\beta}{6}\alpha^2 v_D (C_v^{\text{th}} + C_v^{\text{irr}}) e^{-Q_m/RT},\ C_v^{\text{th}} = e^{-Q_v/RT} \text{ and } C_v^{\text{irr}} \propto \text{dpa}. \quad [1.25a]$$

In general, however, the radiation component of the creep rate is seen to be temperature insensitive and proportional to the flux and stress:

$$\dot{\varepsilon}_{\text{irr}} = B\phi\sigma, \quad [1.25b]$$

and with very little primary creep so that the strain due to radiation creep is given by

$$\varepsilon_{\text{irr}} = B\phi\sigma t. \quad [1.25c]$$

In-reactor results are often sensitive to the neutron spectrum and are very scattered to unequivocally describe the stress and time dependences. However, irradiation creep is not just the thermal creep imposed with high defect density; in the former the interstitial and vacancy loops that form during irradiation play a major role in the creep mechanism while in the latter the creep rate increases with temperature. Two mechanisms are proposed to explain the irradiation creep phenomenon: (a) in a stress-induced preferential absorption (SIPA),[35] extra planes of atoms accumulate on crystal planes so as to produce creep strain in the direction of the applied stress, whereas (b) stress-induced preferential nucleation (SIPN) assumes that nucleation of loops is preferred on planes with a high resolved normal stress.[36,37] Both of these mechanisms assume that the growth or formation of loops occurs at a favourable orientation with respect to applied stress and causes macroscopic strain. Irradiation generated point defects (vacancies and self interstitials) migrate to different sinks like dislocations and grain boundaries, in order to reduce the energy of the system, and do so in a preferential manner due to the anisotropy of the zirconium crystal lattice. Because of the diffusional anisotropy, interstitial atoms tend to migrate to dislocations lying on prism planes and to grain boundaries oriented parallel to prism planes while vacancies drift preferentially to dislocations lying on basal planes and to boundaries parallel to basal planes. This gives rise to elongation in one direction and contraction in the other. The creep rate can be controlled by suitable alloying additions and modifying the texture of the zirconium matrix such that the dislocation glide rate is reduced. The complex behaviour was modelled by Nichols[38] using dislocation climb-glide processes as a function of stress and neutron radiation dose. There is an extensive literature on radiation creep of different materials and the reader is referred to the literature for more details.

1.3.4 Effect of radiation on fatigue

Since LCF and HCF are controlled by ductility and strength respectively, and radiation results in hardening and embrittlement, we expect life in HCF to be improved and that in LCF to be degraded. Murty and Holland[15] examined the fatigue characteristics of Type 304 SS from hexagonal cans of EBR-II guide tubes before and after irradiation to a fast fluence of ~8 × 10^{26} n/m^2 (Fig. 1.12a). Tests were performed in four-point bend mode at constant displacements under symmetrical strain reversal fatigue at 0.1 cps and strains were varied from ~1% to 2.4% with the number of cycles to failure varying from 500 to 40 000. While a slight decrease in fatigue life is noted at high strains or low cycles, the data clearly revealed improved fatigue life at low strains or high cycles (Fig. 1.26) where the model predictions using universal slopes are correlated with experimental results. According to their model,

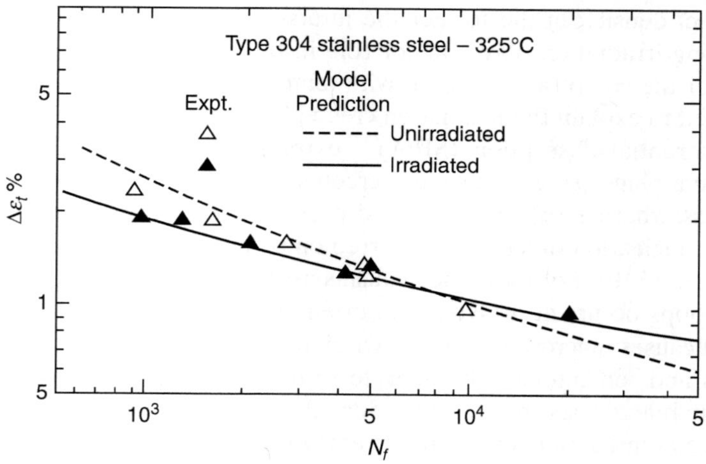

1.26 Strain amplitude versus number of cycles to failure at 325°C for unirradiated and irradiated 304SS depicting improved fatigue life in HCF and degradation in LCF.[15]

$$\Delta\varepsilon = \frac{2f}{\dot{\varepsilon}}\left[\left(\frac{\pi}{2}10^4 D^n \frac{\sigma_u}{E}\right)^2 N_f\right]^{-0.6} + 3.5\frac{\sigma_f}{E}\left(\frac{N_f}{10}\right)^{-0.12}, \qquad [1.26]$$

where f is the frequency of cycling, $\dot{\varepsilon}$ the strain-rate, D the ductility, n the work-hardening parameter, σ_u and σ_f are true tensile and fracture stresses and the remaining factors are as defined earlier.

Fatigue crack growth follows the same kinetics as described in section 1.2.4 but with K_{IC} replaced by K_{IC}^{irr} in Equation [1.14], similar to the case of environmental effects where K_{IC} is replaced by K_{ISCC}.

1.3.5 Corrosion-related problems

Corrosion is a major concern for reactor structures because in their construction many different materials are used which corrode at different rates by electrochemical effect and the corrosion (pitting, cracking, etc.) is accelerated by neutron radiation. More importantly, the corrosion products from steam generators, piping and other components are transported through the core and deposit on the fuel rods leading to formation of crud, in turn leading to increased fuel temperature and fuel failure. Corrosion can also lead to deposition of radioactive corrosion products on out-of-pile surfaces of the primary loop (e.g. heat exchanger) which becomes a safety concern for maintenance personnel. Further, the flow of the medium replenishes the concentration and pH at the corroding site which aggravates the corrosion (flow-assisted corrosion).

An important aspect in the degradation of the Zircaloy clads, the sole barrier between the hot fuel and coolant, is the oxidation and corrosion problem. The various oxidation processes in Zircaloys have led to major degradation phenomena that are described in detail in subsequent chapters. Stable, adherent oxide films form which act as a protective coating and offer resistance to environmental cracking as in the case of stainless steels. However, in the case of Zircaloys long exposures lead to the film flaking – that results in wall thinning – which in some cases may result in through-wall failures. More and specific details can be found in the chapters on Zr-alloys in Part II.

The addition of transition metals (Fe, Cr, Ni) to zirconium was aimed at reducing the severe oxide growth stresses and localized cracking of the grain boundaries that exposes more grains to the corrosive environment. Formation of second phase precipitates (SPPs) containing the transition elements help by aiding uniform oxidation of the grains and preventing localized cracking and spalling as the oxide grows.[37,39] It has been noted in Zircaloy-4 that the fraction of SPPs decrease and correspondingly the oxide thickness increases with fluence (Fig. 1.27).[40]

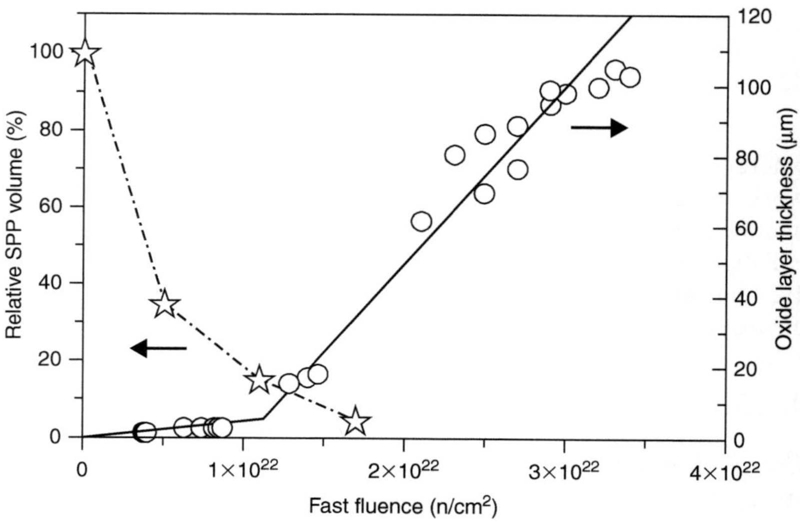

1.27 Dissolution of SPPs with fluence and the increase in oxide layer thickness under BWR condition in Zircaloy-4 at 290°C.[41]

A few factors under irradiation conditions are recognized as affecting the corrosion of clad material: (1) The increase in thickness of the oxide film (water side) decreases the thermal conductivity of the film and increases the temperature of the zirconium matrix which results in a higher corrosion rate;[41] (2) the changes that occur in the microchemistry of the Zircaloy-4 matrix due to irradiation are seen to accelerate weight loss (when compared with unirradiated material);[39,42] (3) accumulation of lithium in the oxide layer reduces the protective nature of the oxide at the metal oxide interface and enhances the corrosion rate;[43] and (4) high concentrations of zirconium-hydride are found in locations where the oxidation of Zircaloy is high.[41,44] The vast research work conducted over the years has led to some understanding of these problems: contrary to the expectation that irradiation-induced defects can cause breakdown of the protective film on Zircaloy clads,[45] the examinations of irradiated clads did not show any evidence of oxide damage;[46,47] though the exact mechanism of lithium induced corrosion of zirconium is not clear, the general understanding of the lithium effect is that Li gets incorporated in solid solution in the ZrO_2 and alters the vacancy concentration and distribution in the oxide film; since zirconium alloy corrosion proceeds by oxygen diffusion through the film, the increased number of vacancies should increase the diffusion and, hence, the corrosion rate; further, it has always been noted in irradiated Zircaloys that hydride density is high in locations where the oxide thickness is high; this clearly indicates that the cathodic hydriding and anodically favoured oxidation occur

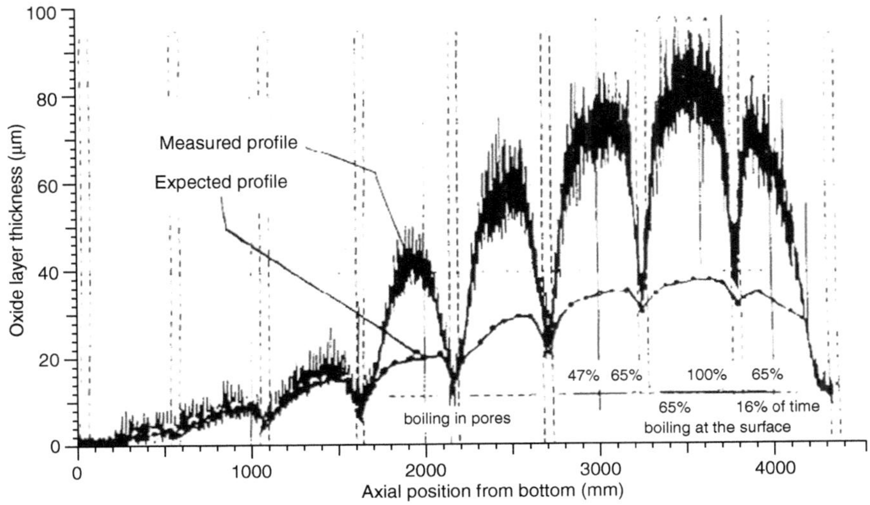

1.28 Temperature profile along the length of a PWR fuel clad results in increased oxide layer thickness.[48]

independent of each other. Change in fabrication route, which can result in the second phase with a different composition, can reduce the susceptibility to nodular corrosion but can lead to increased uniform corrosion.

The most commonly encountered corrosion types in the nuclear reactor are uniform corrosion, nodular corrosion and shadow corrosion. Uniform corrosion, as the name suggests, is associated with uniform oxide thickening and is commonly seen in PWRs and BWRs. Unlike the PWR environment where dissolved hydrogen is present in the coolant and the oxide films remained uniform over a very large thickness, the intermetallics present in Zircaloy, under the BWR environment, promoted nodular corrosion. The mechanism of oxide layer growth on Zircaloy under an irradiation environment is complex. Uniform corrosion starts with low burnup and the thickness of the grey oxide layer increases with burnup and operating temperature. Unlike in BWRs where the outlet temperature and pressure are limited, PWRs can operate at higher outlet temperatures but with the risk of increased corrosion, and this effect is explicitly seen from the increased oxide thickness with the elevation of the fuel rod. Figure 1.28[48] depicts the profile of the oxide thickness layer with the elevation in a typical PWR fuel rod, indicating the increased oxide layer thickness with increase in temperature. The increased turbulence in the coolant close to the spacer grids (which increases the cooling efficiency) and their parasitic absorption of neutrons result in the suppression of clad oxidation at these locations whereas the fuel rod temperature along the length, and hence the oxide thickness, is fairly uniform under a BWR environment.[49]

A more crucial parameter than metal coolant temperature is the metal oxide interface temperature which is difficult to measure but may be calculated with large uncertainty. A further complexity arises as the thermal conductivity reduces with burnup (due to penetration by the coolant into the porosity, cracking and spalling of the oxide films and crud deposition). The measurement of the thermal conductivity of the loose or non-adherent crud layers, which modifies the metal oxide interface temperature, is extremely difficult as the properties of subsequent layers deposited may not be the same. It is known that the thermal conductivity of the crud is higher than the zirconia layer or water or steam[50] which in a way increases the heat transfer characteristics.

Uniform corrosion occurs in both PWRs and BWRs. The oxide that forms is uniform in thickness, consists of several different layers and depends on many factors such as initial SPP size, extent of cold work and irradiation, alloy and water chemistries, temperature, local thermohydraulics, etc. The microstructure of the Zircaloys used in BWRs is continuously evolving, leading to dissolution of the SPP and formation of small and thin patches of white oxide on the otherwise black uniform oxide layer, which thicken at an accelerated rate. The sensitivity of nodular corrosion can be related to the second phase particles present in the alloy, though the number of nodules may not bear a one-to-one relation with the number of particles. Nodular corrosion is encountered in BWRs and starts appearing after a few to 100 days from the start of operation and usually saturates at higher exposure times. Nodules, in general, do not form in Zircaloys with small SPP sizes (<0.1 μm) but initiate early in materials with large SPPs and grow at a decreasing rate with fluence. Figure 1.29[51] shows the appearance of nodular corrosion on the fuel clad of a BWR fuel pin. The shape of nodular corrosion can be lenticular or spherical and growth in Zircaloy-2 decreases at high burnups.[52] The nodular corrosion problem can be eliminated (or delayed) by judiciously controlling the second phase particle sizes through appropriate β quench treatment although this may enhance uniform corrosion.[53] Nodules, whose thickness greatly exceeds the uniformly growing film, are prone to spalling and promote hydrogen pick-up. They can also be a cause of introducing zirconia particles to the coolant. Though PWR and WWER structures are not prone to the nodular corrosion attack, nodular corrosion can be a problem if steam forms at the oxide-coolant interface.[54]

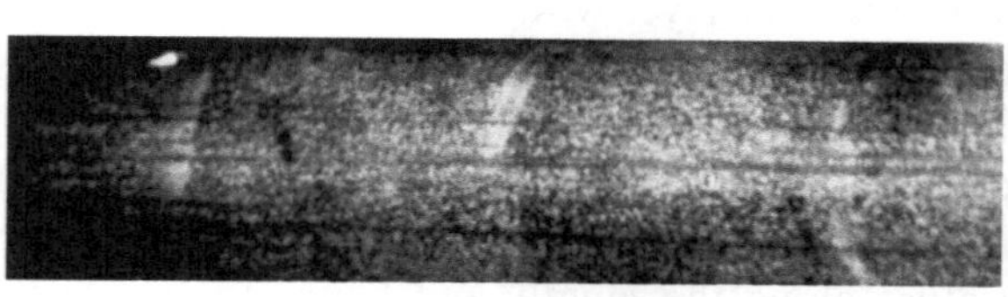

1.29 Nodular corrosion on the fuel clad of a BWR fuel pin.[51]

There is another type of localized corrosion in Zircaloys: the enhanced in-reactor corrosion when Zircaloy is placed close to a noble metal (under BWR conditions it is stainless steel or a nickel alloy), and where Zircaloy 'mimics' the noble metal corrosion. This is termed as 'shadow corrosion'.[55] The oxide thickness is unusually large and often appears to be particularly dense and uncracked. This localized corrosion is a special case of crevice corrosion and is predominantly seen in BWR components, although there is no direct electrical contact between Zircaloy and the material producing the shadow effect. The oxidation of H_2O_2 to HO_2+H^+ at the noble metal surface is balanced by the regeneration of H_2O_2 on the ZrO_2 surface and the coupling between the two metals (Zircaloy and nickel) is maintained by the ionic transport under a concentration gradient. The driving force is the potential difference between the two metals and radiolysis of water is required to sustain this reaction.[55] Shadow corrosion is invariably noted in BWRs and not in PWRs where the coolant is high in hydrogen concentration, which in turn reduces or eliminates galvanic potentials between dissimilar alloy components.

Environmentally assisted cracking is another manifestation of corrosion-related problems and is very often encountered in the steam and feed water piping as well as in condensate systems, RPV feed water nozzles and the secondary circuit of LWRs. This process is accelerated by stress (i.e. SCC) and neutron flux (i.e. IASCC). A typical fracture surface of IASCC is shown in Fig. 1.30.[56] Attempts are being made to reduce IASCC. Figure 1.31 shows the effect of hydrogen injection into the BWR environment on IASCC of 304 SS. The mechanism of crack growth mitigation by hydrogen injection could be explained by analyzing the corrosion potential of the system. The presence of molecules like H_2O_2 and O_2 increases the free corrosion potential which falls into the cracking range and hence the crack velocity is enhanced following the slip dissolution model and Faraday's law. Whereas, when hydrogen is introduced into the environment it helps the recombination of species and thus reduces the corrosion potential far below the cracking range.[57]

Austenitic stainless steels (e.g. blade sheathing in BWR) at high temperature and in a neutron-rich environment (>0.7 dpa), further influenced by higher oxygen levels in the water (BWR environment), exhibit IASCC. Other steels and nickel-base alloys also undergo IASCC at lower stress levels. Another aspect is that IASCC occurs in almost all materials and is known to occur in components at low stress levels. It is an expensive process to detect and repair the affected component. SCC is a major issue of PWR components like steam generator tubes, RPV penetrations, pressurizer nozzles, etc. While SCC can be controlled by modifying the water chemistry and the composition of the alloy, that is by replacing components with those resistant to SCC (e.g. Alloy 690, 52, 152), IASCC is more complex. Though

1.30 Intergranular fracture surface morphology of IASCC (304 SS, 3 dpa). Corrosion debris and cracks along grain boundaries can be seen.[56]

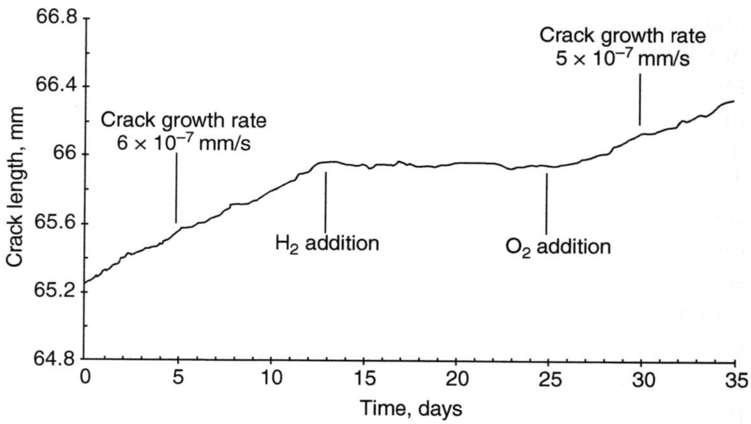

1.31 Typical reduction in crack growth rate by the addition of hydrogen in annealed 304 SS.[57]

both IASCC and IGSCC require external stress, temperature and dissolved oxygen in the water environment, the former is accelerated by neutron radiation. It is recognized that the influence of water chemistry becomes weaker and disappears at high doses (>50 dpa) suggesting that mechanical processes dominate chemical processes in IASCC.[58] The major distinctions between IASCC and other environmental cracking phenomena (e.g. SCC) are that in the former (1) the microstructure is modified by fast neutrons with time and (2) the chemistry of the environment is altered by the ionizing radiation. However, the overall stability of water increases with increasing temperature and the yields of molecular decomposition products (H_2, O_2 and H_2O_2) correspondingly reduced.[59] Austenitic stainless steel is the major material that has been the subject of IASCC investigation as compared to other grades. In the case of in-core structures, radiolysis increases the electrochemical potential in that region where the SCC susceptibility is high. Among the various radiolytic products, H_2O_2 is the most concentrated species present in irradiated, aerated water which gives rise to high corrosion potential for stainless steels. However, the critical potential to mitigate SCC of irradiated materials has not yet been established.[60,61]

Slow strain-rate tests have been carried out on type 304 stainless steel with prior thermal sensitization of the grain boundaries (to produce grain boundaries with chromium depletion) that show that the electrochemical potential of stainless steel increased significantly on irradiation in oxygenated water but decreased slightly in the hydrogen treated water. Though the mechanism is not fully understood, it is now realized that neither Cr depletion near grain boundaries[62] nor RIS (of S or P) at the boundaries[63] alone plays the detrimental role in IASCC. Further, the low stacking fault energy (SFE) of the matrix leads to localized deformation through dislocation channelling and irradiation has been found to accelerate the IASCC process.[64] Studies done under simulated BWR environments to examine the susceptibility of four steels with varying SFE under irradiation showed that the one with the highest SFE exhibited good resistance to cracking whilst that with the lowest SFE was seen to be susceptible to cracking at all of the doses studied[65] (Fig. 1.32). In 316 SS, the initiation and propagation of IASCC in a water environment depends on the dissolved hydrogen and the stress required decreases with (a) increase in dissolved hydrogen and (b) decrease in the rate of straining.[66] Corrosion problems are equally important in storage and disposal of nuclear wastes where long term safety and reversibility act as guidelines in designing the basic layout of a geological repository. Unlike conventional engineering structures, the ageing and degrading clad tubes should not only serve trouble free all through their service life but also maintain their integrity in repository conditions.[67]

1.4 Degradation mechanisms of specific nuclear reactor structures

Following the introduction to various degradation phenomena, fundamentals of radiation damage and radiation effects, we now turn our attention to

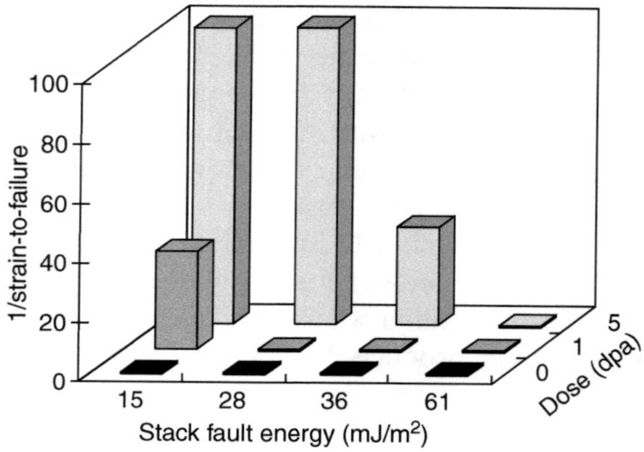

1.32 Effect of stacking fault energy and dose on the strain to failure.[66]

specific structures and materials that experience degradation during reactor operation.

1.4.1 Fuel

The fuel U-235, used in PWRs in its oxide form UO_2, is enriched up to 5% (maximum) and is used as dried pellets of about 10 mm in diameter and 10 mm in height with theoretical density (TD) ~95%. The stoichiometry of UO_2 is kept such that the ratio of O:U is never allowed to go beyond 2:1. The surface temperature of the fuel can reach ~1400°C with the centre temperature still higher. The oxide pellets are enclosed in a Zircaloy clad tube and the tube is capped on both sides to make a fuel rod. The fuel-clad gap is filled with high thermal conductivity helium gas and the conductivity degrades slowly with the fission gases diluting helium. Many such fuel rods (17 × 17) are bundled to form a fuel assembly. Many such fuel assemblies (~200) are immersed in a pool of light water, flowing at a pressure of ~16 MPa, which is the heat transfer fluid in the primary loop of a PWR. During start up, the pellet-clad gap gets reduced due to thermal expansion of the fuel but soon increases as the fuel densifies under irradiation. At high burnups, the gap slowly reduces and eventually an intimate contact is established with the fuel swelling and the clad collapsing under creep due to the coolant pressure (*clad creep-down*). All cladding tubes have some ovality which increases with creep-down and the direct impact of creep-down is the increased gap (between rods) and increased water volume. This increased water volume increases the moderation effect and gives rise to power peaks in the neighbouring fuel pellets.[68] The swelling of the fuel applies a severe hoop stress on the clad which slowly gets embrittled by irradiation and by absorption of hydrogen. The thin oxide layer on the clad breaks and bonding between fuel and clad becomes established, leading to a condition called pellet clad interaction (PCI). In extreme cases, these factors lead to cracking of the clad and to release of the fission gases into the coolant. Such situations may demand reduction of reactor power or the shutdown of the reactor if safe discharge limits are crossed. The strain concentration produced in the cladding by the sharp corners of radially cracked pellet during a power increase is increased by (i) the increasing coolant pressure, (ii) the pellet/cladding friction coefficient, (iii) the pellet radius and (iv) the circumferential temperature gradient. This severity can be marginally reduced by using a stronger clad material and by reducing the number of radial pellet cracks. For burnups greater than 70 MWd/kgU, a high burnup structure (HBS) with bubble size much larger than the grain size forms and traps the fission gases. A rapid rise in temperature can lead to shattering of the HBS and can put severe stress on the cladding.[69]

1.4.2 Cladding

The standard cladding material in a LWR is a dilute zirconium-base alloy containing some other elements such as tin, niobium, iron, nickel, chromium and oxygen (Zircaloy-2 for BWR and Zircaloy-4 for PWR). Being a hexagonally close packed (HCP) crystal structure and hence inherently anisotropic, zirconium acquires further anisotropic properties after fabrication due to induced texture (Fig. 1.25b) with the <c> axis of the HCP crystal oriented at ~30° from the radial direction of the tube. Minor modifications in the chemistry of the alloys are made to reduce the water side corrosion in the clad tubes. Formations of inter-metallic precipitates (which increase the corrosion rate) are avoided by giving the clad material a beta quench (fast cooling from the beta phase). The fuel and the pressure boundary (clad tube) experience time-related ageing and degradation; the former may affect linear power rating while the latter can lead to catastrophic clad failure. The burning of the fuel leads to release of fission gases and to fuel swelling. The fuel makes contact with clad which has picked up hydrogen from the coolant and leads to degradation of the clad tube.[70] In order to improve the structural rigidity, spacer grids made of Inconel/Zircaloy are placed at definite intervals along the length of the assembly which holds the fuel elements in the assembly with spring forces. When these forces relax (due to creep) a gap is created between the grid and rod, and the rod can vibrate. This fretting may result in a breach in the clad integrity.[71]

Hydride related problems in clad

Hydride related problems are viewed as an issue to be considered for clad failure. The main sources of hydrogen for the clad are: the corrosion reaction of metal with water, hydrogen released by radiolysis of water and hydrogen gas that is added in the coolant to keep the oxygen potential low.[72] The defects present in the clad (such as manufacturing defect, PCI crack, debris fretting, etc.) can aid the pick-up of hydrogen and the coolant could surge in through these defects when they grow through-thickness and form steam. The steam reacts with the fuel and hydrogen is released. When the hydrogen-to-steam ratio crosses a critical value (steam starvation), the growing oxide layer on the ID of the tube finally breaks down and the hydrogen diffuses into the matrix of the tube. The hydrogen thus picked up can reduce the toughness of the zirconium matrix in three ways: (i) hydride reorientation, (ii) delayed hydrogen cracking and (iii) formation of a hydride blister.

The solubility limit of hydrogen in zirconium at the reactor operating temperature is about 100 ppm. When the temperature is reduced (for instance, during reactor shutdown), the excess hydrogen precipitates in the form of hydride. The hydride precipitates along the radial direction of the tube owing to the texture and the hoop stress in the tube. The hoop stress required for reorientation (in an unirradiated and recrystallized Zircaloy-2) is about 80 MPa.[73] The differential temperature between ID and OD of the clad wall (either during service or at wet repository) drives the hydrogen to the OD side which is at a lower temperature. The concentration of hydrides found in the tube after irradiation is higher near the water side than at the fuel side which is attributed to the corrosion reaction between the clad OD and the coolant.[74] The threshold stress for failure of irradiated and hydride-reoriented spent fuel cladding is significantly higher than the stress due to the internal pressure of the fuel rod. The degree of oxidation and hydriding in the more advanced fuel claddings commonly used these days in LWRs, such as low-Sn Zircaloy-4 (Sn content around 1.3 wt.%), optimized Zircaloy-4, Zirlo (a Zr-1Sn-1Nb-0.1Fe alloy), M5 (a Zr-1Nb alloy) and optimized Zircaloy-2, is relatively low even at high burnup.

Frequently, a hydride blister is produced when a fuel rod that contains spalled oxide is operated continuously to high burnup. During steam starvation, hydrogen ingress is faster than its diffusion into the tube matrix. This leads to excess amounts of hydrogen getting localized at the inner wall of the clad tube forming a large hydride called a blister. The hydrogen atoms, diffusing down the temperature gradient, form radial hydrides in a sun-burst pattern.

Delayed hydrogen cracking (DHC) is important for spent fuel in either wet or dry repository and is a two-step process. Hydrogen migrates up the stress gradient towards a stressed crack-tip and precipitates as hydride that cracks and extends further. There is an incubation time for the hydrogen to arrive and the concentration to build to the required level, so that the solubility limit is exceeded for the hydrides to form and grow, before the crack extends further. The crack-tip can undergo a corrosion reaction and the hydrogen released can either be absorbed by the matrix close to the crack-tip or the hydrogen can diffuse through the matrix to the crack-tip. The former is called corrosion hydrogen cracking and the latter is known as DHC. Knowing the crack velocity enables prediction of the failure time of the tube. Since there is no way to measure the crack velocity in the reactor, it is assumed that the crack starts at the centre of the tube and proceeds in both directions with a velocity in the range of 2.5×10^{-7}–6.6×10^{-7} m/s which is determined from out-of-pile unirradiated laboratory specimens.[75] It is possible that the velocity may be much higher in reactor as the stress state is more severe. The stress arises partly from the increased pellet volume because of increased temperature (due to reduced thermal conductivity as the pellet cracks up with burnup) and partly from the increased volume of the oxide layer of the Zr-liner on the interior of the tube. The increase in stress due to this volume expansion is faster than the creep relaxation by the clad.

Fuel assembly bow

Various reports of incomplete control rod cluster assembly insertion triggered investigations to identify causes of the sticking problem. The root cause was understood to be excessive deformation (bowing) of fuel assemblies. When the bowing exceeds the limit, it increases friction between the control rod and the guide thimble and can result in the breaking of the control rod cluster assembly.[72] The mechanism of bow, though not clearly understood yet, is believed to be caused by creep affecting the overall assembly and guide thimble. If the fuel assembly experiences a flux gradient, the tubes at the lower flux side will grow less than those in high flux side causing the fuel rod to bow. Provision is made in the design, to accommodate an increase in length of the fuel rod (due to creep and growth) on either end and any restriction in the free movement leads to bowing of the rod. Rods under cold-worked stress-relieved (CWSR) conditions show more elongation than those under fully recrystallized conditions.[72]

Creep of fuel cladding

Fuel failure occurs when the cladding barrier is degraded and breached. The fuel rod failure rate in LWRs has been significantly reduced since 1987.[75] This is due, besides design improvements, to the introduction of many improved variants of Zr-base alloys over the years. With the recent developments in LWR fuels, Zr-1%Nb-Sn-Fe alloys with higher resistance to irradiation-induced growth, creep and corrosion, are being used for guide tubes and for fuel rod cladding with extended residence time (5–6 years).[76] At low burnup, the pellet densifies and the external water pressure causes the clad tube to creep-down. On power ramp, the pellet expands and applies excess strain on the clad. This leads to the pellet touching the clad and results in PCI failure or hydride related cracking. The sheath should have good creep rupture property not to fracture. Further, the creep of the fuel assembly and guide thimble can lead to bowing of the assembly.[77] An analysis performed at Ringhals concluded that the bowing in this reactor had been caused by a large creep deformation due to excessive compressive forces of the hold down spring on the fuel assemblies, with a decrease in lateral stiffness. The proposal to introduce advanced cladding and guide thimble materials with a low growth rate and higher creep resistance to improve the dimensional stability of assemblies is also being considered (M5 by Areva and ZIRLO of Westinghouse);[78] further details can be found in Part II on Zr-alloys.

The creep behaviour of unirradiated material is taken as a benchmark to postulate its performance in reactor. Though these out-of-pile tests may not be representative of their in-reactor behaviour, they have been used successfully to grade various materials during alloy development programmes and to gain a basic understanding of the material behaviour. It has been recognized that hoop strain in a clad tube (in-pile or during spent fuel storage) is a vital parameter in the breach of fuel clad and evaluation of their creep and burst behaviours is very important to assess the integrity of the tube. Biaxial creep[79] and burst[80] characteristics are usually studied using internally pressurized tubing over a range of pressures and temperatures. One may also use ring–creep[81] tests to characterize hoop creep behaviour under uniaxial hoop loading; this might be advantageous in cases where only a limited amount of material is available and also in evaluating radiation effects that require relatively small size samples. It becomes relatively more complex for Zircaloys that exhibit distinct textures leading to anisotropic deformation and creep[82] that need to be accounted for, along with possible radiation effects, in predicting the dimensional changes in reactor. As demonstrated by Murty and co-workers, the relatively weak hoop direction for CWSR material became stronger following recrystallization annealing, illustrating the profound effect of heat treatment on the creep anisotropy of Zircaloys.[83] The observation that under equi-biaxial stress state the secondary slip systems (basal and pyramidal) are also favoured along with the easier prism slip, concur with the observation that the irradiated recrystallized Zircaloy exhibited a creep locus similar to that of isotropic material (texture reduced as all slip systems were favoured). The recent work on the thermal creep of Zr-2.5%Nb alloy by Kishore *et al.*[84] indicates that a microstructure containing a stable phase creeps faster than the one with a meta-stable phase and a phase redistribution is established. The stable β phase (80 wt%Nb) dissolves during creep deformation and re-precipitates as meta-stable β phase (~35 wt%Nb), this phase change adding to the creep strain. Moreover, transients during sudden load drops or changes in temperature need to be considered in proper predictions.[85] The importance and significance of studying the transitions in creep mechanisms along with microstructural characterization following deformation were clearly outlined by Gollapudi *et al.*[86] and the reader is referred to Chapter 3 on Creep for further details.

Role of hydrogen on creep

In Zircaloy-4, the creep rate was reported to depend on the condition of the material – whether in CWSR or annealed condition; CWSR alloy showed a significant strengthening upon addition of hydrogen. The reason for this behaviour is attributed to hydrogen influencing strain-hardening rate and static recovery of the material. Biaxial tests in Zircaloy-4 showed that the presence of hydrides in the cladding will help to prevent the cold work microstructure from being annealed out of dislocations and thereby maintain lower creep rates in the spent fuel cladding.[87] The same alloy in annealed condition showed an increase in creep rate when the hydrogen is in solution and a decrease when part of hydrogen is precipitated as hydrides. This behaviour is attributed to the reduction in the stacking fault energy due to diffusion of hydrogen to the core of the screw dislocations and their increased mobility. On the other hand, when the hydrogen is present in the form of hydrides, it increases the matrix strength and reduces the creep rate. Other researchers noted that the increase in creep due to hydrogen content of around 200 wt ppm may be due to the reduction in the modulus value when hydrogen was added.[88] In a Zr-2.5 wt%Nb alloy, the creep rate at 723 K was reported to increase by 2–2.5 times for a hydrogen content of 160 wt ppm and the stress exponent reduced from 2.41 to 1.59, indicating a change in the creep mechanism.[89]

Since dry storage of spent fuel is gaining importance, it is necessary to assure clad integrity during interim storage. The high burnup rods are likely to contain large amounts of hydrogen (1000 ppm) and with a hoop stress of 100–120 MPa the clad should not creep to failure. In order to reduce the hydride problem the initial level of hydrogen (and other impurities) is kept low and pickup during service is controlled by using new corrosion resistant alloys.

Fuel failure data (PWRs and BWRs)

From the vast reactor operating experience it is noted that the cause of fuel failures in terms of the number of units with leakers has decreased over the years (Fig. 1.33) and the US nuclear industry has been tending toward 100% no-leakers. As of July 2010, more than 90% of units in the

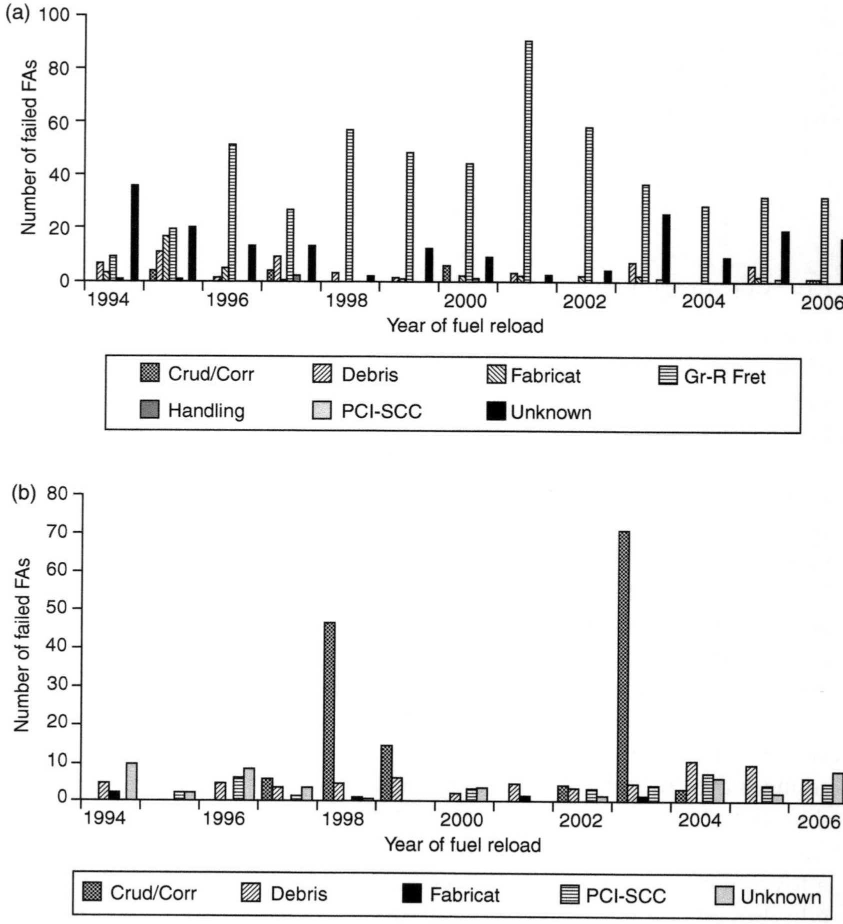

1.33 **Fuel leaker causes in (a) US PWRs and (b) world-wide BWRs.[90]**

United States were failure-free.[90] A recent review of the fuel failures in LWRs by the International Atomic Energy Agency (IAEA) revealed the following causes:[91] crud/corrosion, debris, PCI/SCC, grid-to-rod fretting (GTRF), fuel handling, fabrication with some 'unknown' where the cause could not be pinpointed. Interestingly GTRF was seen to be the major issue confronting the PWR industry, accounting for around 50% of the total fuel failures. Figure 1.34a and 1.34b summarize the fuel leak causes in PWRs in the United States and BWRs across the world, respectively. Debris (31%), crud/corrosion (32%) and unknown (24%) accounted for

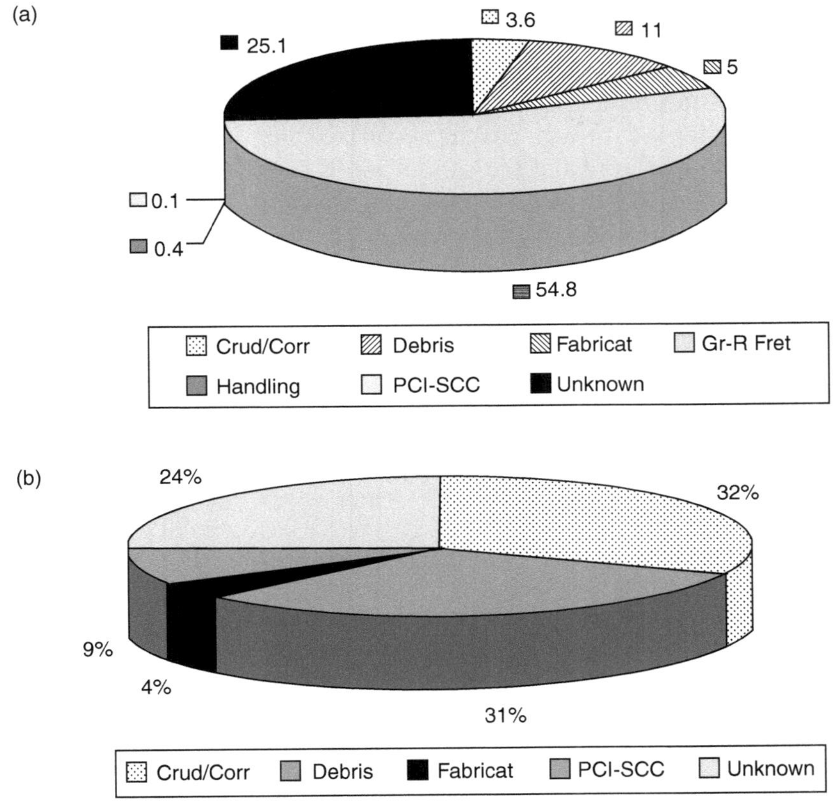

1.34 Fuel leaker causes in (a) PWRs and (b) BWRs.[92]

around 87% of fuel failures in BWRs while a relatively low proportion (9%) was noted to be due to PCI/SCC. Many of these issues are dealt in detail in Part II of this volume.

1.4.3 PWR internals

The internals (flux thimble tube, lock bars, baffle bolts and re-entrant corners) are made from austenitic stainless steels of type SA304, 316 or 347. The insignificantly low swelling behaviour, observed from the limited tests conducted in environments that are more severe than those of a PWR, is believed to be associated with irradiation-induced formation of very fine precipitates (such as G phase, carbides and γ' phase) in high number density whose interfaces act as efficient sinks to irradiation-induced vacancies and thereby the agglomeration of the vacancies is suppressed.[92]

1.4.4 Reactor pressure vessel (RPV)

The RPV of a PWR is a cylindrical vessel with two hemispherical shells – one at the top, bolted with flange, and one at the bottom, welded – which contains the core through which pressurized light water is circulated at an average temperature and pressure of 300°C and 16 MPa, respectively. The inner surface of the cylindrical vessel is lined with a thin layer (thickness around 3–10 mm) of austenitic steel (SAE 308/309) to protect the vessel from corrosion. The vessel is made from plates of low alloy steel (typically ASTM SA302/SA533B or ASME SA508) with a thickness of about 225 mm. Although many materials are acceptable for reactor vessels according to Section III of the ASME Code, the special considerations pertaining to fracture toughness and radiation effects limit the basic materials for most parts of vessels to SA533 Grade B Class 1, SA508 Class 2 and

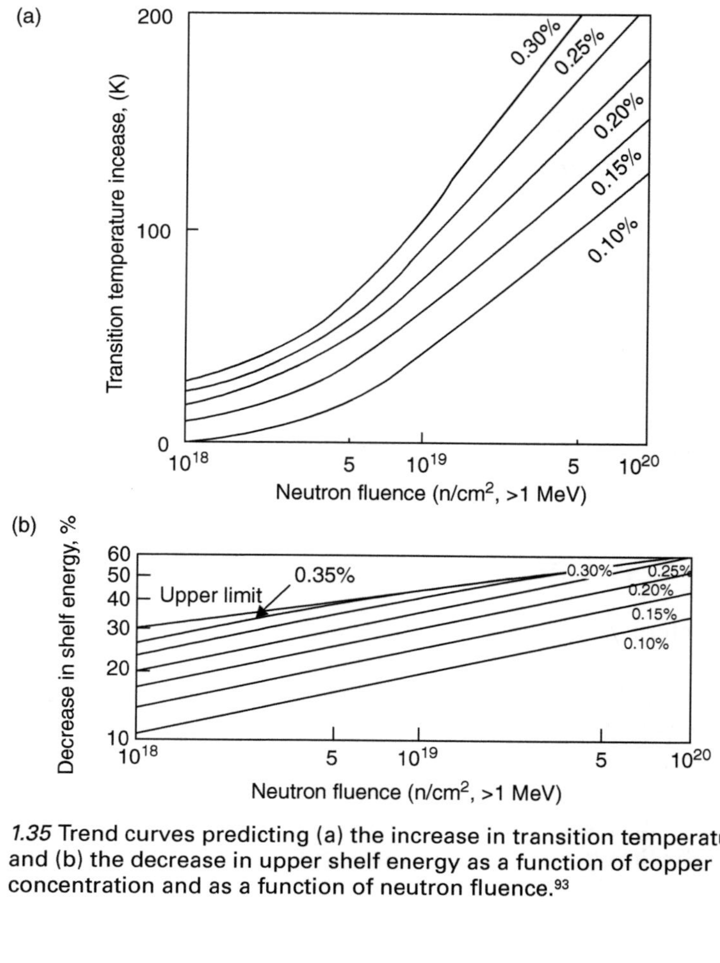

1.35 Trend curves predicting (a) the increase in transition temperature and (b) the decrease in upper shelf energy as a function of copper concentration and as a function of neutron fluence.[93]

SA508 Class 3. Creep *per se* does not pose any safety related problem to RPVs.

A major issue with the ferritic steels used for RPV applications, as described earlier, lies in the increase in DBTT and decrease in the upper shelf energy due to radiation exposure and these factors depend on the concentration of alloying and/or impurity elements. Sensitivity of radiation embrittlement of ferritic steels to the concentration of copper is illustrated in Fig. 1.35a and 1.35b that depict the effects of copper composition on the increase in transition temperature and decrease in the upper shelf energy, respectively, with neutron radiation dose.[93] Welds and HAZ are relatively more sensitive to radiation exposure than the base metal, the reasons being variations in composition, microstructures, etc. Detailed knowledge gathered on the effects of alloying compositions (in particular Cu, S, P and Ni) on radiation embrittlement of RPV steels makes it possible to design materials for new systems to be devoid of these issues that are confronting the currently operating reactors where the fracture characteristics in the weld materials are degraded, mainly by impurities such as Cu. There have been numerous studies to understand the underlying micro-mechanisms responsible for the observed radiation embrittlement of RPV steels and the role of alloying and impurity elements.[94,95] Recent emphasis has been on atomistic modelling along with the characterization of defects using advanced microstructural evaluation techniques such as HRTEM, atom probe microscopy, small angle neutron scattering (SANS), etc.[96]

In the reactor vessel surveillance programmes (RVSPs), samples taken from the base, weld and HAZ of the actual vessel material used during its construction are included in a capsule that is placed closer to the reactor core so that the samples withdrawn after different neutron dose levels can be tested (tensile, Charpy and fracture toughness). The results from these surveillance samples provide information regarding the degradation of the real structure and corrective action can be taken before any major damage occurs. While hardness and tensile tests are routinely performed to get an idea on the effects of neutron irradiation, the effect of radiation dose on RT_{NDT} and upper and lower shelf energies through Charpy tests are significant in evaluating the radiation embrittlement. The RVSP capsules are taken out at intervals during reactor operation and changes in the properties of the samples are monitored to make sure that these changes are less than those prescribed by the NRC regulation guide (10CFR50). In cases where the results reveal degradation greater than the limit prescribed by the regulation guide, the reactor vendor/utility needs to take appropriate actions to demonstrate the safety of continued operation of the reactor so that the RPV does not fail in a brittle mode. It is to be noted that the Charpy tests do not yield fracture toughness (K_I) data which are related to the crack length (see, e.g. Equation [1.9]) and the specimen size required for obtaining valid K_{IC} tests, as in the case of low strength RPV steels, is too large to be practical to investigate radiation effects. Thus, efforts are being put in to correlating the C_V values obtained from standard Charpy tests to fracture toughness evaluated using compact tension (CT) and/or elastic-plastic toughness tests (J_{IC}, etc.).[97] The master curve approach provides an alternative transition temperature index parameter to the RT_{NDT} data measured from Charpy tests. This new parameter, defined as RT_{T_0},[98] is based on a simple addition of 19.4°C (35°F) to the value of T_0 evaluated according to ASTM E 1921. The advantage of this approach is that RT_{T_0} can be measured directly on irradiated samples rather than having to measure initial properties and then add the transition temperature shift.[99] It is also worthwhile considering 'dynamic' values such as dynamic fracture toughness (K_{Id}) which are sensitive to applied strain-rate and which are of importance during accidents such as loss of coolant (LOCA); K_{Id} is generally determined using pre-cracked samples by instrumented Charpy impact tests[100] though these are not routinely considered in RVSP schedules. Further details on the reactor vessel integrity are included in a later chapter in Part II.

Creep of RPV and internals

For RPV steels which undergo a damage of about 0.1 dpa, deterioration due to irradiation creep is much less in comparison to toughness loss. But creep crack growth studies indicate that the HAZ, with a different microstructure and coarser grain size than the base metal, can lead to lower life after prolonged neutron exposure in the temperature range 320–420°C.[101] Many of the components of PWR internals (screws, core barrel and baffle assembly) are made of austenitic stainless steels and undergo an average damage rate of about 1 dpa/year (=5×10^{13} n/cm^{2}s) at a temperature which may reach a maximum of 400°C due to gamma heating. They undergo irradiation-induced creep and stress relaxation.[102]

In some PWRs the core baffle consists of sheets and formers. The sheets are separated by small gaps (0.2–0.4 mm). The connection between the core baffle sheets and the formers, and between the formers and core barrel is completed by a large number of bolts (about 900). During the core baffle manufacturing process the bolts are tightened with well defined pre-stress to guarantee the geometrical and mechanical stability of the structure. During operation the pre-stress of the bolts becomes reduced as a consequence of thermal and mechanical loads aided by neutron irradiation which can possibly affect the flow induced vibrations of fuel rods in the outer fuel elements.[103] Biaxial creep rates measured in solution annealed (SA) 304L (used as baffle plates) and cold-worked 316 (used as bolts) in the temperature range 280°C to ≈380°C and irradiated to a dose level of 120 dpa indicate that SA 304L creeps faster than CW 316.[104] These results suggest that the correct grade of steel and optimum metallurgical conditions can reduce the creep rate and prolong the life of these components.

1.5 Conclusions

Structural components in NPPs undergo ageing with continuous operation and eventually reach the end of life. The rate of degradation depends on their inherent ability to withstand the stress, temperature and service environment. To get the best potential from a material the acumen of the designer, the alacrity of the operator and dexterity of the surveillance personnel should play a non-compromising role. The engineering structures in a NPP can be broadly classified into two categories: (i) the components of steam generators, turbines, etc., which experience thermal and mechanical environment and (ii) in-pile components such as fuel clad, reactor pressure vessel, etc., which are subjected to an added condition of intense neutron irradiation. Materials in both categories also have to face high-temperature flowing water, the energy transfer medium, which corrodes/erodes the material. The feedback data on the performance of materials in these environments help material scientists to modify the materials and to manoeuvre their properties to perform better. This closed cycle needs to be kept active to meet the required technological advancements.

The properties of materials used in LWR power plants are evaluated for the service they have to render: a fluctuating load requires material with good fatigue strength, constant pressure at elevated temperature demands good creep strength and stress relaxation, good toughness is needed even after prolonged neutron irradiation, low tendency to absorb hydrogen so as to minimize hydrogen related problems, etc. It is difficult to have one material endowed with all these properties and hence more than a dozen materials are used inside a reactor – these need to be joined in some way and this adds to corrosion-related problems.

The elastic and plastic deformations of a material, whilst obeying a generic relationship, will show a marginal difference in their properties because of its metallurgical condition. The constants used in these relationships are material- or microstructure-specific. Despite such variability it is still possible to isolate a material with the required microstructure to serve under specified environmental conditions, and above all, for a known life expectancy. An indication of the crack length in a material helps to keep a check on its degradation if its fracture toughness property is known. Charpy impact tests provide an easier alternative to LEFM tests and are used to grade the deterioration of the material. In situations where the initial toughness of a material is unknown for comparison, the master curve technique is convenient to evaluate the irradiation embrittlement of steels. The growth rate of a crack can be estimated from the known relation between the crack length and applied stress. As irradiation is known to benefit HCF and, as the material behaviour under HCF is well understood, a prudent design for longer life becomes possible. Knowledge on the creep rate of a material alerts for corrective measures as the dimensional changes are predictable. The activation energy for creep indicates which metallurgical parameter is crucial in limiting the life. Resolving the yield stress into a source hardening and frictional terms helps understanding of the flow response of the material to nuclear irradiation. It is now known that synergistic effects of neutron irradiation and DSA could lead to beneficial effects on strength and ductility in certain temperature and strain-rate regimes. By making a judicial choice of the temperature and fluence, a steel can be safely used in the blue brittleness range. Understanding the metallurgical treatment and the material response has helped in choosing the right material such as SA 304L instead of CW 316 for better creep resistance for baffle plates. In Zr-2.5%Nb alloy, the stable β phase (80%Nb) is seen to be less creep resistant than the β phase (35% Nb) and the pressure tubes (in Pressurized Heavy Water Reactors (PHWRs)) can have a longer life with this modification.

Corrosion is another major problem in nuclear reactors. Uniform, nodular and shadow corrosion that affects the reactor components, and which are not influenced by any external stress, are controlled by modifying alloy and water chemistries. Routine surveillance test programmes enable better understanding of material behaviour. This has helped to substitute some of the components which suffer from SCC with those having better resistance (e.g. Alloy 690, 52,152). IASCC is known to occur in almost all materials and in components at low stress levels and this phenomenon is yet to be understood well to come out with effective solution.

This chapter serves as an introduction to the various materials degradation phenomena as summarized above while the subsequent chapters dwell on various details with Part I on various fundamental phenomena, Part II on specific and varied components of LWRs while Part III covers management strategies adopted by various nuclear utilities/vendors.

1.6 References

1. 'Light Water Reactor Sustainability Research and Development Program Plan, Fiscal Year 2009–2013', Idaho National Laboratory Idaho Falls, Idaho 83415, p. 7; http://www.inl.gov; Prepared for the U.S. Department of Energy Office of Nuclear Energy Under DOE Idaho Operations Office Contract DE-AC07-05ID14517.

2. C.O. Smith, *Nuclear Reactor Materials*, Addison-Wesley, Reading, MA, USA, 1967.
3. J.T.A. Roberts, *Structural Materials in Nuclear Power Systems*, Plenum Publishing, New York, NY, USA, 1981.
4. H.G. Rickover, *History of Development of Zirconium Alloys for Use in Nuclear Power Reactors*, US ERDA, NR&D,Washington, DC, USA, 1975; see also R. Krishnan and M.K. Asundi, 'Zirconium alloys in nuclear technology', in *Ally Design*, S. Ranganathan, V.S. Arunachalam and R.W. Cahn (eds), Indian Academy of Sciences, Bangalore, India (1981) 139–154.
5. G.S. Was, *Fundamentals of Radiation Materials Science*, Springer, New York, NY, USA, 2007.
6. G.R. Odette and G.E. Lucas, 'Recent progress to understanding reactor pressure vessel embrittlement', *Radiation Effects Defects Solids*, **144** (1998) 189–231.
7. R.G. Carter, N. Soneda, K. Dohi, J.M. Hyde, C.A. English and W.L. Server, 'Microstructural characterization of irradiation-induced Cu-enriched clusters in reactor pressure vessel steels', *J. Nucl. Mater.*, **298** (2001) 211–224.
8. M. Meyers and K. Chawla, *Mechanical Behavior of Materials*, Cambridge University Press, New York, 2009.
9. G.E. Dieter, *Mechanical Metallurgy*, McGraw-Hill, New York, 1988.
10. J. Roester, H. Harders and M. Baeker, *Mechanical Behaviour of Engineering Materials*, Springer, New York, 2006.
11. D.M.R. Taplin and A.L. Collins, Fracture at high temperatures under cyclic loading', *Ann. Rev. Mater. Sci.*, **8** (1978) 235.
12. L.F. Coffin, 'A note on low cycle fatigue laws', *J. Mater.*, **6** (1971) 388–402.
13. S.S. Manson, 'Challenge to unify treatment of high temperature fatigue – partisan proposal based on strain-range partitioning', in *Symposium on 'Fatigue at Elevated Temperatures'*, ASTM STP 520 Storrs, CT, American Society for Testing and Materials (1972) 744–782.
14. S. Majumdar and P.S. Maiya, 'A unified and mechanistic approach to creep-fatigue damage', in *Proceedings of the International Conference on Mechanical Behavior of Materials*, ICM-II, American Society for Metals, Metals Park, Ohio (1976) 924–928.
15. K.L. Murty and J.R. Holland, 'Low-cycle fatigue characteristics of irradiated 304SS', *Nucl. Technol.*, **58** (1982) 530–537.
16. L.K. Mansur, 'Theory and experimental background on dimensional changes in irradiated alloy', *J. Nucl. Mater.*, **216** (1994) 97–123.
17. M. Kiritani, 'Microstructure evolution during irradiation', *J. Nucl. Mater.*, **216** (1994) 220–264.
18. D.R. Olander, 'Fundamentals aspects of nuclear reactor fuel elements', National Technical Information Service, U.S. Department of Commerce, Springfield, VA, 1976.
19. M.J. Makin and F.J. Minter, 'Irradiation hardening in copper and nickel', *Acta Metall.*, **8** (1960) 691–699.
20. K.L. Murty, 'Role and significance of source hardening in radiation embrittlement of iron and ferritic steels', *J. Nucl. Mater.*, **270** (1999) 115–128.
21. T.S. Byun and K. Farrell, 'Plastic instability in polycrystalline metals after low temperature irradiation', *Acta Mater.*, **52** (2004) 1597–1608.

22. I. Charit, C.S. Seok and K.L. Murty, 'Synergistic effects of interstitial impurities and radiation defects on mechanical characteristics of ferritic steels', *J. Nucl. Mater.*, **361** (2007) 262–273.
23. K.L. Murty and D.J. Oh, 'Friction and source hardening in irradiated mild steel', *Scripta Metall.*, **17** (1983) 317–320.
24. K.L. Murty, 'Strain-aging behavior of irradiated and denitrided mild steel', *Mat. Sci. Eng.*, **59** (1983) 207–215.
25. K.L. Murty and I. Charit, 'Static strain aging and dislocation–impurity interactions in irradiated mild steel', *J. Nucl. Mater.*, **382** (2008) 217–222.
26. J.E. McLennan and E.O. Hall, 'The carbon levels in radiation damaged and strain-aged mild steel', *J. Aust. Inst. Met.*, **8** (1963) 191–196.
27. K.L. Murty and E.O. Hall, 'Dynamic strain aging and neutron irradiation in mild steel,' in *Irradiation Effects on the Microstructure and Properties of Metals*, St. Louis, ASTM STP 611 F.R. Shober (Chairman) American Society for Testing and Materials, Philadelphia (1976) 53–71.
28. K.L. Murty, 'Is neutron radiation exposure always detrimental to metals (steels)?' *Nature*, **308** (1984) 51–52.
29. K.L. Murty, 'Beneficial (!) effect of neutron irradiation on mechanical properties of mild steel in blue brittle range', *Scripta Metall.*, **18** (1984) 87–89.
30. S.B. Kass and K.L. Murty, 'Effect of neutron irradiation on mechanical properties of ferritic steels', in *Proceedings of Second International Conference on Microstructures and Mechanical Properties of Aging Materials*, P.K. Liaw *et al.* (eds) TMS (1996) 27.
31. G.R. Odette and G.E. Lucas, 'The effect of nickel on irradiation hardening of pressure vessel steels', in *Effects of Radiation on Materials*, N.H. Packan, R.E. Stoller and A.S. Kumar (eds), ASTM STP 1046, Vol. 2 (1990) 323–347.
32. C. Guionnet, Y. Robin, C. Flavier, A. Lefort, D. Gros, R. Perdreau, 'Radiation embrittlement of a PWR vessel steel: effects of impurities and nickel content', in *Effects of Radiation on Materials*, H.R. Brager and J.S. Perrin (eds), ASTM STP 725 (1981) 20–37.
33. Y.H. Jung and K.L. Murty, 'Effect of interstitial impurities on fracture characteristics of A533B class1 pressure vessel steel,' in *13th International Symposium on Influence of Radiation on Mechanical Properties*, ASTM STP 956 American Society for Testing and Materials (1987) 395–407.
34. J. Hong, R. Zou, J. Britt and K.L. Murty, 'Role of friction and source hardening in DSA and radiation embrittlement of ferritic steels and iron,' in *Proceedings of SMiRT-12*, paper F09/3 (1993).
35. L.K. Mansur, 'Irradiation creep by climb-enabled glide of dislocations resulting from preferred absorption of point defects', *Phil Mag.*, **A39** (1979) 497–506.
36. A.D. Brailsford and R. Bullough, 'Irradiation creep due to the growth of interstitial loops', *Phil. Mag.*, **A27** (1973) 49–64.
37. L.K. Mansur, 'Irradiation creep by climb-enabled glide driven by transient point defect processes', *Mater. Sci. Forum 97–99 Trasn. Tech. Publication*, Switzerland (1992) 489–498.
38. F.A. Nichols, 'Effect of high neutron exposures on properties of materials', *Annu. Rev. Mater. Sci.*, **2** (1972) 463–500.
39. K.L. Murty and J.R. Holland, 'Low cycle fatigue characteristics of irradiated type 304 stainless steel', *Nucl. Tech.*, **58**:3 (1982) 530–537.

40. B. Cox, 'Some thoughts on the mechanisms of in-reactor corrosion of zirconium alloys', *J. Nucl. Mater.* **336** (2005) 331–368.
41. F. Garzarolli, H. Ruhmann, and L. Van Swam, 'Alternative Zr alloys with irradiation resistance properties for high Burnup BWR applications', in *Zirconium in the Nuclear Industry: Thirteenth International Symposium*, ASTM STP 1423, G. Moan and P. Rudling (eds), ASTM International, West Conshohocken, PA (2002) 119–132.
42. B.-C. Cheng, R.M. Kruger and R.B. Adamson, 'Corrosion behavior of irradiated zircaloy', in *ASTM STP 1245 Zirconium in the Nuclear Industry: Tenth International Symposium*, A.M. Garde and E.R. Bradley (eds), American Society for Testing and Materials, Philadelphia, PA (1994) 400–418.
43. D. Pecheur, J. Godlewski, P. Billot and J. Thomazet, 'Microstructure of oxide films formed during the waterside corrosion of the zircaloy-4 cladding in lithiated environment,' in *Zirconium in the Nuclear Industry: Eleventh International Symposium*, ASTM STP 1295, American Society for Testing and Materials, Philadelphia, PA (1996) 94–113.
44. A.M. Garde, 'Enhancement of aqueous corrosion of zircaloy-4 due to hydride precipitation at the metal-oxide interface,' in *Zirconium in the Nuclear Industry: Ninth International Symposium*, ASTM STP 1132, American Society for Testing and Materials, Philadelphia, PA (1991) 566–594.
45. B. Cox, 'Effects of irradiation on the oxidation of zirconium alloys in high temperature aqueous environments: a review', *J. Nucl. Mater.*, **28** (1968) 1–47.
46. E. Hillner, 'Long-term in-reactor corrosion and hydriding of zircaloy-2 tubing,' in *Zirconium in the Nuclear Industry: Fifth Conference*, ASTM STP 754, American Society for Testing and Materials, Philadelphia, PA (1982) 450–478.
47. F. Garzarolli, T. Broy and R.A. Busch, 'Comparison of the long-time corrosion behavior of certain Zr alloys in PWR, BWR, and laboratory tests', in *Proceedings of the 11th International Symposium on Zirconium in the Nuclear Industry*, ASTM-STP-1295, E.R. Bradly and G.P. Sabol (eds), American Society for Testing and Materials, Philadelphia, PA (1996) 850–864.
48. F. Garzarolli, W. Beck W, H.P. Fuchs, E. Steinberg and R.A. Perkins, *Proceedings EPRI-Utility Workshop on Fuel Corrosion*, Washington DC, USA, Electric Power Research Institute Palo Alto, CA (1993).
49. F. Garzarolli, R. Manzel, S. Reschke and E. Tenckhoff E., 'Review of corrosion and dimensional behavior of zircaloy under water reactor conditions', in *Zirconium in the Nuclear Industry, Fourth Conference*, ASTM STP 681, American Society for Testing and Materials, Philadelphia, PA (1979) 91–106.
50. B. Cox and G. Wikmark, ZIRAT-6, 'Impact of water chemistry and CRUD on fuel performance', ANT International, Mölnlycke, Sweden (2001).
51. D.N. Sah, U.K. Viswanathan, E. Ramadasan, K. Unnikrishnan and S. Anantharaman, 'Post irradiation examination of thermal reactor fuels', *J. Nucl. Mater.*, **383** (2008) 45–53.
52. ZIRAT 12-Special Topic Report Corrosion Mechanisms in Zirconium Alloy, 'Corrosion mechanisms in zirconium alloys', Ron Adamson, Friedrich Garzarolli, Brian Cox, Alfred Strasser, Peter Rudling (eds), A.N.T. International, October 2007, ANT International, Skultuna, Sweden.
53. W. Goll and I. Rey, 'The behavior of intermetallic precipitates in highly irradiated BWR LTP cladding', in *Proceedings of the 13th International Symposium*

on Zirconium in Nuclear Industry, ASTM STP 1423, G.D. Moan and P. Rudling (eds), American Society for Testing and Materials, W. Conshohocken, PA, US (2002) 80–95.

54. V. V. Likhanskii and I.A. Evdokimov, 'Effect of additives on the susceptibility of zirconium alloys to nodular corrosion', *J. Nucl. Mater.*, **392** (2009) 447–452.
55. N. Ramasubramanian, 'Shadow corrosion', *J. Nucl. Mater.*, **328** (2004) 249–252.
56. H.M. Chung and W.J. Shack, 'Irradiation-assisted stress corrosion cracking behavior of austenitic stainless steels applicable to LWR core internals', Chapter 9 (NUREG/CR-6892, ANL-04/10) http://www.nrc.gov/reading-rm/doc-collections/nuregs/contract/cr6892/.
57. T.M. Karlsen and C. Vitanza, *Proceedings of the International Symposium on Plant Aging and Life Predictions of Corrodible Structures*, Sapporo, Japan, May 1995, p. 741; see also, Tetsuo Shoji, Shun-ichi Suzuki and K.S. Raja, 'Current status and future of IASCC research', *J. Nucl. Mater.*, **258–263** (1998) 241–251.
58. Fukuya K, Nakano M, Fujii K, Torimaru T, 'IASCC susceptibility and slow tensile properties of highly-irradiated 316 stainless steels', *J. Nucl. Sci. Tech.*, **41** (2004) 673–681.
59. P. Scott, 'A review of irradiation assisted stress corrosion cracking', *J. Nucl. Mater.*, **211** (1994) 101–122.
60. R.S. Glass, G.E. Overturf, R.A. Van Konynenburg, R.D. McCright, 'Gamma radiation effects on corrosion – I. Electrochemical mechanisms for the aqueous corrosion processes of austenitic stainless steels relevant to nuclear waste disposal in tuff', *Corr. Sci.*, **26**:8 (1986) 577–590.
61. I. Aho-Mantila, 'Corrosion in the primary coolant systems of water cooled reactors', Coolant Technology of Water Cooled Reactors, Volume 2 IAEA, Vienna 1992, IAEA-TECDOC-667, ISSN 1011–4289; Austria (1992).
62. A.J. Jacobs, 'The relationship of grain boundary composition in irradiated type 304SS to neutron fluence and IASCC', in *16th International Symposium on Radiation on Materials*, ASTM-STP 1175, A.S. Kumar, D.S. Gelles, R.K. Nanstad, E.A. Little (eds), American Society for Testing and Materials, Philadelphia, PA (1993) 902–918.
63. J.T. Busby and G.S. Was, *Proceedings of 11th International Conference on Environmental Degradation of Materials in Nuclear Power Systems*, Water Reactors, American Nuclear Society, La Grange Park, IL (2003) 995.
64. Z. Jiao, J.T. Busby, R. Obata, G.S. Was, *Proceedings of 12th International Conference on Degradation of Materials in Nuclear Power Systems*, Water Reactors, The Minerals, Metals and Materials Society, Warrendale PA (2005) 379; (see also) Z. Jiao, J.T. Busby, G.S. Was, 'Deformation microstructure of proton irradiated stainless steel', *J. Nucl. Mater.*, **361** (2007) 218–227.
65. Z. Jiao and G.S. Was, 'Localized deformation and IASCC initiation in austenitic stainless steels', *J. Nucl. Mater.*, **382** (2008) 203–209.
66. K. Fukuya, H. Nishioka, K. Fujii and T. Torimaru, 'Effects of dissolved hydrogen and strain rate on IASCC behavior in highly irradiated stainless steels', *J. Nucl. Sci. Tech.*, **45** (2008) 452–458.
67. F. Cattant, D. Crusset and D. Feron, 'Corrosion issues in nuclear industry today', *Mater. Today*, **11** (2008) 32–37.
68. D.G. Franklin and R.B. Adamson, 'Implications of Zircaloy creep and growth to light water reactor performance', *J. Nucl. Mater.*, **159** (1988) 12–21.

69. D.R. Olander, 'Light water fuel designs and performances,' in *Concise Encyclopedia of Materials for Energy Systems*', J.W. Martin (ed.), Elsevier (2000), 31.
70. P.L. Anderson, 'Irradiation assisted stress corrosion cracking,' in *Material Performance and Evaluation*, R.H. Jones (ed.), ASM International, Materials Park, OH (1998) 181.
71. D.G. Franklin and R.B. Adamson, 'Implications of Zircaloy creep and growth to light water reactor performance', *J. Nucl. Mater.*, **159** (1988) 12–21.
72. K. Edsinger, 'Review of fuel degradation in BWRs', in *Proceedings of International Topical Meeting on LWR Fuel Performance*', ANS (2000) 523.
73. K. Sakamoto and M. Nakatsuka, 'Stress reorientation of hydrides', *J. Nucl. Sci. Tech.*, **43** (2006) 1136–1141.
74. A.M. Garde, 'Effects of irradiation and hydriding on the mechanical properties of Zircaloy-4 at high fluence', in *Zirconium in the Nuclear Industry, 8th International Symposium*, ASTM STP 1023, L.P. Van Swam and C.M. Eucken (eds), American Society for Testing and Materials, Philadelphia, PA (1989) 548–569.
75. K. Edsinger, J.H. Davies and R.B. Adamson, 'Degraded fuel cladding fractography and fracture behavior', in *12th International Symposium on 'Zirconium in the Nuclear Industry*, ASTM STP 1354, G.P. Sabol and G.D. Moan (eds), American Society for Testing and Materials, Philadelphia, PA (1999) 316–339.
76. J.A. Kuszyk *et al.*, Interim Report, Zion Unit 1 Cycle 6, Fuel Performance, EPRI Project RP611-1, WCAP-10280.
77. T.C. Rowland and S. Gehl, BWR Fuel Rod Performance Evaluation Program, EPRI Report NP-4602 (1986).
78. 'Review of fuel failures in water cooled reactors', IAEA Nuclear Energy Series No. NF-T-2.1, International Atomic Energy Agency, Vienna (2010).
79. G. Srikant, B. Marple, I. Charit and K.L. Murty, 'Characterization of stress rupture behavior of cp-Ti via burst testing', *Mat Sci Eng.*, **A463** (2007) 203–207.
80. Y. Zhou, B. Devarajan and K.L. Murty, 'Short-term rupture studies of Zircaloy-4 and Nb-modified Zircaloy-4 tubing using closed-end internal pressurization', *Nucl. Eng. Design*, **228** (2004) 3–13.
81. C.S. Seok, B. Marple, Y.J. Song, S. Gollapudi, I. Charit and K.L. Murty, 'High temperature deformation characteristics of Zirlo™ tubing via ring-creep and burst tests', *Nucl. Eng. Design*, **241** (2011) 599–602.
82. K.L. Murty, 'Applications of crystallographic textures of zirconium alloys in nuclear industry', in *Zirconium in the Nuclear Industry: Eighth Symposium*, ASTM STP 1023, Swam and C.M. Eucken (eds), American Society for Testing and Materials, Philadelphia, PA (1989) 570–595.
83. K.L. Murty, 'Creep studies for life-prediction in water reactors', *J. Metals*, Oct. (1999) 32–39.
84. R. Kishore, S. Banerjee and P. Rama Rao, 'First report on observation of abnormal creep in Zr-2.5%wt%Nb alloy at low stress', *J. Mater. Sci.*, **44** (2009) 2247–2256.
85. K.L. Murty, 'Deformation mechanisms and transients in creep of zircaloys: applications to nuclear technology', *Trans. IIM*, **53** (2000) 107–120.
86. S. Gollapudi, I. Charit and K.L. Murty, 'Creep mechanisms in Ti–3Al–2.5V alloy tubing deformed under closed-end internal gas pressurization', *Acta Mat.*, **56** (2008) 2406–2419.

87. N. Rupa, M. Clavel, P. Boufloux, C. Domain and A. Legris, 'Impact of hydrogen on plasticity and creep of unirradiated Zircaloy-4 cladding tubes', in *Zirconium in Nuclear Industry: 13th International Symposium*, ASTM STP 1423, G.D. Moan and P. Rudling (eds), American Society for Testing and Materials (2000) 399–424.
88. D. Setoyama and S. Yamanaka, 'Indentation creep study of zirconium hydrogen solid solution', *J. Alloy Compds*, **379** (2004) 193–197.
89. R. Kishore, 'Effect of hydrogen on the creep behavior of Zr–2.5%Nb alloy at 723 K', *J. Nucl. Mater.*, **385** (2009) 591–594.
90. K. Edsinger, 'EPRI and the zero fuel failures program', The Nuclear News Interview, The Nuclear News, Dec 2010, 40.
91. 'Review of fuel failures in water cooled reactors,' IAEA Nuclear Energy Series No. NF-T-2.1, International Atomic Energy Agency (2010).
92. H.M. Chung, 'Assessment of void swelling in austenitic stainless steel core internals', NUREG/CR-6897, ANL-04/28, U.S. Nuclear Regulatory Commission Office of Nuclear Regulatory Research, Washington, DC 20555–0001.
93. K.L. Murty, 'Interstitial-impurity radiation-defect interactions in ferritic steels', *J. Metals* (1985) 34–39.
94. M.S. Wechsler, 'Impurity-defect interactions on radiation hardening and embrittlement', *J. Engg. Mater. Tech.* (Transactions ASME), **101** (1979) 114–121.
95. G.R. Odette, 'On the dominant mechanism of irradiation embrittlement of reactor pressure vessel steels', *Scripta Metall.* **17** (1983) 1183–1188; see also G.R. Odette and R.K. Nanstad, 'Predictive Reactor Pressure Vessel Steel Irradiation Embrittlement Models: Issues and Opportunities', *J. Metals* **61** (2009) 17–23.
96. G.R. Odette and G.E. Lucas, 'Embrittlement of nuclear reactor pressure vessels', *J. Metals* **53** (2001) 18–22.
97. N. Rupa, H. Churier-Bossennec and G. Bezdikian, 'Materials and NDE aspects in the RPV operating condition behavior,' in *Contribution of Materials Investigations to Improve the Safety and Performance of LWRs*, Fontevraud, France, SFEN, Paris, France (2006) 715.
98. 'Use of fracture toughness test data to establish reference temperature for pressure retaining materials, Section XI, Division 1', ASME Boiler and Pressure Vessel Code Case N-629, ASME, New York (1999).
99. 'Guidelines for application of the master curve approach to reactor pressure integrity in nuclear plants', Technical Report Series No. 429, International Atomic Energy Agency, Vienna (2005).
100. K.L. Murty, R.P. Shogan and W.H. Bamford, 'Dynamic fracture toughness of irradiated A533 Grade B Class1 pressure vessel steel', *Nucl. Techn.*, **64** (1984) 268–274.
101. R. Wu, R. Sandstorm, F. Seitisleam, 'Low temperature creep crack growth in low alloy reactor pressure vessel steel', *J. Nucl. Mater.*, **336** (2005) 279–290.
102. J.C. Van Duysen, P. Todeschini and G. Zacharie, 'Effects of neutron irradiation at temperatures below 500 C on the properties of cold-worked 316 stainless steels: a review', in *Effects of Radiation on Materials: 16th International Symposium*, ASTM STP 1175, A.S. Kumar, D.S. Gelles, R.K. Nanstad and

E.A. Little (eds), American Society for Testing and Materials, Philadelphia, PA (1994) 747–776.

103. E. Altstadt, H. Kumpf, F.-P. Weiss, E. Fischer, G. Nagel and G. Sgarz, 'Analysis of a PWR core baffle considering irradiation induced creep', *Ann. Nucl. Energy*, **31** (2004) 723–736.

104. A.J. Garnier, Y. Bréchet, M. Delnondedieu, C. Pokor, P. Dubuisson, A. Renault, X. Averty and J.P. Massoud, 'Irradiation creep of SA 304L and CW 316 stainless steels: mechanical behaviour and microstructural aspects. Part II: Numerical simulation and test of SIPA model', *J. Nucl. Mater.*, **413** (2011) 70–75.

2 Corrosion in pressurized water reactors (PWRs)

T. COUVANT, EDF R&D, France

DOI: 10.1533/9780857097453.1.70

Abstract: Corrosion is one of the major obstacles to extending the lifetime of nuclear power plants within agreed safety requirements. A large variety of the structural metals present in primary and secondary circuits of pressurized water reactors (PWRs) suffer corrosion. Uniform corrosion, flow-accelerated corrosion (FAC), pitting, stress corrosion cracking (SCC), environmentally assisted fatigue and hydrogen embrittlement can all affect the major components of PWRs, despite stringent selection of materials for component manufacture. Remedies can vary: adjusting water chemistry, reducing superficial strains and stresses, replacing materials or changing microstructures. Experience in the field has demonstrated that increasing chromium content is an efficient strategy: to date nickel alloys containing 30% chromium exhibit very good resistance to corrosion such as SCC. It can be shown that tendency to corrosion can largely depend on manufacturing conditions.

Key words: corrosion, austenitic alloys, pressurized water reactors, primary water, cracking.

2.1 Introduction

We begin the chapter with an outline of the history and fundamental principles of corrosion.

2.1.1 History

Corrosion and its effects have been observed since the first steps in metallurgy. Corrosion damage increased with the use of iron over the centuries. In 1830, de la Rive (1801–1873) showed that bimetallic junctions suffered fast corrosion due to impurities present in zinc. Later, Faraday (1791–1867) correlated the current flow with the associated rate of corrosion. In the 1930s, Wagner (1901–1977) showed that the uniform dissolution of metals did not require separate anodic and cathodic sites but that metal dissolution and the accompanying cathodic reaction can occur randomly with respect to space and time over the surface. In the 1950s, Pourbaix (1904–1998) edited a series of major diagrams giving the domain of stability of many elements as a function of potential and pH.

From the 1950s the importance of corrosion to the economy became increasingly evident. Today, corrosion is one of the major degradations to overcome in order to extend the lifetime of nuclear power plants in agreement with safety requirements.

2.1.2 Fundamental principles of corrosion

The second law of thermodynamics is an expression of the tendency over time differences in temperature, pressure and chemical potential will equilibrate in an isolated physical system. In other words, every material tends to reach the maximum of disorder, in order to minimize its potential energy. With regard to corrosion, it means that leaving the crystalline network under the action of an electric field, metal ions yields energy. According to thermodynamics, almost all metals have negative free energy, suggesting their reactivity in environments where they are exposed.

Corrosion reactions are electrochemical in nature, based on mass and charge transfers. Reactions can be split into partial oxidation and reduction reactions. The potential is the propensity to exchange electrons: the metal donating electrons is oxidized, while the metal receiving electrons is reduced. The stability of elements in a given medium is predicted by the corresponding Pourbaix diagram, where predominant phases are defined in agreement with thermodynamics. However, reaction kinetics play a major role in the evolution of the system (such as changes in pH, potential or temperature).

For example, if iron is introduced into hydrogenated water at 300°C (without any dissolved oxygen) at pH 7 and a potential of –700 mV_{SHE}, cations Fe^{2+} are dissolved in the water (Reaction [2.1]). This anodic reaction (oxidation of metal) is coupled to the cathodic reaction (reduction of water) described by Reaction [2.2]. Then, dissolved cations Fe^{2+} can join oxidant ions OH^- (Reaction [2.3]), to form ferrous hydroxide $Fe(OH)_2$.

$$Fe \rightarrow Fe^{2+} + 2e' \tag{2.1}$$

$$2H_2O + 2e' \rightarrow H_2 + 2OH^- \tag{2.2}$$

$$Fe^{2+} + 2OH^- \rightarrow Fe(OH)_2 \tag{2.3}$$

Cations can be released in the water due to dissolution. When the saturation in ferrous hydroxide $Fe(OH)_2$ or ferrous cation Fe^{2+} is reached, according to the Pourbaix diagram, magnetite Fe_3O_4 forms based on Schikorr Reaction [2.4].

$$3Fe(OH)_2 \rightarrow Fe_3O_4 + H_2 + 2H_2O \tag{2.4}$$

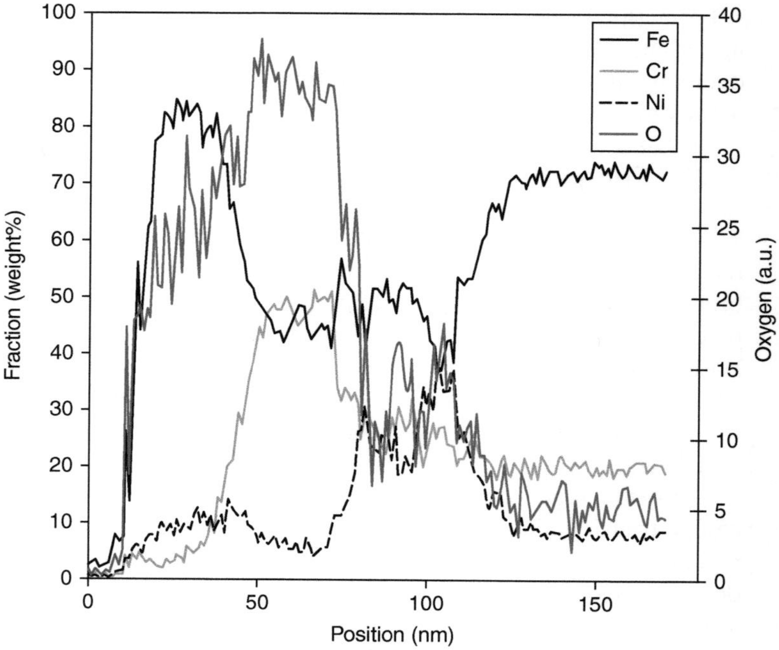

2.1 EDX profile on 304 L exposed to water (360°C, $pH_{325°C} = 7.2$).

Finally, passivation is the process of building a protective layer of oxide isolating the surface of the material from the aggressive environment. Some corrosion inhibitors help the formation of such layers. Figure 2.1 shows energy-dispersive x-ray (EDX) analysis of the oxide formed on stainless steel exposed to water at 360°C (–600 mV_{SHE}, $pH_{325°C} = 7.2$). The passive film is the 50 nm-thick layer containing a significant level of chromium at the surface of the metal.

2.2 Pressurized water reactors and the main types of corrosion

2.2.1 PWRs

A large variety of structural metals present in primary and secondary circuits of PWRs suffer corrosion:

- **Carbon steels** are cheap iron-base metals with less than 1% of alloying element present. These materials exhibit a poor resistance to corrosion but their forming, machining and welding are superior.
- **Low-alloy steels** are iron-base metals containing a few percent of, for example, nickel, chromium, molybdenum, vanadium, which are usually

used to achieve better hardenability. These alloys have a limited resistance to corrosion.

- **Stainless steels** are iron-base alloys containing more than 13% chromium. An austenitic structure is obtained when they contain a large nickel content. Austenitic stainless steels have excellent resistance to corrosion despite the fact that they can suffer pitting and SCC under some conditions. A martensitic structure is obtained when the nickel content is low. Martensitic stainless steels have high mechanical strength but they are less resistant to corrosion than austenitic structures.
- **Ni alloys** have 15–30% Cr, with a high resistance to uniform and pitting corrosion, but they are susceptible to SCC, except for those with the highest range of chromium content. Ni alloys are expensive, especially with 30% chromium metals.

In the **primary circuit**, the water is at a temperature ranging from 270°C to 345°C. Boron is introduced as boric acid to absorb and to control the core reactivity. In order to counteract the general corrosion of materials, lithium hydroxide is added to the water, in order to reach a slightly alkaline pH of 7.2 at 300°C. Finally, hydrogen is dissolved to counteract the radiolytic decomposition of water into oxidizing compounds that may lead to SCC of stainless steels. In addition, a few ppb of zinc can be introduced to mitigate the activation of cobalt.

In the **secondary circuit**, the pH of the boiling water is also made slightly alkaline in both liquid and steam phases to limit corrosion. The original operating chemistry combined ammonia (for its slightly alkaline pH) with phosphate (to buffer the various potential contaminants that may enter the system through the condenser). Current chemistries involve all volatile treatment (AVT) without any phosphate addition in high quality water. Ammonia is added to get a $pH_{25°C}$ higher than 9.8 in plants without any copper alloys to avoid FAC of carbon steel. In other situations, an amine is preferred such as ammonia, morpholine or ethanolamine which exhibit a high thermal stability. Last, hydrazine is introduced to obtain a reducing environment and to limit the SCC of Alloy 600 tubing.

2.2.2 Main types of corrosion observed in PWRs

Uniform corrosion proceeds over the entire surface area of the material exposed to the environment leading to a general slow thinning accompanied by a release of corrosion products. Uniform corrosion is usually relatively easy to measure and predict. In the primary circuit of PWRs, released ions are transported and may be activated when they reach the reactor pressure vessel (RPV), which is a major problem to be overcome.

Therefore, uniform corrosion is minimized combining appropriate materials (nickel alloys and stainless steels), surface finish, passivating treatments and alkaline pH.

Flow-accelerated corrosion or flow-assisted corrosion (FAC) is a mechanism in which the passive layer dissolves in fast flowing water, without any mechanical erosion. As a consequence, the underlying metal continuously corrodes to recreate the protective oxide. FAC rate decreases when the flow velocity decreases and when the pH increases. FAC stops as soon as oxygen is dissolved in water. FAC affects carbon steel piping of the secondary circuit where water or wet steam circulates.

Pitting is a localized corrosion forming holes at the surface of the metal, induced by the local depassivation of an area, which becomes anodic while a large area becomes cathodic. The acidity inside the pit is sustained by the spatial separation of the cathodic and anodic half-reactions, which creates a potential gradient and the transport of anions into the pit. The presence of surface defects, such as scratches and local changes in chemical composition promote pitting which causes little loss of material but it may lead to deep corrosion in a component. Pitting mainly affects materials such as austenitic stainless steels exhibiting a good resistance to uniform corrosion thanks to their good passivation. However, the presence of chlorides and oxygen at relatively low temperature (typically 80°C) may weaken the passive layers and enhance pitting via an autocatalytic process: Cl^- ions start to concentrate in the pits for charge neutrality and promote the reaction of positive metal ions with water to form a hydroxide corrosion product and H^+ ions. The increasing acidity within the pits accelerates the process.

Stress corrosion cracking (SCC) is a progressive failure affecting metals subjected to a tensile stress (residual or applied) while they are exposed to a corrosive environment. SCC occurs in specific and limited conditions in terms of water chemistry, material and loading. SCC usually involves a long incubation period prior to initiation, followed by a slow crack extension stage and transition in a fast crack propagation stage leading to failure. Stress concentrations, cold work and irradiation promote SCC. The material most susceptible to SCC is Alloy 600 (in both primary and secondary waters). However, stainless steel becomes susceptible to SCC in primary water under specific conditions: polluted environments (oxygen plus chlorides), high level of cold work and irradiation. The susceptibility of nickel alloys to SCC strongly decreases when the chromium content of the material increases, especially above 20%. Basically, SCC results in the oxide ingress, usually at grain boundaries, which locally weaken the material. The oxide penetration is enhanced by the presence of strain and is affected by precipitation. If local stresses are sufficient to fail weakened grain boundaries, a crack extension occurs.

Environmentally assisted fatigue occurs under the combined actions of low frequency cyclic loading and oxidation. Therefore, ingredients are very similar to those involved in SCC mechanisms in the sense that a synergy operates locally between oxidation and mechanics. The major difference with SCC is the nature of the loading: cyclic loading strongly promotes strain localization in shear bands. The movement of dislocations (defects allowing the non-reversible deformation of the metal) enhances oxide ingress and failure, especially when shear bands emerge at the surface, breaking the passive layer. Therefore, strain rate is one of the key controlling parameters of the mechanism.

Last, it should be noted that one of the corrosion products is able to embrittle metals. Indeed, as reported before, the cathodic reaction (reduction of water) produces hydrogen which partly enters into the metal and interacts with the microstructure. The entry of hydrogen is limited by the growth of the passive layer; also its transport and interactions with the metal strongly depend on temperature. **Hydrogen embrittlement** is the process by which a localized accumulation of a sufficient level of hydrogen can eventually lead the metal to fracture under residual or applied stress. Therefore, situations limiting the transport of hydrogen (low temperature, presence of traps) promote such embrittlement. Other mechanisms of introducing hydrogen into metals exist, such as manufacturing (welding) and irradiation.

2.3 Major components experiencing corrosion

We continue the chapter by describing the major components within the reactor which are subject to corrosive damage.

2.3.1 Reactor pressure vessel (RPV)

Reactor vessel heads (RVH) can experience different types of corrosion. In 2002 boric acid crystal deposits and iron oxide were found to have flowed out from several openings in the lower service structure support skirt after removal of insulation from the Davis-Besse RVH, after an accumulated ≈16 effective full power years (EFPYs) of operation. A large corrosion cavity was found on the downhill side of the low-alloy steel RVH.[1] Boric acid corrosion wastage occurred on the RPV head surface and lead to a total low-alloy steel loss of ~4.3 cm^3. Boric acid corrosion was not the only mechanism involved in the degradation: it was supposed that erosion–corrosion may have played a role in the initial cavity formation; galvanic corrosion between the low-alloy steel and the stainless steel occurred around the perimeter of the exposed cladding; and axial stress corrosion cracks were observed in five control rod drive mechanism (CRDM) nozzles adjacent to the J-groove weld.[1] Both Alloy 600 and 182 weld metal failed by primary water stress corrosion cracking (PWSCC). There was no conclusive evidence that hot cracking contributed to the J-groove weld cracking. The Davis-Besse event illustrates the severe consequence of in-service cracking of RVH-penetration components fabricated from Ni-base Alloy 600 and 182 weld metal in which PWR water leaking from the cracked nozzles severely corroded the RPV head low-alloy steel material down to the 308 stainless steel cladding material.[2]

In the vessel, internals are exposed to irradiation. Under neutron flux the microstructure of the material can evolve: segregation at the grain boundaries associated with dechromization and hardening induced by the recombination of point defects. The first cracked baffle-former bolts were observed in 1988 in Bugey Unit 2 (PWR, France), during ultrasonic testing (UT) controls. Several bolts were examined[3–6] and the failure was attributed to a particular case of SCC: irradiation assisted stress corrosion cracking (IASCC). Periodic inspections and a replacement program were set up in the affected reactor types. The assessment of the damage affecting the bolts revealed that significant differences in cracking behaviors exist between the various reactors. For instance, taking into account the number of cracked bolts, Bugey Unit 2 (100 cracked bolts in 140 000 h) and Fessenheim Unit 2 (46 cracked bolts in 140 000 h) are the most affected reactors (the remaining reactors were mostly less than 30 cracked bolts). Additionally, their bolts were made from the same heat, suggesting the influence of initial composition and microstructure.

2.3.2 Steam generators (SGs)

In 2004, a failure occurred at Mihama 3, in the pipe of a loop condensate system between the fourth feedwater heater and the deaerator, on the secondary side of the PWR.[7] The accident resulted in five deaths among the workers preparing for periodic inspections at the time of the piping rupture. The rupture opening in the carbon steel pipe measured as follows: 51.5 cm (axial direction) by 93.0 cm (circumferential direction). At the time of the initial plant service, the nominal wall thickness of the pipe was 10 mm, with the thinnest section only 0.4 mm. Designed with a maximum service temperature of 195°C and a maximum service pressure of 1.27 MPa, the pipe ruptured when the temperature was only 140°C with a pressure of 0.93 MPa; the flow rate through the pipe was 1700 $m^3 h^{-1}$. There were no precursor indicators before the accident or special operations shown on the review of the plant parameters which could have caused the pipe to rupture. An investigation concluded that water quality had been maintained since the commissioning of the plant. A microscopic inspection was then conducted, which revealed that a fish-like pattern covered almost the entire inner surface of the ruptured pipe downstream of the orifice. The bottom of the pipe on the inside was also covered with a thick surface film. These findings are characteristic of FAC.

Intergranular SCC (IGSCC) and **intergranular attack** (IGA) are the most serious degradation processes affecting SG tubes, on the secondary side. This degradation commonly occurs in crevice regions at tube support plate and tube sheet locations or under sludge piles,[8,9] although intergranular SCC has also been observed in the free span of the tubes. The presence of lead in the secondary circuit was supposed to enhance IGSCC[10,11]: Pb ions would influence passivity of the Alloy 600 surface, being incorporated into the Alloy 600 specimen surface and enhancing electronic conductance. Lead may cover a significant fraction of the Alloy and shift equilibria for the Ni oxide formation. IGA was observed at the secondary side, at the roll transition zone underneath crud deposits of SG tubes from McGuire Unit 1, for example. A wide variety of elements were present in the crud deposits, including Fe, Ni, Cr, Al, Si, Mg, Cu, Ti, Mn, Ca, K, and S. The copper was present in the deposit as metallic copper. The presence of metallic Cu indicates that the electrochemical potential was below the Cu/Cu oxide equilibrium. The SG unit had operated initially with Ni-Cu moisture separator reheaters. On the primary side, IGSCC occurs at locations of high stress, typically at regions where substantial plastic strain has occurred within the tube, during the SG manufacturing process and from in-service straining. Thus, IGSCC has been observed at the apex and at the transition from bent to straight portions of small radius U-bends.[12] Examinations of SG tubes revealed the presence of axial cracks mainly at regions of transitions from expanded to non-expanded portions of the tube/tubesheet joint and circumferential cracks at the end of the transition.[13]

More recently, **primary water SCC** was found (2004) at the surface of the warm side of the divider plate of the SG #171 at Chinon Unit B4,[14,15] exposed to the primary environment at 325°C. Cracks initiated in a area which had been subject to grinding, on the hot side of the partition stub made of Alloy 600, close to the welds (Alloy 182), where a significant cold work was present and where a limited intergranular precipitation was observed. Examination showed intergranular and intragranular precipitates in the materials. The divider plate exhibited large non-recrystallized grains close to the surface. Cross-sections indicated the presence of IGSCC perpendicular to the surface. The maximal crack depth was 1.2 mm (<4% of the total thickness). After neutron diffraction examinations, the mean plastic strain present in the Alloy 600 stub was estimated as 5.3%, while the maximal strain in the heat affected zone reached 11%. In addition, the deformation was higher at the surface than in the bulk of the material. R&D studies[16–18] on representative hot rolled Alloy 600 have shown that no significant crack growth is expected as soon as the deformation is lower than 7–10%. Therefore, it was concluded that observed cracks could not significantly propagate in the bulk of the plate, where the deformation is low, even after 60 years.[19]

2.3.3 Pressurizers

Since 1997, cracking incidences of pressurizer heaters were encountered in French PWRs.[20,21] The heater sheath (outer diameter of 22 mm, 2 mm thick) is made of 316 L stainless steel. The lower part of the heater is attached to electrical connectors, out of the pressurizer, while the upper part, introduced into the pressurizer, consists of a coaxial heating element coiled round a copper mandrel. After the final assembly of the heater, the sheath is cold swaged to reduce the gaps between the heating element and the sheath and therefore, to improve the thermal exchanges. Under operating conditions, the heaters are exposed to hydrogenated and non-polluted primary environment at 345°C. Nevertheless, the temperature at the outer surface of the sheath, in nominal condition, could reach 360°C. Nine cases of SCC were found after 12 destructive examinations out of the 1200 failures observed on heaters. When SCC leads to primary leakage (boron traces on connectors), the heater has to be replaced no later than the next outage. Elemental analyses (energy-dispersive X-ray spectroscopy) did not reveal any trace of pollutant at the surface of the retired heaters. Therefore, it was concluded that SCC occurred in the nominal hydrogenated primary environment. No chromium depletion was present at the grain boundaries of the stainless steel. Such depletion can result in the precipitation of chromium carbides at grain boundaries and promotes IGSCC in an oxidizing environment.

A surface annealing heat treatment was developed (induction heating) to counteract the initiation of SCC on the original cold-worked outer layer exposed to the primary water. The goal of the heat treatment is to anneal the surface of the material, decreasing strain-hardening and residual stresses without any damage of the electrical properties of the heater element. As a result, the Vickers micro hardness decreased from 320 HV_1 (higher than the threshold necessary to initiate SCC)[22] down to 200 HV_1 (below the threshold necessary to initiate SCC) at the surface of heat treated heaters, and residual stresses were removed as shown by corrosion tests in $MgCl_2$ medium.

2.4 Conclusion

Despite the original stringent selection of the materials used to manufacture the components, uniform and localized corrosion occurs in PWR environments. Remedies can be of different types: adjusting the water chemistry, reducing superficial strains and stresses, replacing materials or changing microstructures. In particular, experience in the field demonstrated that an increase in chromium content is an efficient strategy: to date nickel alloys containing 30% chromium, used to replace 16% chromium nickel alloys, have exhibited very good resistance to localized corrosion, such as SCC. The history of the degradations shows that for a given type of material, the tendency to corrode can largely depend on the manufacturing conditions.

2.5 References

1. H. Xu, S. Fyfitch, J.W. Hyres, T.A. Lang and T.T. Pleune, 'Laboratory investigation of PWSCC of CRDM nozzle 3 and its J-Groove weld on the Davis-Besse reactor vessel head,' *Proceedings of 12th International Conference of Environmental Degradation of Materials in Nuclear Power Systems – Water Reactors*, The Minerals, Metals and Materials Society, 2005.
2. H. Xu, S. Fyfitch and J.W. Hyres, 'Laboratory investigation of the stainless steel cladding on the Davis-Besse reactor vessel head,' *Proceedings 12th International Conference of Environmental Degradation of Materials in Nuclear Power Systems – Water Reactors*, The Minerals, Metals and Materials Society, 2005.
3. R. Cauvin, O. Goltrant, Y. Rouillon, E. Verzaux, A. Cazus, P. Dubuisson, P. Poitrenaud and S. Bellet, 'Endommagement des structures internes inférieures soumises à fortes fluences: apports de l'expertise,' International Symposium, Fontevraud III (France), 1994.
4. O. Goltrant, R. Cauvin, D. Deydier and A. Trenty, 'Eléments internes inférieurs: apports de l'expertise d'une cornière de Chooz A,' International Symposium, Fontevraud III (France), 1994.
5. I. Monnet, G.M. Decroix, P. Dubuisson, J. Reuchet and O. Morlent, 'Investigation of the Chooz A nuclear power plant bolts,' International Symposium, Fontevraud IV (France), 1998.
6. C. Pokor, G. Courtemanche, JL. Fléjou, M. Tommy-Martin, I. Rupp, B. Tanguy, J.-P. Massoud and N. Monteil, 'IASCC of core internals of PWRs: EDF R&D and engineering program to assess internals lifetime management,' International Symposium, Fontevraud VII (France), 2010.
7. NRC information notice 2006–08: secondary piping rupture at the Mihama power station in Japan, March 2006.
8. J.M. Boursier, P. Jardet, F. de Keroulas, P. Lemaire and Y. Rouillon, 'Contribution des expertises metallurgiques de tubes extraits à la comprehension de la corrosion en milieu secondaire des générateurs de vapeur REP d'EDF,' International symposium, Fontevraud IV (France), 1998.
9. J.M. Boursier, M. Dupin, P. Gosset and Y. Rouillon, 'Secondary side corrosion of French PWR steam generator tubing: contribution of the surface analyses to the understanding of the degradation process,' *Proceedings of 9th International Symposium on Environmental Degradation of Materials in Nuclear Power Systems-Water Reactors*, 1999.
10. A. McIlree, K. Fruzzetti, J. Gorman, W. Shack, R. Staehle and J. Stevens, 'A case of the major influence of lead and sulfur in the secondary side degradation of PWR steam generators tubing,' International symposium, Fontevraud VI (France), 2006.

11. J. Lumsden and A. McIlree, 'Factors affecting PBSCC in alloy 600 and alloy 690 steam generator tubing,' International symposium, Fontevraud VI (France), 2006.
12. R.T. Begley, A.W. Klein and R.E. Gold, 'Investigations of ID stress corrosion cracking in field and laboratory programs,' International symposium, Fontevraud (France), 1985.
13. F. Cattant and F. de Keroulas, 'Examens des tubes en alliage 600 extraits des générateurs de vapeur des centrales EDF,' International symposium, Fontevraud (France), 1985.
14. C. Bibollet, C. Beroni, S. Raimbault, M. Stindel, N. Verdiere and J.-P. Gauchet, 'EDF field experience on steam generator divider plate examinations,' Fontevraud 6, 2006.
15. D. Deforge, L. Duisabeau, S. Miloudi, Y. Thebault, T. Couvant, F. Vaillant and E. Lemaire, 'Learning from EDF investigations on SG divider plates and vessel head nozzles. Prior deformation effect on stress corrosion cracking,' Fontevraud 7, Avignon (France), 2010.
16. F. Vaillant, S. Le Hong, C. Amzallag and C. Bosch, 'Crack growth rates on vessel penetrations in alloy 600 in primary water,' Fontevraud 4, 1998.
17. F. Vaillant *et al.*, 'Crack growth rates in thick materials of alloy 600 and weld metals of alloy 182 in laboratory primary water. Comparison with field experience,' Fontevraud 5, 2002.
18. T. Couvant, F. Vaillant and E. Lemaire, 'Stress corrosion crack growth rate in rolled alloy 600 exposed to primary PWR environment,' *Proceedings of 14th International Symposium on Environmental Degradation of Materials in Nuclear Power Systems-Water Reactors*, Virginia Beach (Virginia), 2009.
19. T. Couvant, S. Miloudi, F. vaillant, D. Déforge, Y. Thébault, 'PWSCC of steam generator divider plates in alloy 600: coupling field characterizations with R&D studies,' *Proceedings Fontevraud 7, International Symposium Contribution of Materials Investigations to the Resolution of Problems Encountered in PWR*, SFEN, 2010.
20. T. Couvant, P. Moulart, L. Legras, P. Bordes, J. Capelle, Y. Rouillon, T. Balon, 'PWSCC of austenitic stainless steels of heaters of pressurizers,' Fontevraud 6, 2006.
21. Y. Thébault, P. Moulart, K. Dubourgnoux, J. Champredonde, T. Couvant, Y. Neau, J.M. Fageon, D. Lecharpentier, A. Breuil, V. Derouet, 'PWSCC of Thermocoax Pressurizer Heaters in Austenitic Stainless Steel and Remedial Actions to preventing SCC,' *15th International Conference on Environmental Degradation of Materials in Nuclear Systems-Water Reactors*, Colorado Springs (Colorado), 2011.
22. T. Couvant, L. Legras, F. Vaillant, J.M. Boursier, Y. Rouillon, 'Effect of strain-hardening on stress corrosion cracking of AISI 304 L stainless steel in PWR environment at 360°C,' *12th International Conference on Environmental Degradation of Materials in Nuclear Systems-Water Reactors*, Snowbird (Utah), 2005.

3 Creep deformation of materials in light water reactors (LWRs)

K. L. MURTY, North Carolina State University, USA, S. GOLLAPUDI, Massachusetts Institute of Technology, USA, K. RAMASWAMY, Bhabha Atomic Research Center, India, M. D. MATHEW, Indira Gandhi Center for Atomic Research, India and I. CHARIT, University of Idaho, USA

DOI: 10.1533/9780857097453.1.81

Abstract: The time-dependent deformation of materials or creep governs the useful life of many engineering structures. It assumes even higher significance in the case of structures constituting a nuclear reactor, wherein materials bombarded with neutrons develop defects that assist faster diffusion leading to greater plastic deformation. As a result, an understanding of the creep deformation process and factors controlling it is necessary for gauging the usefulness of materials in a nuclear reactor as well as for predicting life-times of various structures. Thus in this work we discuss the various mechanisms of creep, the rate controlling factors, deformation mechanism maps and useful life prediction methodologies. We also identify a few cases where direct application of simple creep correlations might not be feasible. Finally, we discuss the various factors that control the creep behavior of materials in light water reactors.

Key words: creep, diffusion creep, dislocation creep, deformation mechanism maps, modeling, zirconium alloys, stainless steels, irradiation creep.

3.1 Introduction

Creep is time-dependent plastic strain under a constant load/stress at a given temperature and often becomes the life limiting criterion for many structures that experience loads and temperatures, and becomes significant for materials in light water reactors (LWRs) due to imposed radiation effects. A thorough understanding of the plastic deformation behavior of materials is essential for the sound design of engineering structures. Fail-safe designs are based on the ability to predict the response of a structure to applied loads and ensuing plastic deformation. While brittle materials such as ceramics fail after relatively low plastic strains, a significant number of engineering materials such as metals and alloys are characterized by large scale plastic deformation leading to failure. The extent of deformation is controlled by intrinsic factors such as bond strength, presence of secondary phases and defect concentration. At the same time extrinsic factors such as applied loads, temperature, deformation rates and geometry of the structure also determine the amount of plastic deformation. It has been well established that high applied loads and temperatures generally accelerate the rate of plastic deformation. This is because high temperatures and stresses provide the necessary activation energy required for defects to overcome barriers to plastic deformation. While plastic deformation at room temperature or low homologous temperatures (T/T_m) occurs when the applied stress exceeds the yield stress σ_y, deformation at high temperatures can occur at stresses significantly smaller in comparison to the yield stress. The branch of metallurgy which attempts at understanding material deformability at high homologous temperatures and small applied stress has come to be known as creep. The kinetics of deformation processes become important with increasing temperatures and hence creep is defined as the time dependent plastic deformation of a material under constant load or stress.

The earliest studies on time dependence of plastic strain were carried out by Andrade.[1] The time dependence of elongation under tensile loads was investigated at constant temperature. Andrade observed that the total deformation could be divided into three periods: (a) immediate extension upon loading (mainly elastic with relatively small instantaneous plastic), (b) an initial flow which gradually disappears and (c) a constant flow which takes place throughout the elongation. Subsequent studies by Hanson and Wheeler[2] showed the presence of a period where the extension increases continuously until fracture. This period was found to occur following the period of constant flow and was understood to be due to decreased cross-sectional area accompanying the elongation. At constant loads, the cross-sectional area decrease leads to the increase in effective stress and a corresponding increase in strain rate.

3.1.1 Creep curve

The time dependence of plastic strain is described by plots of strain against time, also known as creep curves. A typical creep curve is shown in Fig. 3.1,[3] and consists of three different regions: the primary, secondary and tertiary creep regions. Usually the primary creep region commences only after the material has experienced an instantaneous strain, ε_0 which is a result of sudden loading of the material and corresponds to period 'a' observed by Andrade.[1] The instantaneous strain is composed of elastic (recoverable on release of load), anelastic (recovers with time) and plastic (non-recoverable) components. Though the applied stresses for creep are smaller than

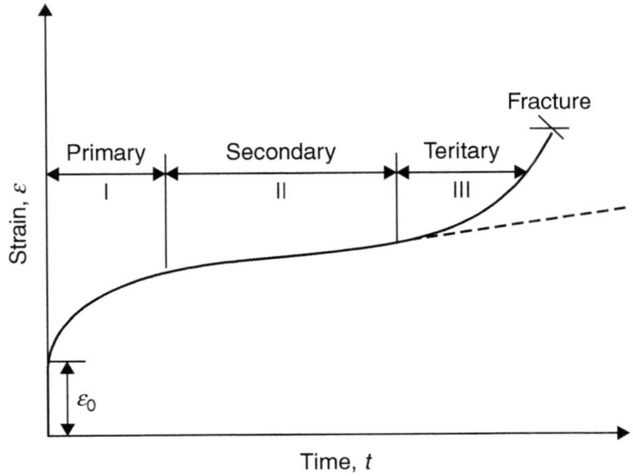

3.1 A typical creep curve.[3]

the yield strength of the material, the instantaneous strain is composed of a plastic strain component.

The primary creep (also known as transient creep) region, as the name suggests, describes the first or initial stage of creep deformation and corresponds to period ‘b’ in Andrade’s work.[1] Such a region is characterized by a strain rate decreasing with time. The decrease in strain rate continues until the secondary stage (also known as steady-state creep) is attained. In the secondary creep region (period ‘c’ in Andrade’s study) the strain rate of deformation remains constant. This is evident in Fig. 3.1 with the secondary creep region described by a straight line indicating a constant slope. The secondary stage strain rate is the minimum strain rate of the creep curve. The useful creep life of most engineering materials is generally estimated from secondary stage creep strain rate values. However such a methodology might not be applicable for materials which have a large primary creep region or where the tertiary creep region completely dominates the primary and secondary creep stages. The tertiary creep region is the last stage of creep deformation and concludes with the failure of the material. In the tertiary creep regime (as identified by Hanson and Wheeler[2]) the material undergoes deformation at very high strain rates. The tertiary stage of the creep curve usually occurs over significantly smaller time periods in comparison to the primary and the secondary stage, and is often regarded as ‘fracture’ mode.

Nature of the creep curve

The previous section described the different regions of a creep curve and their corresponding characteristics. Mechanistically, the creep curve is a result of the changes occurring in a material at a microstructural level.

The creep curve is basically the outcome of the competition between the processes of strain hardening and recovery. Materials usually strain harden during plastic deformation due to dislocation multiplication. The strain hardening is a kind of 'defense' mechanism in response to an applied stress. Further plastic deformation can occur only if the applied stress exceeds the increase in flow stress of the material due to strain hardening. Alternatively, deformation can proceed at the initial applied stress if the material softens. The mechanism of recovery acts to soften a deformed specimen thus allowing further plastic deformation. In the primary stage, the rate of strain hardening is greater than the rate of recovery. This is due to the formation of a more resistant creep substructure. The substructure could be the formation of dislocation networks or the arrangements leading to the formation of subgrains. In the secondary stage of creep, the rate of strain hardening is balanced by the rate of recovery due to dislocation annihilation and deformation occurs at a constant strain rate. In the tertiary stage of creep, the increase in applied stress due to a reduction in specimen cross-sectional area surpasses the increase in flow stress due to strain hardening. The reduction in specimen cross-sectional area can be due to *necking* or *internal void formation*. The tertiary creep is often associated with metallurgical changes such as the coarsening of precipitate particles, recrystallization or diffusional changes in phases present, void formation and so forth.

Figure 3.2 depicts four types of creep curves that have been generally observed.[4] The shape of the creep curve is dependent on the initial condition

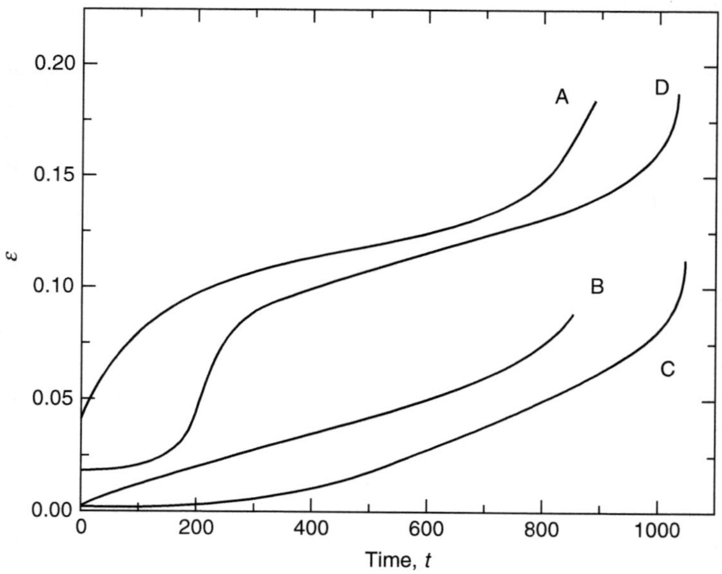

3.2 Illustration of different types of creep curves.[4]

of the material prior to deformation. Curve A is a typical creep curve observed in several materials. The curve consists of a normal primary stage characterized by a decreasing strain rate, a secondary stage where deformation proceeds at a constant rate and a tertiary stage where the material deforms at increasing strain rates with time/strain leading to eventual failure. Such creep curves are usually exhibited by annealed metals and certain alloys (known as class-M or class-II type). In comparison to curve A, curve B depicts a very small primary creep stage. In fact, it appears as if the material enters the steady state immediately. Such a type of curve is obtained when the substructure pertaining to creep remains constant such as in some alloys (known as class-A or class-I type). Curves of type C are obtained from materials that have been previously crept at a higher stress. The increasing creep rate over the primary creep stage is due to the recovery of the substructure corresponding to the previous steady-state condition. The sigmoidal type of creep curve (curve D) suggests the nucleation and spread of slip zones until a steady state is achieved. Such creep behavior has been exhibited by certain dispersed phase alloys.

In certain cases, it is possible that the total creep curve is in the primary stage, and the secondary and tertiary creep stages are not attained at all. Such a curve has been seen for materials tested at low temperatures ($T < 0.3\ T_M$) where effects due to diffusion are suppressed (no annealing or recovery) and the entire deformation is due to work hardening (dislocation controlled). The primary creep strain rate tends towards a value of zero at long periods. The strain hardening due to long range dislocation interactions precludes a constant rate of creep deformation. The absence of recovery processes due to the low test temperatures allows the strain hardening process to be the creep-controlling mechanism. This eventually leads to an increase in strength of the material to a value greater than the applied stress value. Further deformation can only occur under the application of a higher stress or in the presence of a higher temperature. Such a behavior is described as an exhaustion creep behavior and the creep curve can be described by a logarithmic creep equation.[4]

3.2 Standard creep equations

On the basis of his studies, Andrade[1] proposed that the creep curve could be described by an equation of the form

$$\varepsilon = \varepsilon_0 \left(1 + \beta t^{1/3}\right) e^{\kappa t}, \tag{3.1}$$

where ε is the strain, t is the time and β and κ are constants. The transient creep, that is the primary creep stage, is described by β; the constant κ representing the secondary stage describes an extension per unit length that proceeds at a constant rate. Elsewhere, the dependence of creep rate on time was described by a power series[5]

$$\dot{\varepsilon} = \sum_i a_i t_i^{-n_i}, \tag{3.2}$$

where $\dot{\varepsilon}$ is the creep rate, a_i and n_i are functions of both temperature and stress. The primary stage of a normal creep curve, namely curve A, can be described by Equation [3.2] when n attains a value of 2/3. In such a case, the time dependence of creep strain (ε) is described by

$$\varepsilon = \varepsilon_0 + \beta t^{1/3}, \tag{3.3}$$

where ε_0 is the instantaneous strain, β is a constant and t is time. Equation [3.3] is in accordance with the time law of creep proposed by Andrade,[1] known as Andrade's β-flow.

For steady-state creep, $n = 0$. The creep curve is then described by

$$\varepsilon = \varepsilon_0 + \dot{\varepsilon}_s t, \tag{3.4}$$

where $\dot{\varepsilon}_s$ is the steady-state creep rate. The total creep curve consisting of the primary, secondary and tertiary region can be then described by

$$\varepsilon = \varepsilon_0 + \beta t^{1/3} + \dot{\varepsilon}_s t + \gamma t^3, \tag{3.5}$$

where γt^3 describes the tertiary component of the creep curve. There were several other time-creep law equations proposed to describe creep data. A major objection to the Andrade's β-flow equation is that it predicts infinite creep-rate at the instant of loading (i.e. as t approaches 0) which is considered to be unrealistic.[4] Garofalo[6] proposed the following equation:

$$\varepsilon = \varepsilon_0 + \varepsilon_t \left(1 - e^{-rt}\right) + \dot{\varepsilon}_s t, \tag{3.6}$$

where ε_t is the limit for transient creep and r is the rate of exhaustion of the transient creep that is a function of the ratio of initial creep strain rate $(\dot{\varepsilon}_{t \to 0})$.[4] An extension of the Garofalo equation that further includes the tertiary creep regime, would be[7]:

$$\varepsilon = \varepsilon_0 + \varepsilon_t \left(1 - e^{-rt}\right) + \dot{\varepsilon}_s t + \varepsilon_L e^{p(t - t_{ot})}, \tag{3.7}$$

where ε_L is a constant equal to the smallest strain deviation from steady state at the onset of tertiary creep, p is a constant and t_{ot} is the time for onset of tertiary creep.

In comparison to the normal creep equation, the logarithmic creep behavior is usually described by

$$\varepsilon = \varepsilon_0 + \alpha \ln(1+\gamma t), \qquad [3.8]$$

where α and γ are constants. This equation indicates that over a long period of time, the strain rate of deformation tends to become zero. Such an equation, as discussed in the previous section, would be useful for describing exhaustion creep.

While Andrade's equation is an empirical correlation borne out of his experimental observations, a number of researchers have derived similar relations on the basis of physically based mechanisms. For example, the Garofalo equation has been derived by considering sub-structural changes during deformation and by modeling the whole phenomenon as a first order reaction. Webster *et al.*[8] and Amin *et al.*[9] have correlated the stress and temperature dependence of ε_T, r and $\dot{\varepsilon}_s$ to the rate controlling mechanisms of high temperature creep using first order reaction rate concepts. They found that the Garofalo equation can be derived by assuming that the transient creep follows a first order kinetic reaction rate theory with a rate constant $1/\tau = K\dot{\varepsilon}_s$ that depends on stress and temperature. Here $1/\tau$ is the relaxation frequency which is similar to r in Equation [3.6] and τ is the relaxation time for rearrangement of dislocations during transient creep controlled by dislocation climb. The physical mechanisms, for example, dislocation climb, that have been suggested to control or govern creep will be discussed in a subsequent section.

During their analyses of the creep results on Zr-based alloys, Murty[10] found the following equation better describes the primary creep compared to the Garofalo equation:

$$\varepsilon = \frac{A\dot{\varepsilon}_s t}{1 + AB\dot{\varepsilon}_s t}, \quad A = \frac{\dot{\varepsilon}_i}{\dot{\varepsilon}_s} \text{ and } B = \frac{1}{\varepsilon_{tr}}, \qquad [3.9]$$

where A is the ratio of the initial strain rate to steady-state value and B is the inverse of the extent of the primary creep. A was found to be around 10 as reported earlier by Dorn and co-workers[4,9] for many materials that behave like pure metals and class-M alloys.

A distinctly different approach was used by incorporating anelastic[11] strain in the description of the primary creep regime and especially in predicting the transients in creep strain due to sudden stress changes.[12] Accordingly, the total strain at any given time is given by

$$\varepsilon = \varepsilon_E + \varepsilon_a + \varepsilon_p, \qquad [3.10]$$

where the subscripts E, a and p correspond to the elastic, anelastic and plastic strains; the plastic strain (ε_p) is given by Equation [3.9]. The anelastic strain (ε_a) is time dependent, completely recoverable strain in contrast to the permanent plastic strain which is not recoverable and elastic strain which is recoverable but instantaneous (time-independent). The anelastic strain at the loading is given by

$$\varepsilon_a = \frac{\sigma}{\mu} - \left\{\left(\frac{\sigma}{\mu}\right)^{\nu} + \nu\alpha^{*}t\right\}^{\frac{1}{\nu}}. \qquad [3.11]$$

In the above equation, $\nu \sim 7$ and α^* ($\sim 2.5 \times 10^{22}$) are material constants in Hart's equation of state and μ is the anelastic modulus.

In the following section we discuss the importance of the different parameters, namely stress temperature and microstructure. The strain rate of deformation, $\dot{\varepsilon}$ can be expressed as

$$\dot{\varepsilon} = f(\sigma, T, \text{microstructure}), \qquad [3.12]$$

where σ is the applied stress and T is the test temperature.

3.2.1 Effect of stress and temperature

The steady-state strain rate of creep deformation, at a given temperature, has been found to be directly dependent on the applied stress. The functional dependence of strain rate on stress can be expressed by Norton's law[13]

$$\dot{\varepsilon}_s = K\sigma^n, \qquad [3.13]$$

where K is a constant and n is the stress exponent. Similarly, for a constant applied stress, the rate of creep deformation increases with increasing temperature. The effect of temperature can be understood by including an extra term indicated in the following equation

$$\dot{\varepsilon}_s = K_1\sigma^n \exp\left(\frac{-Q_c}{RT}\right), \qquad [3.14]$$

where K_1 is another constant and Q_c is the activation energy of creep deformation. The activation energy term is included due to the fact that the creep deformation is considered to be a first order reaction rate process. The magnitude of the activation energy is dependent upon the physical mechanism governing the deformation process.

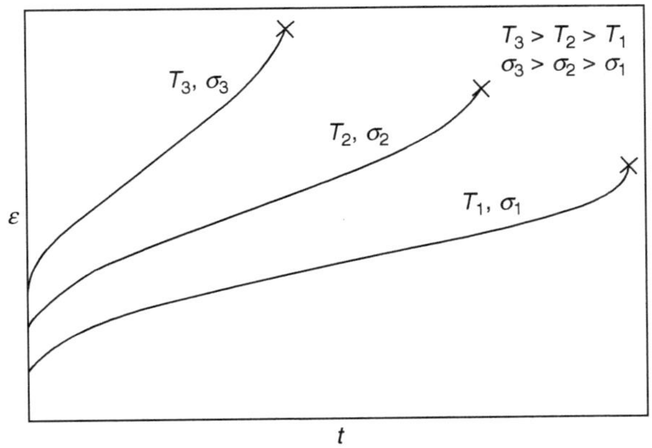

3.3 Illustration of the effect of stress and temperature on creep behavior of a material.

The effect of stress and temperature is clearly illustrated by Fig. 3.3. With increasing stress and temperature, the instantaneous strain at the time of stress application increases, the steady-state creep rate is increased and the rupture lifetime is diminished.

3.2.2 Effect of microstructure

The analysis of any creep data is made by assuming the microstructure to be constant. Some of the microstructural features that could change during the course of a test are phase composition, precipitate size and distribution, and grain size. Thus to estimate the different creep parameters and to determine the mechanism of creep, it is necessary to keep the microstructure constant. To this end, materials are usually heat treated at temperatures higher than the test temperature. Even though thermal stabilization establishes a constant microstructure during the course of a test, stress-assisted processes altering the microstructure cannot be ruled out. Non-equilibrium structures, namely nanocrystalline materials undergo stress-assisted microstructural changes that prevent the attainment of a constant creep microstructure.[14] Creep tests on such materials should be carried out at stresses lower than the critical stress at which microstructural changes could be initiated.[15]

The most important microstructural parameter that plays a major role in controlling the creep properties of a material is the grain size. The dependence of the strength of a material on its grain size can be understood through the Hall–Petch relationship which states that materials with finer grain sizes possess greater strength than materials with larger grain size[3]:

$$\sigma_y = \sigma_0 + \frac{k_y}{\sqrt{d}}, \qquad [3.15]$$

where σ_y is the yield strength, σ_0 the friction hardening (stress felt by dislocations while moving through the lattice), d the grain size and k_y is known as the Petch-unpinning coefficient that frees the dislocations locked by interstitial solute atoms. This is true at low temperatures where grain boundary sliding is not dominant.

On the contrary, under creep conditions the reverse is true. Materials with finer grain size creep faster than coarse grained materials at higher temperatures and at lower stresses. There are certain creep mechanisms that operate faster in finer grained materials in comparison to coarse grained materials. Hence, it is necessary to have knowledge of the grain size of a material. The dependence of the steady-state creep rate on the grain size is understood through the following equation:

$$\dot{\varepsilon}_s = K_2 d^{-p} \sigma^n \exp\left(\frac{-Q_c}{RT}\right), \qquad [3.16]$$

where K_2 is a constant, d is the grain size and p is the grain size exponent. As is clear from this equation at a given stress and temperature, finer grain sized materials are expected to creep faster than coarser grained materials. However for dislocation-based mechanisms which are not grain size dependent, the strain rate of deformation would be the same for both fine grained and coarse grained materials.

3.3 Identifying the mechanisms of creep

It is possible to identify a particular micromechanism of creep through knowledge of the stress exponent (n), the activation energy (Q_c) and the grain size exponent (p). Table 3.1 describes the different mechanisms of creep and their relationship to the creep parameters n, Q_c and p. In addition to these three parameters, the relevant mechanism of creep can be identified through knowledge of the creep constant A given by K_2 in Equation [3.16]. Each mechanism of creep possesses a distinct value of A.

The mechanisms of creep can be broadly classified into two types: diffusion-based processes and dislocation-based processes. Coble creep and Nabarro–Herring (N–H) creep are mechanisms of deformation that fall under the category of diffusion-based processes. Harper–Dorn (H–D), viscous glide and dislocation climb are mechanisms of creep that fall under the category of dislocation-based processes. Grain boundary sliding (GBS) appears to proceed by a combination of diffusion- and dislocation-based

Table 3.1 Identification of the particular mechanism of creep from parameters n, p, Q_c and A (Equation [3.16])

Creep mechanism	n	p	Q_c	A
Nabarro–Herring (N–H)	1	2	Q_L	12
Coble	1	3	Q_{gb}	150
Harper–Dorn (H–D)	1	0	Q_L	3×10^{-10}
Spingarn–Nix (S–N)	1	3	Q_{gb}	75
Grain boundary sliding (GBS)	2	2	Q_{gb}	200
Viscous glide	3	0	Q_s	6
Dislocation climb	4–7	0	Q_L	6×10^7
Power-law breakdown	>7	–	Q_L	–

processes. Power-law breakdown (PLB) occurs at relatively high stresses equal to or greater than around $10^{-3}E$ (E is modulus of elasticity); this has also been correlated with strain rates at or greater than $10^{-9}D$ (D is diffusivity).[16] In this high stress and high strain rate regime, the creep-rates vary with stress via an exponential function. While this region is observed in all materials under appropriate conditions, the underlying mechanism of PLB is still a moot factor.

In this table, Q_{gb} is the grain boundary diffusion activation energy, Q_L the activation energy for lattice diffusion, and Q_s the activation energy for solute diffusion. As Table 3.1 suggests, a stress exponent value of unity (Newtonian viscous) imply that the deformation mechanism could be Coble, N–H or H–D creep. However knowledge of the grain size exponent or the activation energy would establish the right mechanism of creep. For example, a stress exponent of 1 and activation energy equal to that for lattice diffusion would suggest the mechanism of creep to be either N–H or H–D. But if the grain size exponent is equal to 2, it would establish that the mechanism of deformation is N–H. On the other hand, if the steady-state strain rate is found to be independent of the grain size ($p = 0$), the mechanism of creep is H–D. The fact that the Coble creep mechanism is more sensitive to the grain size and is controlled by the grain boundary diffusivity, it becomes dominant at lower temperatures and/or smaller grain sizes while N–H creep becomes predominant at relatively larger grain sizes and higher temperatures. H–D creep becomes significant at large grain sizes and bulk single crystals. Thus, knowledge of the creep parameters would help in identifying the exact mechanism of creep.

3.3.1 The $n = 1$ regime

A stress exponent value of 1 suggests the mechanism of creep to be Coble, N–H or H–D controlled. Recent studies have shown that the Spingarn-Nix

(S–N) mechanism with a stress exponent value of 1 could also be a viable creep mechanism.[17] This mechanism could be identifiable through microstructural evaluation following creep deformation revealing the limitation of only considering the creep parameters. In this section, the mechanism of Coble, N–H, H–D and S–N creep will be discussed.

N–H and Coble creep

The possibility of creep occurring by stress-assisted diffusional mass transport through the lattice was first considered by Nabarro[18] in 1948 and Herring[19] in 1950. A few years later, Coble[20] proposed that grain boundaries could also provide an alternative path for stress-directed diffusional mass transport to take place. Figure 3.4 provides schematics of N–H and Coble creep mechanisms.

As Fig. 3.4 indicates, under the application of a stress, grain boundaries normal to the applied stress will develop a higher concentration of vacancies. On the other hand, grain boundaries parallel to the applied stress (lateral grain boundaries) will experience compressive stresses and will have a reduction in vacancy concentration. This causes a concentration difference between the two boundaries leading to a flux of vacancies diffusing from the normal grain boundaries to the parallel or lateral grain boundaries (atoms diffuse in the opposite direction). The diffusion of vacancies can occur through the lattice (N–H) or via grain boundaries (Coble creep). The diffusion of vacancies or the motion of atoms from one grain boundary to another leads to a crystal strain which in turn contributes to the deformation of the grains and consequently the material. The calculations of the steady-state flux of vacancies and the corresponding steady-state creep rate lead to the following relationships for N–H and Coble creep, respectively,

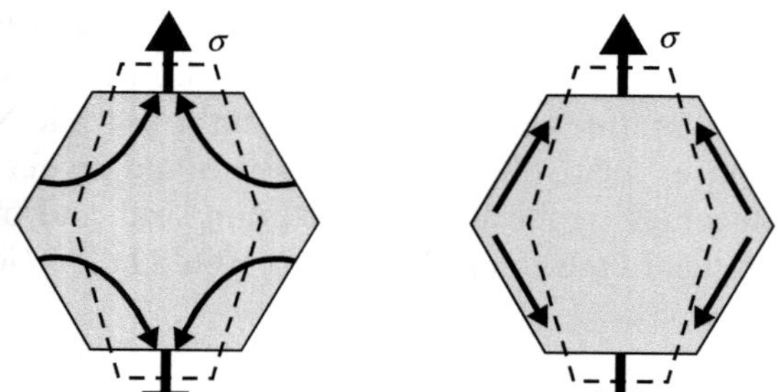

3.4 (a) Schematic of N–H creep. Mass transport occurs through the lattice. (b) Schematic of Coble creep. Mass transport occurs along grain boundaries.

$$\dot{\varepsilon} = B_H \frac{D_L \sigma \Omega}{d^2 kT} \qquad [3.17]$$

$$\dot{\varepsilon} = \frac{B_c}{\pi} \frac{D_B \delta_B \sigma \Omega}{d^3 kT}. \qquad [3.18]$$

In Equation [3.17], B_H is the N–H constant and has a value of about 12–15, D_L is the lattice diffusivity, Ω is the atomic volume and k is the Boltzmann constant. In Equation [3.18], B_c is the Coble constant and has a value of 150, D_B is the grain boundary diffusivity, and δ_B is the grain boundary thickness. From the above relationships it is understood that the creep strain rate varies linearly with stress and is inversely proportional to the grain size. Usually with decreasing grain size, it is observed that Coble creep dominates N–H creep and *vice versa*. But when both the mechanisms operate in parallel the strain rate can be expressed by

$$\dot{\varepsilon} = \frac{B \sigma \Omega}{d^2 kT} D_{\mathrm{eff}}, \qquad [3.19]$$

where D_{eff} is the effective diffusion coefficient and is given by

$$D_{\mathrm{eff}} = D_L \left(1 + \frac{\pi}{d} \frac{D_B \delta_B}{D_L} \right). \qquad [3.20]$$

From Equations [3.19] and [3.20], it is clear that the grain boundary diffusion will contribute more to the creep rate for larger D_B/D_L ratios and for smaller grain sizes. In the derivation of the N–H and Coble creep equations, the following assumptions were made:

i. The grain boundaries are perfect sources and sinks of vacancies, and
ii. The initial dislocation density of the crystal is low.

This implies that the only sources and sinks for vacancies are the grain boundaries. Since their discovery, both N–H and Coble creep have been found to occur in a variety of materials and experimental results have agreed well with the proposed theory.[21–26]

Harper-Dorn creep

Through their classic experiments on high purity aluminum (99.95%), Harper and Dorn[22] came across a rate controlling mechanism that was seemingly independent of the grain size but still displayed characteristics generally associated with Newtonian viscous creep. The creep experiments carried out at $0.99T_m$ provided stress exponent and activation energy values ($n = 1$ and $Q = Q_L$) considered unique to N–H creep. However the grain size-independent behavior of the material combined with experimental strain rates around 1400 times larger than theoretical N–H creep predictions were suggestive of a new mechanism of creep. When the results obtained by Harper and Dorn were compared with theoretical N–H creep predictions, a large discrepancy was noted. In addition, by using markers Harper and Dorn[22] found that the strains in the center of the grain are equal to the macroscopic strains noted in the creep experiments. The steady-state strain rate of deformation of this creep mechanism, now known as Harper–Dorn (H–D) creep mechanism, is given by

$$\dot{\varepsilon} = A_{\mathrm{HD}} \frac{\sigma b D_L}{kT}. \qquad [3.21]$$

Studies over the years, on a host of other materials have led to a belief that H–D creep is seen only in large grained materials (studies carried out by Harper and Dorn were on Al with a grain size of 3.3 mm) and at very low stresses and high temperatures. The primary characteristics of high temperature H–D creep are summarized below:[5]

- The stress exponent is equal to one.
- The creep rate is independent of grain size and similar creep rates are observed both in polycrystals and single crystals.
- The activation energy for creep is equal to that for lattice diffusion.
- The creep curves show a distinct primary stage which is followed by a steady-state stage.
- There is a random and reasonably uniform distribution of dislocations in specimens crept to the steady state.
- The dislocation density is low, of the order of 5×10^7 m^{-2}, and is independent of stress.
- Very similar results are obtained in pure metals and solid solution alloys revealing that solute concentration has no effect on the creep behavior at these conditions.

While the initial studies were confined to very high temperatures (>0.95 T_M), recent studies show that H–D creep can be rate controlling at intermediate temperatures as well. Creep studies in alpha titanium,[23] beta cobalt,[24] alpha iron[25] and alpha zirconium[26] have shown the presence of a H–D regime at homologous temperatures of around 0.35 to 0.6 for applied stresses around 9×10^{-5}G (G is shear modulus) and grain sizes of around 500 μm.

Several models were proposed to understand the mechanism of H–D creep. The models of high temperature H–D creep were discussed in detail by Langdon and Yavari.[27] Barrett *et al.*[28] proposed a model based on the creep strain resulting from dislocation glide with dislocation multiplication through climb. Murty[29] suggested that H–D creep in Al-Mg solid solution arises from a modified viscous creep glide process (described later) with stress-independent dislocation density. More recently Kumar *et al.*[30] summarized the experimental results obtained on ceramic single crystals. Purity of crystals and a low initial dislocation density were cited as necessary conditions to unequivocally establish the presence of H–D creep as a viable mechanism of deformation. A review of the viscous creep with $n = 1$ was recently made by Lingamurty *et al.*[31]

Spingarn–Nix slip-band model

The S–N model[32] is based on the fact that dislocation climb, when assisted by grain boundaries, can occur at activation energies smaller than those of lattice diffusion. Since grain boundary diffusion is much faster than lattice diffusion, climb rates are increased in the proximity of a grain boundary. The S–N model thus relates to the ideas of diffusional creep and dislocation climb at grain boundaries. This model, also known as the slip-band model, provides a physical mechanism to explain the observation of activation energies equal to the grain boundary self diffusion and a stress exponent equal to 1. According to this model, creep occurs by shearing along slip-bands blocked by grain boundaries. The creep strain at the boundary is in turn accommodated by diffusional flow. A schematic of the slip-band model is shown in Fig. 3.5. Under the application of a shear stress, the slip-band/

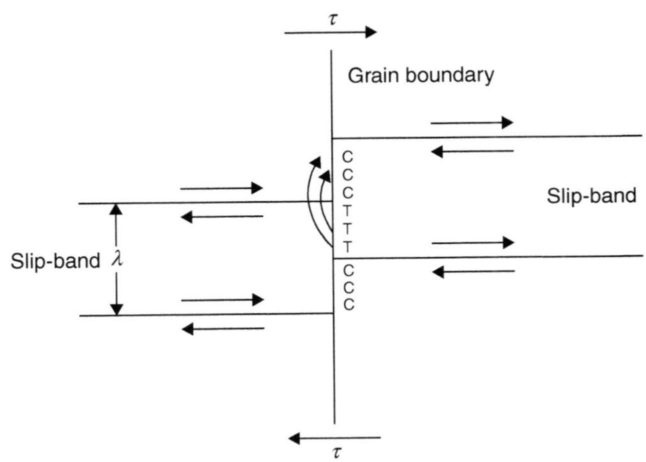

3.5 Schematic of the slip-band model.

grain interfaces slide, generating compressive (C) and tensile (T) tractions at the grain boundary. In order to relieve these tractions, atoms flow from regions under compression to regions experiencing tensile stresses. This atomic flow is accompanied by grain boundary sliding that causes the shear of a continuous slip-band. A mathematical analysis using Fick's law for diffusion of vacancies yields the following expression for the strain rate due to the absorption of dislocations into the boundary:[33]

$$\dot{\varepsilon}_{SB} = \frac{50l}{\lambda^4}\frac{\sigma\Omega}{kT}D_b\delta, \qquad [3.22]$$

where λ is the slip-band width, l is the slip-band length, Ω is the atomic volume, δ is the grain boundary thickness and the rest of the terms are as defined before. The slip-band length, l, can be considered equal to the mean linear intercept grain size, d. Recently Gollapudi *et al.*[17] have studied the feasibility of the slip-band model as a viable creep mechanism in a titanium based alloy.

Microstructural features

The first microstructural evidence for diffusional creep was provided by Squires *et al.*[34] who carried out creep studies on Mg-0.5Zr at 723 and 773K. The initial microstructure had a uniform distribution of inert ZrH_2 particles and investigation of the microstructure of the crept specimen, as shown in Fig. 3.6a, depicts the presence of regions denuded of the inert particles. These denuded zones mostly formed near transverse grain boundaries. Squires *et al.*[34] attributed the formation of denuded zones to diffusional creep of the Mg alloy. Under the application of a stress, Mg atoms diffuse from parallel grain boundaries to the transverse grain boundaries causing a slight elongation of the grains. The inert particles do not travel along with the Mg particles and their absence adjacent to the transverse grain boundaries causes the formation of denuded zones. In subsequent years, denuded zones have been observed by other groups in Mg-Zr[35] and Mg-Mn[36] alloys. Even though the formation of denuded zones as a consequence of diffusional creep appears reasonable, it has been a matter of regular debate.[37–40] Jaeger and Gleiter[41] carried out experiments on a bamboo structured copper coated with Al_2O_3 film. Diffusional creep experiments were carried out on copper at a temperature of around 1348 K. At the conclusion of the creep experiment, it was observed that the alumina film fractured in a few places. The fracturing of the alumina film was ascribed to the deformation incompatibility of the alumina film and the copper beneath. The copper grains elongated under the application of the stress whereas the alumina film did not deform to the same extent causing fracturing of the film. The

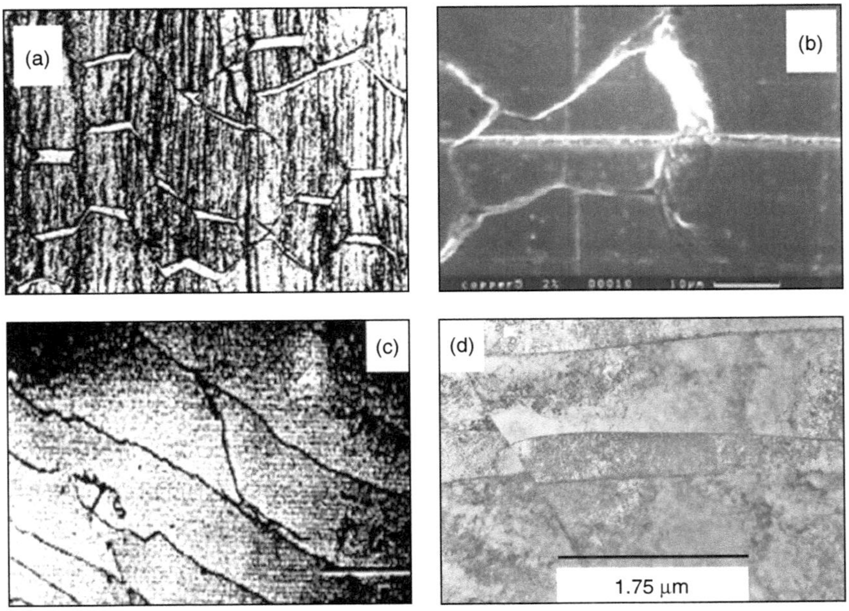

3.6 (a) Denuded zones in Mg-0.5Zr alloy,[38] (b) grain boundary grooves in copper,[42] (c) dislocation cross slip in Al[43] (d) slip-bands in a Ti alloy.[17]

elongation of the copper grains was ascribed to diffusional creep deformation. Recently McNee *et al.*[42] carried out detailed experiments on OFHC grade copper in the diffusional creep regime. A surface scratch technique was employed to establish the operation of a diffusional creep mechanism. In addition to measurements of the surface scratch displacements, grain boundary grooves were identified and subsequently quantified through an atomic force microscopy (AFM). Grooves, as shown in Fig. 3.6b, were formed predominantly on boundaries transverse to the applied stress. Grain boundary grooves were thus suggested as a microstructural feature characteristic of diffusional creep. While denuded zones, elongated grains and grain boundary grooves are essentially features developed due to N–H or Coble creep, the features associated with H–D and S–N creep are different. Dislocations cross slipping[43] as shown in Fig. 3.6c and slip-bands sheared by grain boundaries,[17] Fig. 3.6d, are suggested to be evidence of H–D and S–N creep, respectively.

3.3.2 The $n = 2$ regime: grain boundary sliding

Grain boundary sliding as a mechanism of creep is usually observed at temperatures higher than $0.4T_M$. GBS is typically a response of grain boundaries (usually high angle) to an applied shear stress and is supposed to occur by the relative motion of grains along a common boundary or along a narrow zone immediately adjacent to the boundary. The relative motion of grains along a common boundary is known as Lifshitz GBS when the accommodating process is diffusional creep.[44] On the other hand, when the process of accommodation occurs by glide and climb of dislocations, the GBS process is termed as Rachinger sliding.[45] GBS with the deformation limited to a zone around the boundary comes under the category of Rachinger sliding. Accommodation is necessary to avoid the formation of voids at the grain boundary. Strain compatibility and relaxation of stress concentration are only possible through the process of accommodation, usually by diffusional flow.

Accommodation through diffusional flow

Ashby and Verall[46] proposed a model to describe the process of GBS accommodated by diffusional flow. According to this model

$$\dot{\varepsilon}_s = K\left(\frac{b}{d}\right)^2\left(\sigma - \frac{\sigma_0}{E}\right)D_{\text{eff}} \qquad [3.23]$$

where

$$D_{\text{eff}} = 9D_L\left(1 + 3.3\frac{\delta_B}{D}\frac{D_B}{D_L}\right) \qquad [3.24]$$

and σ_0 is the threshold stress. As shown in Fig. 3.7, a natural outcome of diffusional flow during GBS is grain switching. The grains change their neighbors during the process of sliding and such a change is assisted by diffusional flow. The threshold stress term present in Equation [3.23] appears

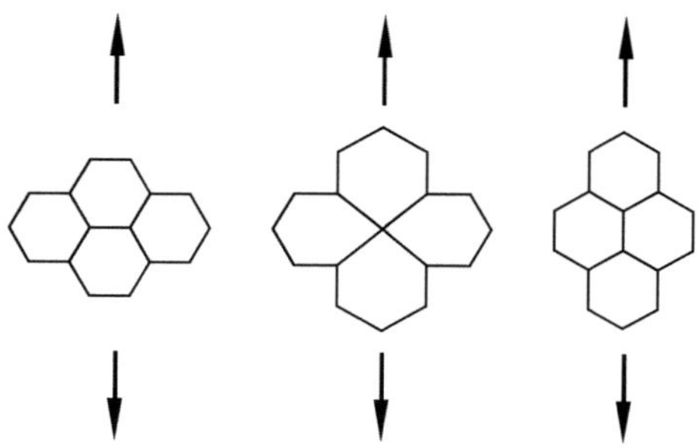

3.7 Illustration of the process of GBS accommodated by diffusional flow.[52]

due to an increase in grain boundary area, a resultant of the grain switching process. Though this model was successful in explaining the experimentally observed switching of grains during deformation, it failed to predict the stress dependence of strain rate. Moreover, Spingarn and Nix[47] suggested that grain switching cannot be entirely attributed to diffusional flow as the diffusion paths are physically incorrect.

Accommodation through dislocation movement

The earliest model to explain GBS accommodated by dislocation movement was proposed by Ball and Hutchison.[48] Later modifications to this model were brought about by Langdon,[49] Mukherjee[50] and Arieli and Mukherjee.[51] The Ball-Hutchison model is well illustrated by Fig. 3.8.[52] As shown in the figure, when the grains tend to slide under the application of a shear stress, strain incompatibilities and stress concentrations are developed at triple points[53] and grain boundary ledges.[54] Dislocation emission from these ledges and triple points is a natural consequence of the stress concentration. The emitted dislocations traverse the grain diameter until they encounter the opposite grain boundary at which point the dislocations start piling up and generate a back stress that prevents the further emission of dislocations. To enable further deformation, the lead dislocation at the pile-up climbs into

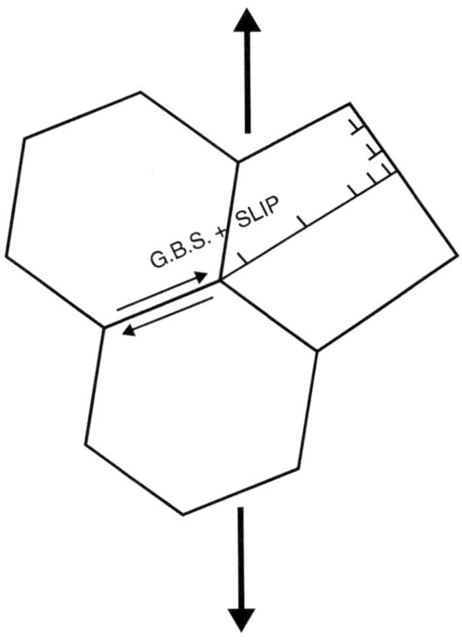

3.8 Illustration of the Ball–Hutchison model of GBS accommodated by dislocation movement.[58]

or along the grain boundary resulting in the rate controlling step being the climb of dislocations at the grain boundary.

Gifkins[56] presented a similar but slightly different model to explain the mechanism of GBS known as the 'core and mantle' model and considered the grain as the core and the regions adjacent to the grain boundary as the mantle. All deformation was assumed to occur only in the mantle region of the grain. This model and the rest of the models predicted strain rates which had an $n = 2$ dependence of the applied stress. According to this model

$$\dot{\varepsilon}_s = K\left(\frac{b}{d}\right)^p \left(\frac{\sigma}{E}\right)^2 D, \qquad [3.25]$$

where $p = 2$ or 3 and $D = D_L$ or D_B – depending on whether the motion of the dislocations is along the lattice or along the grain boundary respectively. Superplasticity – the ability of a material to exhibit high tensile elongations before failing – is primarily attributed to GBS. The mechanism of deformation in superplastic materials is supposed to be in accordance with the mechanisms discussed in this section.

Microstructural features

Fiducial markers are generally employed to study the contribution of grain boundary sliding to the total creep strain.[53] GBS leads to shearing of the fiducial markers and the shear offset provides a measure of the strain contribution. Since the stress concentrations developed during sliding are relieved by dislocation emission, dislocation activity can be expected in the vicinity of the grain boundary. Recent work by Gollapudi *et al.*[57] shows increased dislocation activity close to the grain boundary during deformation controlled by GBS. At the same time, dislocations emitted from a grain boundary are expected to travel across the grain until they encounter a grain boundary. These dislocations subsequently pile up which is relieved by dislocation climb. Dislocation pile-up close to the grain boundary was also observed by Gollapudi *et al.*[57] Figure 3.9a and 3.9b provide microstructural features associated with creep in the GBS regime.

3.3.3 The $n = 3$ regime: viscous glide (class-A alloys)

The $n = 3$ regime, though in principle corresponding to the power-law controlled ($n = 4$–7) creep mechanism, differs from it at a mechanistic level. The power law controlled creep mechanism (as will be discussed in the following section) is mostly dislocation climb-controlled commonly noted in pure metals and class-II or metal-class alloys. In contrast the $n = 3$ regime is dislocation glide-controlled creep usually exhibited by alloys known as

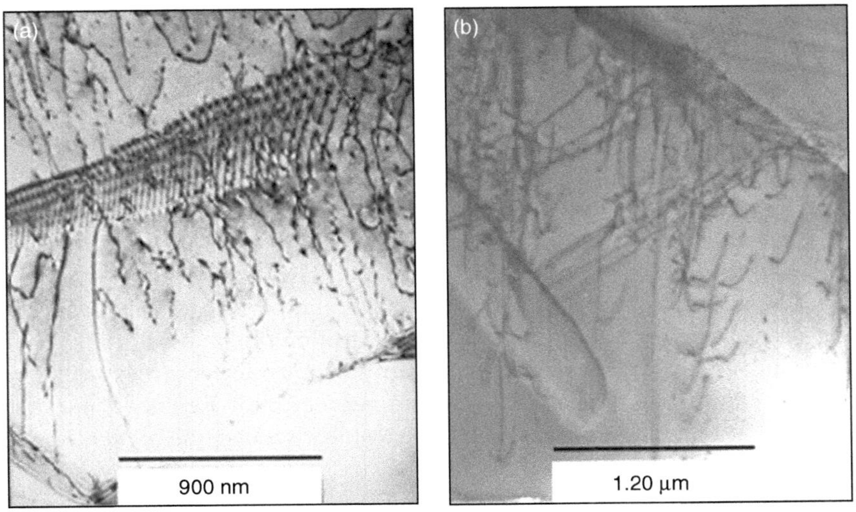

3.9 (a) Dislocation pile up and (b) enhanced dislocation activity in the vicinity of a grain boundary.[57]

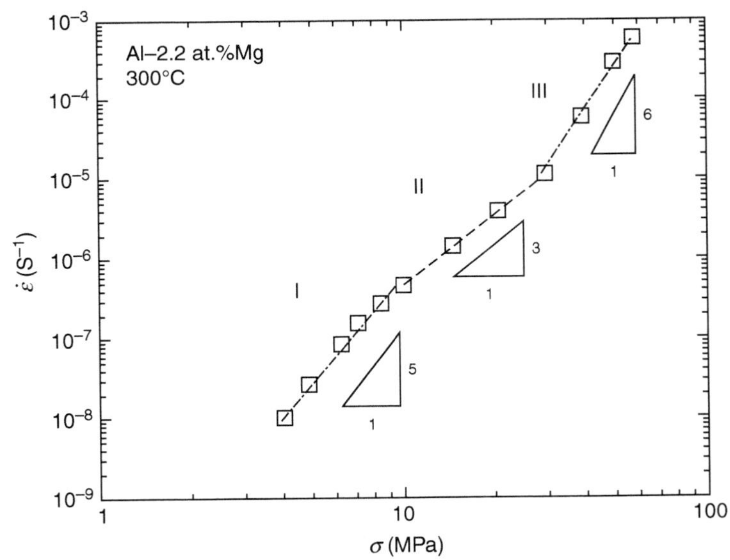

3.10 Illustration of the creep behavior of class-A type materials.

class-I or class-A, and hence the $n = 3$ regime at times is referred to as alloy type creep behavior. The creep behavior of materials can thus be classified into two groups: class-A and class-M. The creep behavior of solid solutions or class-A alloys at intermediate stresses and for specific material parameters consists of 3 regimes. As shown in Fig. 3.10, as stress increases the stress exponent changes from 5 (region I) to 3 (region II) and back to 5 (region III).

The creep behavior illustrated in Fig. 3.10 is a consequence of a competition between two rate controlling mechanisms: dislocation climb and dislocation glide. Once the dislocations are generated from Frank-Read (FR) sources on parallel glide planes, the leading edge dislocations first glide and then climb to annihilation. In pure metals dislocation glide is relatively faster compared to the diffusion-controlled climb and thus climb becomes the rate controlling process resulting in $n = 5$. In class-A alloys, the rate of glide is controlled by the diffusion of the solute atoms, thereby leading to a relatively slower rate of glide compared to that of climb whereby the viscous glide of dislocations becomes the rate controlling process with $n = 3$; this mechanism is known as Weertman microcreep.[54] Region II, the three power-law creep regime, is also known as the viscous glide regime. Viscous glide is described by

$$\dot{\varepsilon}_s = \frac{0.35}{A} D_s \left(\frac{\sigma}{E}\right)^3, \qquad [3.26]$$

where A is an interaction parameter that depends upon the viscous process controlling dislocation glide and D_s is the solute diffusivity.

The viscous process can be of different types. According to Cottrell and Jaswon,[58] the dragging force could be due to the segregation of solute atmospheres to moving dislocations. The dislocation speed in this case is controlled by the rate of migration of the solute atoms. Fisher[55] suggested that the viscous process had its origin in the destruction of the short range order in solid solution alloys. The disorder created by dislocation motion would result in the formation of a new interface thereby the interfacial energy becomes the rate controlling process. Suzuki[59] suggested that the dragging force was an outcome of solute atoms segregating to stacking faults. There are suggestions that the obstacle to dislocation motion could be the stress-induced local ordering of solute atoms. The ordering of the region surrounding a dislocation reduces the total energy of the crystal pinning the dislocation.

The three power-law creep region has usually been observed to occur in solid solutions with a large atom size mismatch. Alloys with higher concentrations of the solute atoms seem to prefer the three power-law creep regime as a viable creep mechanism. In fact, for very high concentrations of the solute atoms, regime II could be suppressed. In addition, class-A alloys usually exhibit either no or little primary creep or a region characterized by an increasing slope (increasing strain rate). This is in sharp contrast to pure metals and class-M alloys that exhibit a distinct primary creep curve with a decreasing strain rate; distinguishing features of class-A and class-M alloys

3.11 Deformation microstructure in Nb-modified Zr-alloy crept in the three power-law regime.[61]

were summarized by Murty.[60] Some of the different alloys that exhibit three power-law creep behavior are Al-Zn, Al-Ag, and Ni-Fe alloys.

Microstructural features

Since the $n = 3$ region is dislocation glide-controlled, recovery based processes (such as climb) are considered to be less important. The deformation microstructures are found to consist of a large number of dislocations as shown in Fig. 3.11. In comparison to class-M alloys, the deformation microstructures of class-A alloys are devoid of subgrains.[61]

3.3.4 The n = 4–7 creep regime: five power-law creep

The five power-law creep regime is observed at higher stresses and lower temperatures. This creep regime is generally controlled by dislocation climb but describing it as dislocation climb-controlled creep is not appropriate. This is because some of the other creep regimes such as GBS could also be dislocation climb-controlled.

The five power-law creep regime is commonly displayed by class-M alloys and can be described by

$$\dot{\varepsilon} = A\,\sigma^{4-7} \exp\left(\frac{-Q_c}{RT}\right), \qquad [3.27]$$

where A is a constant and Q_c is the activation energy corresponding to the rate controlling mechanism. The activation energy indicated in Equation [3.27] is the apparent activation energy. This is because the effect of temperature on the elastic modulus is not included here and that could have a substantial effect.[62] The true activation energy can be obtained from the following equation

$$\dot{\varepsilon} = A\left(\frac{\sigma}{E}\right)^{4-7} \exp\left(\frac{-Q_c'}{RT}\right). \quad [3.28]$$

Here E is temperature-dependent modulus of elasticity,

$$E = E_0 - \frac{\mathrm{d}E}{\mathrm{d}T}(T - T_0). \quad [3.29]$$

The temperature-normalized stress (σ/E) term includes the effect of temperature, and thus the value of Q_c' obtained from Equation [3.28] is the true activation energy. The activation energy thus obtained has been found to be equivalent to the lattice self-diffusion activation energy.

Mechanisms of five power-law creep

There are several models that have been proposed to describe the rate controlling mechanism in the five power-law creep regime. The general consensus is that the five power-law creep regime is diffusion-controlled. This is evident from the equivalence between the activation energy for creep and that for self diffusion. In addition, factors affecting self-diffusion such as phase transformation, superimposed hydrostatic pressure, etc., similarly influence creep-rates thereby rendering support to the fact that the creep-rate is proportional to self-diffusivity (D_L). Thus, all models that have been proposed to explain the five power-law creep regime are built around the concept of a dislocation climb-controlled creep mechanism.

The earliest model to describe creep by dislocation climb was proposed by Weertman[56,63] who considered the creep processes to be a result of the glide and climb of dislocations, with climb being the rate controlling process. The glide motion of dislocations is impeded by long range stresses due to dislocation interactions and the stresses are relieved by dislocation climb and subsequent annihilation. The rate of dislocation climb is determined by the concentration gradient existing between the equilibrium vacancy concentration and the concentration in the region surrounding the climbing dislocation. Creep strain, however, arises mainly through the glide of dislocations. In the glide-climb model (Fig. 3.12) a dislocation produced by the Frank-Read source glides a distance 'L' until a barrier of height 'h' is

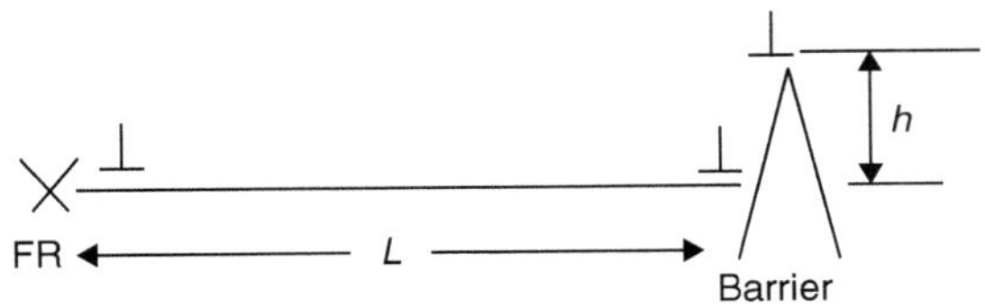

3.12 Schematic of dislocation glide-climb event.

encountered which it has to climb so that another dislocation can be generated by the source.

A rather simple way of deriving the relation between the strain rate and the applied stress and temperature is given by referring to Fig. 3.12: thus the total strain is given by,

$$\Delta\gamma = \text{strain during glide-climb event} = \Delta\gamma_g + \Delta\gamma_c \approx \Delta\gamma_g = \rho\ b\ L$$

$$t = \text{time of glide-climb event} = t_g + t_c \approx t_c = \frac{h}{v_c},\ v_c = \text{climb velocity}$$

$$\dot{\gamma} = \frac{\Delta\gamma}{t} = \frac{\rho b L}{h / v_c} = \rho b \frac{L}{h} v_c,$$

where $v_c \propto \Delta C_v\, e^{-E_m/kT}$, E_m = activation energy for vacancy migration

$$\dot{\gamma} = \rho b \frac{L}{h} v_c, \text{where}\ \Delta C_v = C_v^+ - C_v^- = C_v^0\, e^{-\sigma V/kT} = C_v^0\, 2\sinh\left(\frac{\sigma V}{kT}\right)$$

$$\text{so that}\ \dot{\varepsilon} = \alpha\rho b \frac{L}{h} v_c = \alpha\rho b \frac{L}{h} C_v^0\, e^{-E_m/kT}\, 2\sinh\left(\frac{\sigma V}{kT}\right). \qquad [3.30]$$

At low stresses, sinh $(\sigma V/kT) \approx \sigma V/kT$ so that $\dot{\varepsilon} = A_1 \rho b (L/h) C_v^o e^{-E_m/kT} (\sigma V/kT)$

$$\dot{\varepsilon} = A_1 \rho b \frac{L}{h} D_L \frac{\sigma V}{kT} \approx A_2 \rho\sigma \frac{L}{h} D_L. \qquad [3.31]$$

Assuming that the dislocation density (ρ) varies as stress is raised to the power 2 (σ^2), we find that

$$\dot{\varepsilon} = A\sigma^3 D_L. \qquad [3.32]$$

This is known as 'natural creep law' and Weertman showed that L/h in Equation [3.31] varied as $\sigma^{1.5}$ so that

$$\dot{\varepsilon} = AD_L\sigma^{4.5}. \quad [3.33]$$

This equation with $n = 4.5$ and $D = D_L$ agreed closely with the experimental results on pure aluminum. Subsequently, this has been generalized with n close to 5 and is referred to as the five power-law creep.

At high stresses, Equation [3.30] predicts an exponential stress dependence:

$$\sinh\left(\frac{\sigma V}{kT}\right) \approx \exp\left(\frac{\sigma V}{kT}\right) \text{ so that } \dot{\varepsilon} = A_1\rho b\frac{L}{h}D_L e^{\sigma V/kT}$$
$$\text{and } \dot{\varepsilon} = A\sigma^2 D_L e^{\sigma V/kT} \sim A' D_L e^{\sigma V/kT}, \quad [3.34]$$

as is commonly noted in the PLB regime. Both the power-law and exponential stress regimes can be combined into a single equation as proposed by Garofalo[6]

$$\dot{\varepsilon} = AD_L(\sinh B\sigma)^n, \quad [3.35]$$

which describes both the power-law creep regime at low stresses and exponential stress dependence at high stresses.

Another model that considered the non-conservative motion of dislocations was proposed by Barrett and Nix.[64] This model came to be known as the 'jogged screw dislocation' model. The rate controlling mechanism is the motion of screw dislocations containing edge jogs. The edge jogs impede the motion of the screw dislocations and the non-conservative motion of the edge jogs becomes the rate controlling mechanism. This model is similar to the Weertman model in the sense that the rate of climb of the edge jogs is dependent on the concentration gradient established by the climbing jogs. In the original model Barrett and Nix assumed the jogs to be of atomic height, but recently Viswanathan *et al.*[65] have shown that these jogs could be several times larger than atomic dimensions. The modified jogged screw model proposed by Viswanathan *et al.* has been used to satisfactorily explain the creep behavior of titanium aluminides[65] and some titanium and zirconium alloys.[66,67]

Ivanov and Yanushkevich[68] were the first to identify subgrain boundaries as important rate controlling features. The subgrain walls were suggested as obstacles to the motion of dislocations emitted within the subgrain. Subsequent plastic deformation could occur only when the dislocations were annihilated at the subgrain boundaries. This annihilation process is climb controlled.

There are a few other network and recovery based models that appear attractive. These models consider the dislocation networks (Frank networks) present inside the subgrains to explain the hardening and recovery during creep.

Microstructural features

Subgrain formation is widely believed to accompany deformation during creep in the five power-law regime. The formation of subgrains and other dislocation networks is a natural consequence of plastic deformation during creep. During plastic deformation the total dislocation density increases. An increase in dislocation density is concurrent with the increased work hardening particularly due to long range stresses acting on the dislocations. In the presence of the long range stresses, the dislocations tend to arrange themselves into low energy configurations. These low energy configurations are basically walls of dislocations inside a grain. The grain is thus divided into many smaller sections resulting in the formation of subgrains constituting both low angle tilt and low angle twist boundaries. Figure 3.13a shows the presence of distinct subgrains formed in a near α-Ti alloy.[69]

The subgrain size is an important microstructural feature of the five power-law creep regime. In fact, it has been empirically determined that the subgrain size (λ) varies with applied stress σ according to:[4]

$$\frac{\lambda}{b} = 20\left(\frac{\sigma}{E}\right)^{-1}. \quad [3.36]$$

The dislocations within a subgrain usually form three-dimensional networks known as Frank networks. These networks impede the movement of dislocations and can cause strengthening known as network strengthening. All the dislocations that are not associated with subgrain boundaries usually form Frank networks.

Jogged screw dislocations have also been considered as important features characteristic of deformation in the five power-law regime. Figure 3.13b provides the deformation microstructure of Ti-48Al crept in the five power-law creep regime.[65] Long dislocations with a screw orientation can be observed in Fig. 3.13b. Mills and co-workers[65–67] outlined the conditions under which jogs significantly larger than atomic height can be expected on screw dislocations. These conditions are: (a) screws are compact such that cross slip is relatively easy, (b) screw orientation is preferred due to strong lattice friction, and (c) jog pair or kink pair expansion is sluggish due to

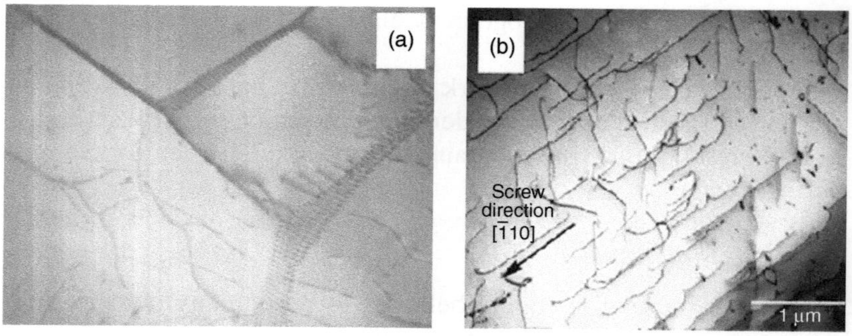

3.13 (a) Subgrain formation in Ti-3Al-2.5V in a five power-law creep regime[69] and (b) deformation microstructure in Ti-48Al showing the presence of jogged screw dislocations.[65]

lattice or solute friction. Accordingly Mills and co-workers suggest that the creep behavior of bcc solid solutions can be expected to be controlled by the jogged screw model.

3.3.5 The $n > 7$ creep regime: power-law breakdown

Creep regimes with a stress exponent greater than 7 are sometimes described as the PLB regime. There are certain dispersion strengthened alloys that exhibit high stress exponents, but such a behavior is rationalized by invoking a threshold stress for the operation of the mechanism of five power-law creep. The PLB, also known as the exponential creep regime, is described by

$$\dot{\varepsilon} = A\exp\left(\frac{-Q_c}{RT}\right)\exp\left(\frac{B\sigma}{kT}\right), \qquad [3.37]$$

where A and B are material constants. The constant B is related to the activation volume given by the area swept by dislocation and its Burgers vector (b). It has been observed that the PLB regime occurs at $\sigma > 10^{-3}E$. At such high stresses, the dislocation density increases more than that predicted by the Taylor equation (σ^2) and thus PLB could be controlled by the motion of dislocations. However the mechanism of steady-state creep deformation in the PLB regime is not clearly resolved.

The activation energy, Q_c, for PLB has been found to be smaller than the activation energy for self diffusion, Q_{sd}. In some cases, the activation energy was found to be equal to the activation energy for pipe diffusion suggesting that the mechanism of deformation could be dislocation climb but facilitated by short circuit diffusion of vacancies through the large

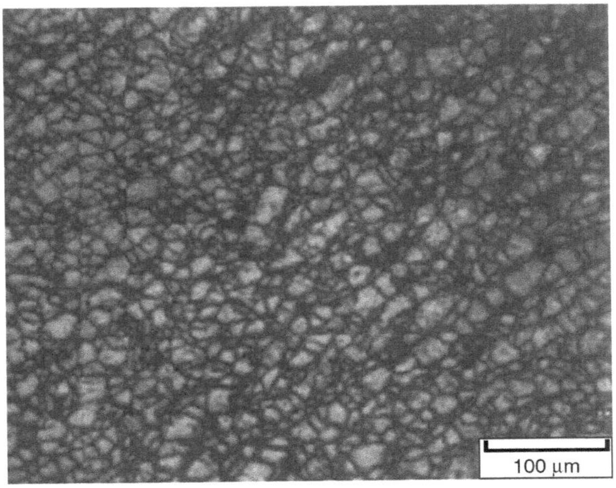

3.14 Microstructure of NaCl corresponding to the power-law breakdown regime.[72]

number of dislocation cores generated at high applied stresses. Sherby and Burke[16] suggest that vacancy diffusion due to vacancy supersaturation at high applied stresses can be associated with the rate controlling mechanism. Others have considered breaking down of subgrain walls[70] and cross slip or cutting of forest dislocations[71] instead of dislocation climb as the rate controlling mechanism. In any case PLB remains poorly understood and this is due to the relatively small number of studies that have been carried out in this regime.

Microstructural features

The PLB regime has been described as similar to a normal tension test. This is primarily due to the high strain rates of deformation experienced in the PLB regime. Naturally the microstructural features associated with such a regime should have similarities to those observed under normal tensile test conditions. TEM studies have shown the presence of high dislocation density. There is a tendency to develop dislocation cell structures. Figure 3.14 provides the deformation microstructure of NaCl crept in the PLB creep regime.[72]

3.4 Rate controlling mechanisms and activation energy

The activation energy of deformation is dependent on the rate controlling mechanism of creep. As shown in Table 3.1, the activation energy changes with the underlying creep mechanism. For example, the activation energy of creep is equal to that for grain boundary diffusion in the case of Coble creep and equal to lattice diffusion activation energy with N–H creep. Usually the activation energy of deformation is constant if a single thermally activated process is rate controlling. The Arrhenius plot – log of strain rate of deformation against reciprocal of temperature (in K) – is a straight line in such a case. However in certain cases more than one mechanism of creep, each with different activation energies, could be rate controlling. The Arrhenius plot in such a case is curved in the temperature range where the activity of the mechanisms is comparable. There are two cases which should be considered.

In case 1, the mechanisms of creep are independent of each other and hence occur simultaneously or in parallel. Each mechanism contributes a strain ε_i and the strain rates of deformation are additive. The total strain rate of deformation in such a scenario is given by

$$\dot{\varepsilon} = \sum_i \dot{\varepsilon}_i. \qquad [3.38]$$

For example, for the case of two mechanisms occurring simultaneously, the temperature dependence of strain rate is given by

$$\dot{\varepsilon}(T) = \dot{\varepsilon}_{01} \exp\left(-\frac{Q_1}{RT}\right) + \dot{\varepsilon}_{02} \exp\left(-\frac{Q_2}{RT}\right). \qquad [3.39]$$

The Arrhenius plot for such a scenario is shown in Fig. 3.15a, and if $Q_1 > Q_2$, mechanism 1 makes the dominant contribution to the creep rate at high temperatures and mechanism 2 becomes dominant at low temperatures. In the temperature range where the activity of both mechanisms is comparable, the Arrhenius plot is curved. At any given temperature, the *faster* mechanism is expected to control the rate of deformation.

In case 2, the mechanisms of creep occur sequentially and are known as series or sequential mechanisms. One mechanism cannot operate unless the other has taken place and *vice versa*. Here instead of the deformation strains, the time periods over which each mechanism has occurred are additive. Thus the total strain rate of deformation, assuming each mechanism contributes to the total strain, is given by

$$\frac{1}{\dot{\varepsilon}} = \sum_i \frac{1}{\dot{\varepsilon}_i}. \qquad [3.40]$$

For the case of two mechanisms occurring sequentially, the temperature dependence of the creep-rate is given by

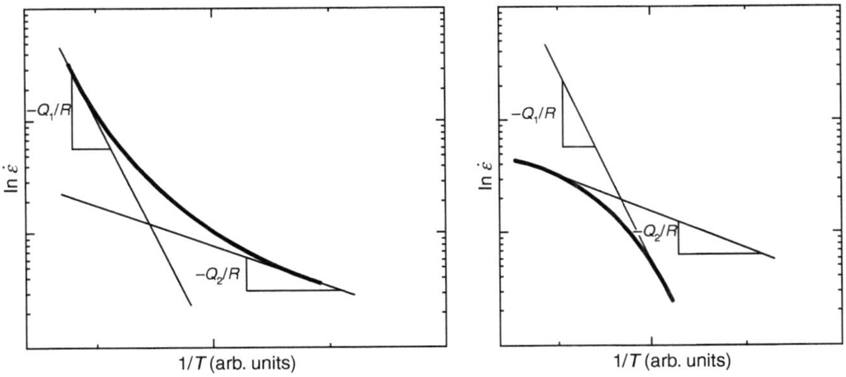

3.15 Arrhenius plots for (a) parallel mechanisms and (b) sequential mechanisms of creep.

$$\dot{\varepsilon}^{-1}(T) = \dot{\varepsilon}_{01}^{-1} \exp\left(\frac{Q_1}{RT}\right) + \dot{\varepsilon}_{02}^{-1} \exp\left(\frac{Q_2}{RT}\right). \quad [3.41]$$

The Arrhenius plot for such a scenario is illustrated in Fig. 3.15b. Here, in any given temperature range, the *slower* process dominates the creep rate. However the amount of creep strain may not necessarily be controlled by the slower process. It could be possible that the slower mechanism contributes little strain but allows the other mechanism with a greater strain contribution to operate. The dislocation glide-climb creep mechanism described earlier (Fig. 3.12) and Equations [3.30–3.34] is an example for this type of series mechanism while the simultaneous occurrence of N–H and Coble creep (Equations [3.17] and [3.18]) falls under parallel mechanisms.

3.5 Transitions in creep mechanisms

A material experiences transitions in mechanisms when the applied stress or the test temperature is varied.

3.5.1 The Bird–Mukherjee–Dorn equation[4]

As discussed earlier, for the same temperature and stress combinations, a material can creep via different mechanisms if the grain size is different. In fact, as Equation [3.25] would suggest, a material creeps with higher strain rates for smaller grain sizes. For relatively smaller grain sizes, creep could occur by diffusion of vacancies through the grain boundaries. But larger grain sized materials, under the same stress and temperature conditions, could creep by dislocation-based processes or by lattice diffusion processes.

In order to illustrate the effect of stress and temperature on transitions in mechanisms it is necessary to suitably modify the creep equation. Sherby analyzed steady-state creep-rate results using strain rate compensated by diffusivity versus stress normalized by temperature-dependent modulus of elasticity[16]

$$\frac{\dot{\varepsilon}}{D} = A\left(\frac{\sigma}{E}\right)^n \qquad [3.42]$$

so that different materials can be compared with each other. While this equation seems to work well, it would be more appropriate to use dimensionless strain rate as well, and Dorn and co-workers[4] proposed a dimensionless equation that can appropriately describe the effect of changes in stress, temperature and microstructure on mechanisms of creep. This equation known as the Bird–Mukherjee–Dorn (BMD) equation is given by

$$\frac{\dot{\varepsilon}kT}{DEb} = A\left(\frac{b}{d}\right)^p\left(\frac{\sigma}{E}\right)^n. \qquad [3.43]$$

As shown in Fig. 3.16a, changes in stress and temperature for a given constant microstructure of the material can reveal changes in the stress exponent value.[60] At low normalized stress values, the deformation mechanism appears to proceed with a stress exponent value of 1. At intermediate stress values a stress exponent value of 2 corresponding to GBS is obtained. At the highest normalized stress values, the mechanism of deformation operates with a stress exponent value of 5 corresponding to power-law creep. The diffusivity value utilized for constructing the plot corresponds to the lattice diffusion activation energy of titanium, and thus data at different temperatures follow different curves in the GBS and viscous creep regimes where the appropriate activation energy is that for grain boundary diffusion. On the contrary, if one chooses to use the activation energy for grain boundary diffusion, the data at high stresses will lead to different lines for different temperatures. The BMD plot thus allows an easy understanding of the transitions in creep mechanisms following changes in stress and temperature. Such an analysis was found to be very useful in delineating various creep mechanisms in Zr-based alloys as depicted in Fig. 3.16b.[73] Moreover, such plots made for different materials would show the material behaviors at equivalent loading conditions.[4]

Transitional creep mechanisms in class-A alloys

It is instructive to examine the transitions in creep mechanisms in solid solutions of class-A type such as the results depicted in Fig. 3.10 where we

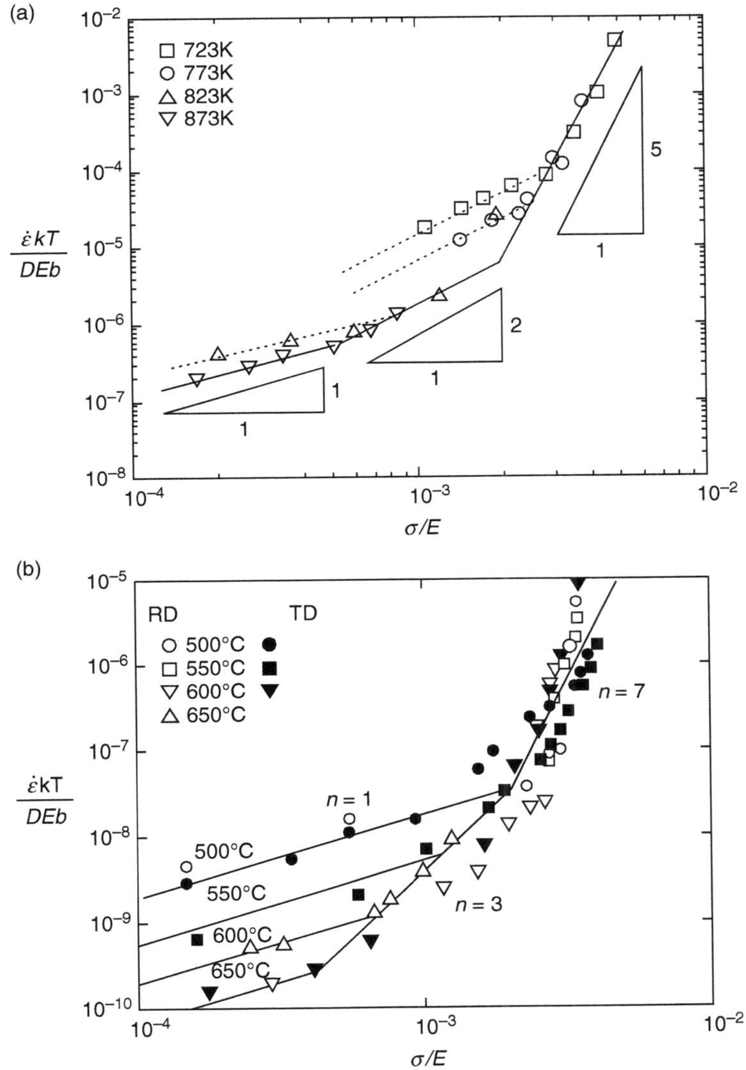

3.16 The BMD plot exhibiting the transitions in creep mechanisms in (a) Ti-3Al-2.5V[57] and (b) Nb-modified Zircaloy[73] as a function of stress and temperature.

note that the stress exponent in the intermediate stress region is 3.5 corresponding to viscous glide of dislocations or Weertman microcreep mechanism. Since glide and climb occur in sequence, when lower temperatures are approached the climb-controlled creep becomes dominant with five power law thus depicting the fact that the slower climb process controls creep. In fact this low stress regime is associated with similar characteristics as the climb-controlled creep with distinct subgrain formation and relatively large primary creep region. On the other hand, dislocations may break away from the solute atmospheres at high stresses, thus entering a climb-controlled regime again as noted at higher stresses with higher n value. Following Murty's work, this breakaway stress can be calculated from the equation[74]

$$\sigma_b = \frac{W_m^2 c_o}{2^\beta kTb^3}, \quad [3.44]$$

where W_m is the binding energy between solute atom and the dislocation, c_o is the solute concentration, and β typically ranges between 2 and 4 depending on the shape of the solute atmosphere. Later, Langdon and co-workers[75] showed that this relation is valid for a number of solid solution alloys. Assuming 0.23 eV as a reasonable value for W_m, the critical stress for breakaway is estimated to be ~7.5×10^{-4}E, which is in agreement with the experimental results obtained from various class-A alloys.[65] At even higher stresses, another regime may appear involving low temperature climb-controlled creep with a stress exponent value of $n + 2$ (i.e. 7). This mechanism is associated with the climb processes involving dominance of dislocation core diffusion (Fig. 3.16b). However, this is often masked because the PLB regime starts in the near vicinity.

As lower stresses are approached, one expects to note viscous creep with $n = 1$ (Fig. 3.16b) either due to N–H or Coble creep mechanisms. Depending on the test temperature, one of the regions such as with $n = 3$ for viscous glide may completely disappear as noted in Fig. 3.16b. This could get further complicated if an intervening GBS regime with $n = 2$ appears between viscous creep and dislocation creep regimes.

3.5.2 Deformation mechanism maps

The concept of deformation mechanism maps was proposed by Ashby.[76] Since different creep mechanisms operate or dominate in different stress, temperature and grain size regimes, Ashby envisioned that a deformation mechanism map would be an ideal representation of the materials constitutive behavior. Over the years, this concept has been extended to describe a variety of other physical phenomena such as sintering,[77] wear[78] and fracture.[79] Figure 3.17 is a deformation mechanism map first reported by Ashby in 1972. The map was plotted as normalized stress (σ/G) against homologous temperature (T/T_m) for a constant grain size. The map was then constructed by determining the stress or temperature boundaries where one mechanism would dominate others. To this end, the creep constitutive relations of different mechanisms were compared and stress and temperature values where

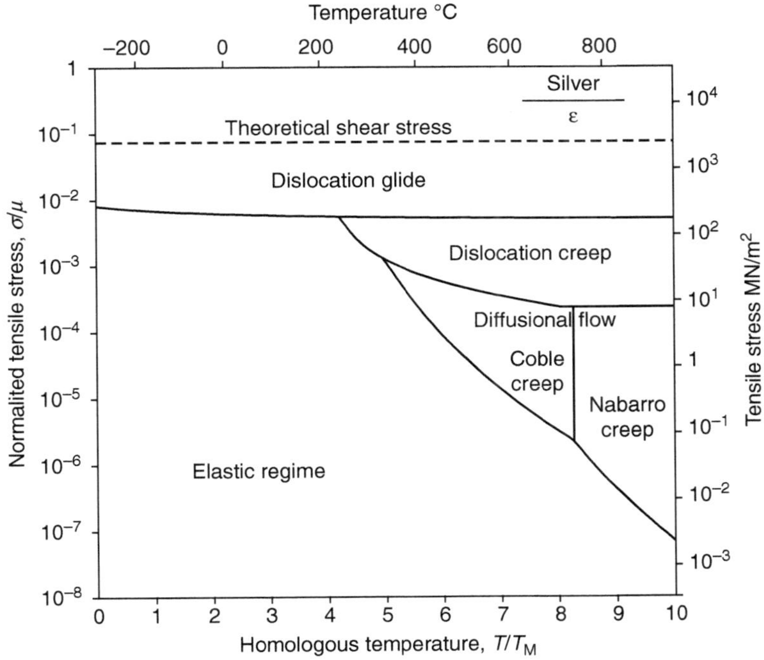

3.17 Deformation mechanism map for pure silver with a grain size of 32 μm and a critical strain rate of 10^{-8} s^{-1}.[76]

transitions from one mechanism to another would occur were determined. For example in Fig. 3.17, at low temperatures and low stresses, the material would resist plastic deformation and the material would behave elastically while in the later modifications this low stress regime was considered to be due to Coble creep.

However, as we continue to increase the temperature and approach higher homologous temperatures, diffusional processes become dominant. Also the applied stresses are sufficient to overcome the flow stress corresponding to that temperature and the material deforms plastically. Since diffusional creep can either be governed by Coble or N–H creep we find the map outlining the regions where these mechanisms are dominant. As Coble creep is controlled by grain boundary diffusion, it is dominant at lower temperatures and the Coble creep field lies to the left of N–H creep on the map. Also, if we increase the stress at a given temperature, dislocation-based mechanisms come into play. Depending upon the homologous temperature, the deformation can be controlled by dislocation climb or glide. At low homologous temperatures dislocation climb is suppressed and hence dislocation glide becomes the dominant deformation mechanism; this is not to be confused with the viscous glide creep discussed earlier which occurs along with climb creep in class-A alloys. For the sake of the reader, we present a small example of how the temperature and stress boundaries of different mechanisms can be determined in a given material. If we assume Coble creep and N–H creep as competing mechanisms for a given grain size, then Coble creep will be dominant when

$$\dot{\varepsilon}_{\text{Coble}} > \dot{\varepsilon}_{\text{N-H}}. \qquad [3.45]$$

From the relevant equations for Coble and N–H creep mechanisms, this would imply

$$\frac{B_c}{\pi}\frac{D_B \delta_B \sigma\Omega}{d^3 kT} > B_H \frac{D_L}{d^2}\frac{\sigma\Omega}{kT}. \qquad [3.46]$$

Cancelling the common terms we obtain

$$\frac{D_B}{D_L} > \frac{K_5}{d}, \qquad [3.47]$$

where K_5 is a constant. At a constant grain size, and after expanding D_B and D_L, the above equation will turn out to be

$$\frac{D_{0B}\exp(-Q_B/RT)}{D_{0L}\exp(-Q_L/RT)} > K_5, \qquad [3.48]$$

where K_5 is a constant. Clearly the transition from Coble to N–H creep is temperature dependent and independent of stress. The transition is only dependent on the activation energies for grain boundary and lattice diffusivities. The temperature dependence of this cross-over is captured by the map where we can observe that a line parallel to the stress axis separates the Coble creep and N–H creep fields.

An alternate way of representing the deformation mechanism maps was proposed by Mohamed and Langdon.[80] Since grain size is an important factor which governs the deformation behavior of materials, the mechanism map can also be plotted for normalized grain size (*d/b*) against normalized stress (σ/G) for a given temperature (Fig. 3.18). As the plot shows, smaller grain sizes are favorable for Coble creep and as we increase the grain size N–H and H–D creep mechanisms become dominant. Since dislocation creep is independent of grain size, transitions from dislocation creep to other mechanisms are represented by lines parallel to the grain size axis. The climb-glide mechanisms are noted for larger grain sizes with

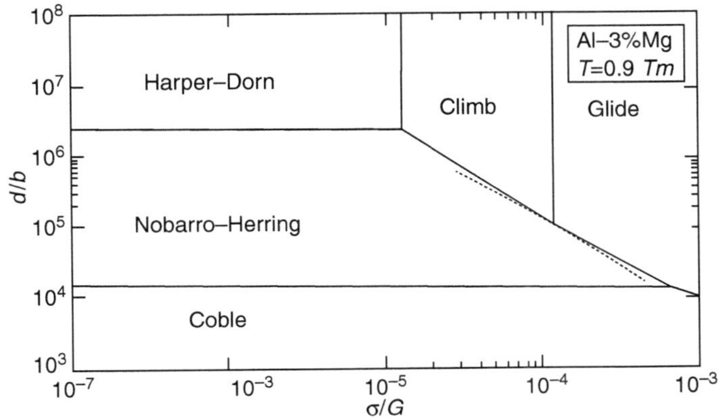

3.18 Mohamed-Langdon deformation mechanism map describing transitions in creep mechanisms as functions of grain size and stress.[80]

climb occurring at lower stresses; this plot did not consider the climb region at the higher stress-end as described earlier. Also missing is the GBS that is expected between viscous and dislocation creep mechanisms.

3.6 Modeling creep life: extrapolation of strain and rupture data

In the previous sections, we discussed the different mechanisms of creep that have been observed in various materials. The stress, temperature and microstructural dependence of each mechanism was described and the steady-state strain rate of deformation of each mechanism was correlated to these parameters. We also outlined the different regions, through deformation mechanism maps, where a given mechanism would be dominant over others. In all these sections, more emphasis was laid on the secondary creep region and the mechanism maps were also constructed taking into account these steady-state creep-rates. However, such a methodology would be based on the premise that the secondary creep stage accounts for a significant fraction of the useful creep life. While such a method is not entirely wrong, it is unsuitable for several materials which tend to have larger primary or tertiary creep regimes. For example, Ni-based superalloys have been found to exhibit primary creep strains of the order of 1% or more.[81] These alloys are used as materials for fan and compressor blades of aero-engines. The dimensional tolerance for these components is very small and plastic strains in the order of 1% are sufficient to wreck the stability of the engine. Hence under such conditions, modeling by considering only the steady-state creep rates will grossly overestimate the useful creep life of the material. Furthermore some of the mechanisms of creep, for example in the power-law creep regime, have been proposed following microstructural studies on the crept specimens. For instance observation of subgrains in the crept microstructure is considered evidence for creep controlled by climb of edge dislocations. Similarly observation of jogged screw dislocations is believed to indicate deformation controlled by the Barrett-Nix model or its recent modification proposed by Mills and co-workers.

In most cases the deformation microstructures are investigated through TEM studies. Hence the sample studied, due to its very small volume, may not be a real representation of the condition of the material. Thus there is some uncertainty associated with the rate controlling mechanism. While the physically based mechanisms discussed in the previous sections are important for understanding and predicting deformation rates, an equally large number of studies has been carried out to predict creep life using mathematical models and empirical correlations. The Larson-Miller parameter (LMP), Monkman-Grant constant, *θ-projection* concept and a host of other graphical and mathematical methods have been utilized to predict the creep life of various engineering materials.

Generally engineering components are designed for a stress level below which there is no danger of rupture or excess deformation during the service life of the component. The stress level is decided by one of the following two criteria: (a) stress level at which rupture/failure would be caused in 100 000 or 200 000 h, whichever period is appropriate and (b) stress level which produces a nominal strain of 0.1%, 0.2% or 0.5% in a certain period, say 100 000 h.[82] However there are not many tests carried out till 100 000 h even for established materials and hence it is necessary to extrapolate data from much shorter tests, say 10^3–10^4 h. This is especially important for new materials where it is necessary to understand their long term behavior within a short span of time. Hence the extrapolation techniques become important and in this section we discuss some of the existing extrapolation techniques for predicting long term creep behaviors. Penny and Marriott[82] provide an excellent review of the various extrapolation methods and also the advantages and disadvantages associated with each method. They divide the extrapolation techniques into three main groups:

1 Parametric methods
2 Graphical methods
3 Algebraic methods.

Equations correlating time-temperature or stress-time fall under the parametric method. Functional relationships between time, temperature and stress are established and it is believed that when stress is plotted against a function of time and temperature, a single master curve will be obtained. This master curve can be constructed by performing short term tests at

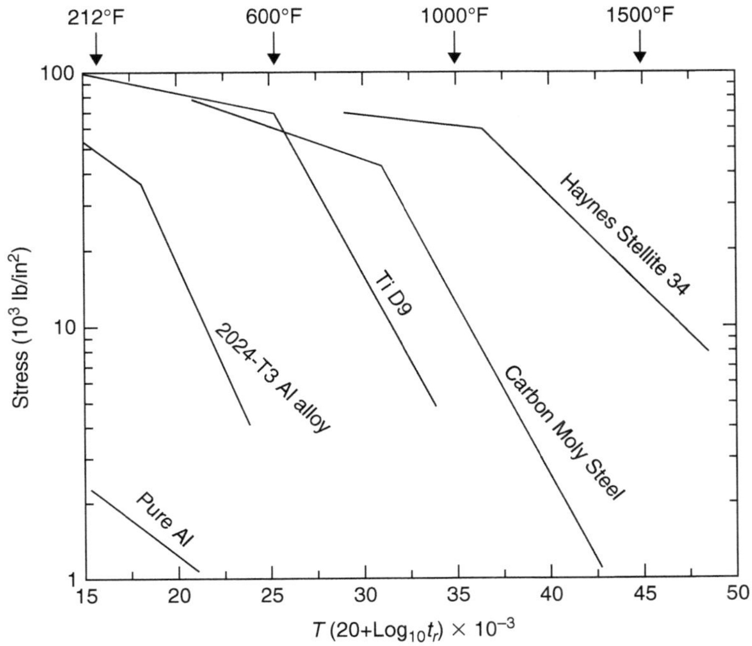

3.19 Larson-Miller plot for various materials.[88]

higher temperatures. It is then assumed to be equally valid for longer times and lower temperatures thus allowing for extrapolation. The Larson–Miller method[83] is based on this logic. The original Larson–Miller equation is given by the following:

$$\text{LMP} = T(C + \log_{10} t_r), \qquad [3.49]$$

where LMP is the Larson–Miller parameter and C is a constant which was assumed to be equal to 20 and was found to be reasonably accurate for many materials. Plots of applied stress versus the LMP would then allow extrapolation of short term data for long term predictions. Figure 3.19 shows a LMP obtained from short term tests for a variety of materials. It is interesting to note the change in slope as lower stresses are approached. Some of the other parameters which fall under the category of parametric methods are by Dorn and Shepherd,[84] Manson and Haferd,[85] Murry,[86] etc. However, LMP is quite commonly used in creep life predictions and extrapolations.

Under graphical methods, there are procedures which seek to extrapolate rupture curves by direct manipulation of the plotted data. Grant and Bucklin,[87] Glen,[88] Mendelsohn and Manson[89] and others proposed methods

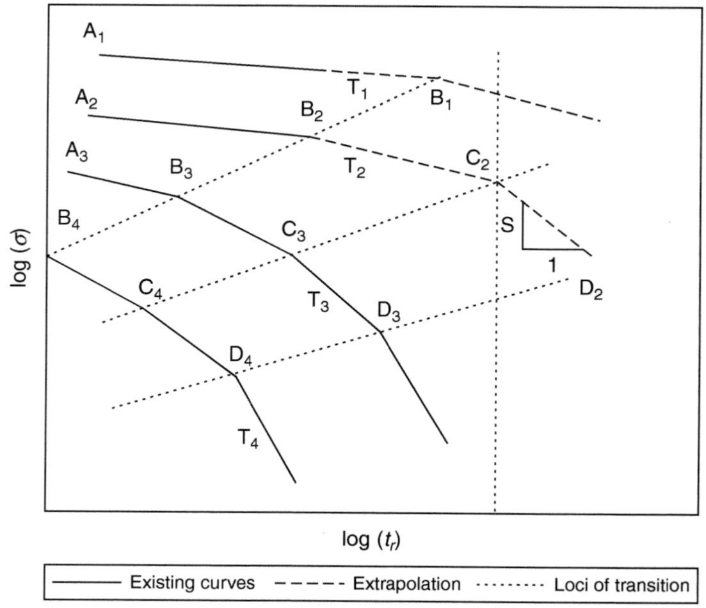

3.20 **Grant-Bucklin methodology for determining creep life.**

that fall under this category. Here we provide a brief description of the Grant-Bucklin method. Grant and Bucklin considered the fact that creep rupture would be influenced by several time- and temperature-dependent effects and hence mode of failure might not be uniform over the whole range of time and temperature. They identified distinct segments of the rupture curve where one mode of failure might be dominant. These segments were later described by linear relations (Fig. 3.20). By plotting the slopes of like segments against temperature, it is possible to extrapolate to temperatures outside the experimental range. Secondly the positions of the transition points may be plotted on axes of temperature versus t_r for extrapolation. However Penny and Marriott[87] indicate that such extensions are subjective and sensitive to the ability or judgment of the analyst, albeit Grant and Bucklin imply that reliable extrapolations of the rupture curves are not critically dependent on the accurate determination of either slopes or transition points.

The algebraic methods are similar in a way to the parametric methods. The difference lies in finding functions which can combine the effects of stress, temperature and time into a single relation such as

$$f(\sigma, t_r, T)^c = \text{constant.} \quad [3.50]$$

Any function $f(\sigma, t_r, T)^c$ which can be separated into two functions such as

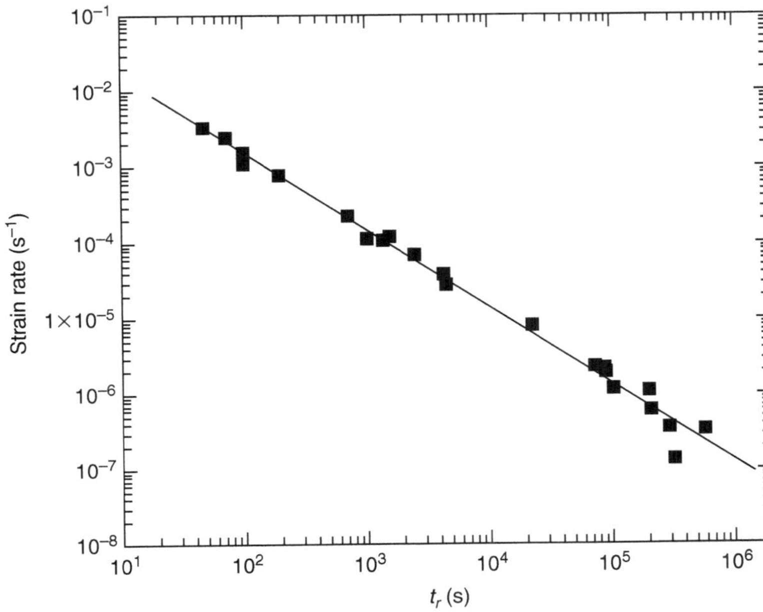

3.21 Monkman–Grant plot for cp-Ti tubing.[91]

$$f(\sigma,t_r,T)^c = f_1(\sigma)f_2(t_r,T) = \text{constant} \quad [3.51]$$

is similar to the time-temperature method of parametric types. In addition to these methods, there are several other methods which have been proposed and found to provide reasonable predictions. Monkman and Grant[90] proposed a relationship between the steady-state strain rate and rupture time:

$$\dot{\varepsilon}_s t_r = \kappa, \quad [3.52]$$

where κ is a material constant known as the Monkman–Grant constant. Figure 3.21 depicts such a plot for internally pressurized cp-Ti tubing.[91]

Other methods of extrapolation include the *θ-projection* method advocated by Wilshire and co-workers[92,93] where the total creep curve is described by a series of θ parameters. Wilshire[92] suggests that for most materials the secondary creep region is only an *inflection* that appears to be a constant over a limited strain range. Hence it was emphasized that creep life modeling should take into account the total creep curve including the tertiary creep regime rather than just focusing on the secondary creep rates. On this premise, Wilshire and co-workers advocated the *θ-projection* concept where the total creep curve would be described as

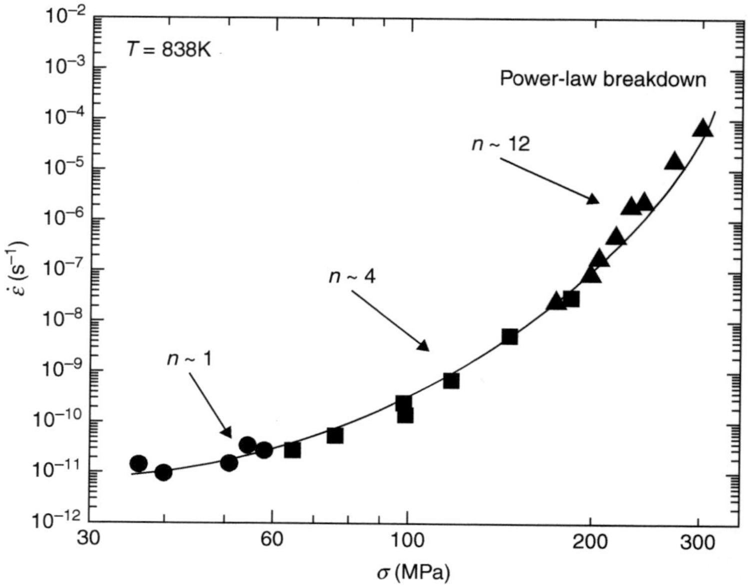

3.22 Experimental creep rates obtained in 0.5Cr-0.5Mo-0.25V steel 838 K. The solid line corresponds to the predictions of the θ-projection concept and the plot shows the sound agreement between the experimental and theoretical predictions.[92]

$$\varepsilon = \theta_1(1 - \exp(-\theta_2 t)) + \theta_3(\exp(\theta_4 t) - 1), \qquad [3.53]$$

where θ_1 scales the primary creep regime, θ_2 is a rate parameter governing the curvature of the primary stage, θ_3 scales the tertiary creep regime and θ_4 is a rate parameter quantifying the shape of the tertiary curve. These parameters are found to change with stress and temperature conditions and accordingly influence a change in the shape of the creep curve. A determination of the stress and temperature dependencies of the θ parameters would allow the prediction of long term creep properties. Furthermore Wilshire counters the widely accepted view of transitions in creep mechanisms with changing stress and temperature conditions. The creep characteristics of a 0.5Cr-0.5Mo-0.25V ferritic steel could thus be described by the θ-projection over a wide range of stress values based on a single dislocation-based mechanism. However, as shown in Fig. 3.22, there are definite changes in stress exponent values with changing stress. Wilshire argues that if different mechanisms operate in different stress and temperature regimes, data collected in one mechanism regime should not be able to predict the creep behavior in a different mechanism regime. Furthermore Wilshire contends that the θ-projection approach can be utilized to quantify material behavior in complex, non-steady stress-temperature conditions encountered in service conditions.

In addition to these methods, creep life predictions are also guided by damage mechanics. The irreversible material damage caused by mechanical loading and environmental features during creep eventually leads to very high strain rates of deformation and failure. Damage could be due to cavity formation, microcracks and gross deformation such as strain- or ageing-induced. A materials scientist viewpoint on micromechanical causes of damage is given by Le May.[94] In addition to creep damage, other mechanisms of damage such as fatigue, surface oxidation and internal corrosion are also important. Although some of these phenomena are not temperature dependent, their interactions with creep, such as creep-fatigue interaction, can have significant effects on high temperature damage accumulation. The different damage processes constitute ductile creep rupture, intergranular cavitation during creep, continuum creep rupture, continuum fatigue damage, environmental damage and age- and strain-induced hardening and softening. In contrast to creep life predictions based on mechanistic models, continuum damage mechanics (CDM) attempts to provide a holistic view of the damage process and accordingly models the useful creep life of a material. By accepting the fact that damage is a result of the complex interactions between different mechanisms, CDM provides greater accuracy in creep life estimation in comparison to models based on a single mechanism of creep, namely grain boundary sliding or dislocation creep. While there have been many continuum damage mechanics models advocated over the years, a unique model is the one proposed by Kachanov,[95] later elaborated by Rabotnov[96] and commonly referred to as the Kachanov–Rabotnov model. A brief review of the Kachanov-Rabotnov model is presented below.

3.6.1 The Kachanov–Rabotnov CDM model

Kachanov represented continuum damage as an effective loss in material cross-section due to the formation and growth of internal voids. Consequently the internal stress corresponding to a nominal externally applied load increases with increasing damage. Kachanov assumed that damage could be represented by a quantity which he called the 'continuity.' The continuity is essentially the ratio of the remaining effective area A to the original area A_0. With accumulation of damage, the resulting internal stress (σ_i) increases from initial value σ_0 to a value given by

$$\sigma_i = \sigma_0 \frac{A_0}{A}. \quad [3.54]$$

The continuity term was later modified by Rabotnov and was called the damage parameter ω, where

$$\omega = \left(1 - \frac{A}{A_0}\right). \quad [3.55]$$

By assuming a power-law dependence of stress, the creep rate at constant temperature was described as

$$\dot{\varepsilon}_c = \frac{k\sigma_0^m}{(1-\omega)^p}, \quad [3.56]$$

where m and p are material parameters. At time $t = 0$, $\omega = 0$ and the above equation assumes the power-law form. As ω increases, the creep rate increases and when it achieves a critical value, the creep rate tends towards infinity and failure follows.

In order to describe the evolution of damage, Kachanov assumed that damage is a function of the initial stress σ_0. This was later generalized by Rabotnov who assumed that the damage is instead a function of the instantaneous stress and described the rate of change of damage through the following:

$$\frac{\mathrm{d}\omega}{\mathrm{d}t} = \frac{B\sigma_0^k}{(1-\omega)^r}. \quad [3.57]$$

Solving the above two equations gives the creep strain in the following form:

$$\frac{\varepsilon_c}{\varepsilon_R} = \left[1 - \left(1 - \frac{t}{t_R}\right)^{\frac{1}{\lambda}}\right] \quad [3.58]$$

where ε_c is the instantaneous creep strain, ε_R is the rupture strain, t is the time and t_R is the time to rupture. The shape of the creep curve described by Equation [3.58] is as shown in Fig. 3.23.

The damage tolerance parameter λ is given by the following equation:

$$\lambda = \frac{1+r}{1+r-p}. \quad [3.59]$$

The material fails in the steady-state creep regime when $\lambda = 1$. Ashby and Dyson[97] have demonstrated that each damage micromechanism has a characteristic λ and a characteristic shape of the creep curve. This implies that

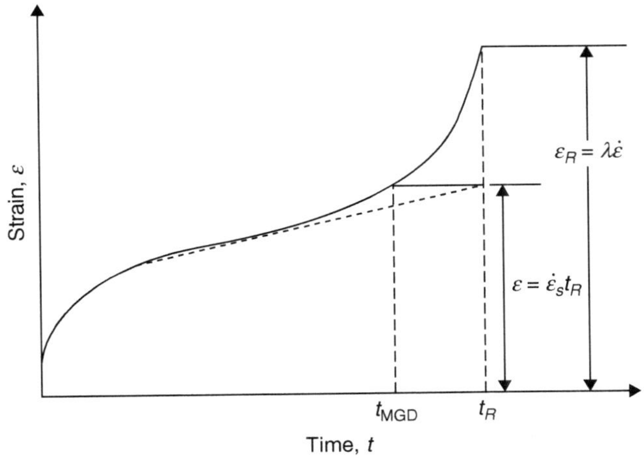

3.23 Representation of creep strain growth following the Kachanov–Rabotnov model.

the creep curve would assume different shapes for different values of λ. Phaniraj *et al.*[98] have established a correlation between the ratio of time to Monkman-Grant ductility (t_{MGD}) and time to rupture (t_R) and the damage tolerance parameter as given by

$$\frac{t_{MGD}}{t_R} = 1 - \left[\frac{\lambda - 1}{\lambda}\right]^{\lambda}. \qquad [3.60]$$

Figure 3.24 is based on this Equation [3.60] and shows that t_{MGD}/t_R is essentially constant for $\lambda > 4$. The t_{MGD} was suggested as time for onset of true tertiary creep damage and was considered to be an important parameter in identifying the useful creep life of a material. It also describes the time for which minimum creep ductility is ensured. Hence Phaniraj *et al.* contend that the stress to cause t_{MGD} in 10^5 h can be used as a useful design criterion for creep of elevated temperature components.

Before concluding we present a few examples where the concepts discussed in the previous sections may not be directly applied. Rather subtle modifications to the models are necessary in order to simulate the actual behavior of the material.

3.7 Case studies illustrating the role of other factors

In the following section, the effects of impurities, second phases and multiaxial loadings on creep of materials are discussed with examples taken from various classes of materials including ionic solids.

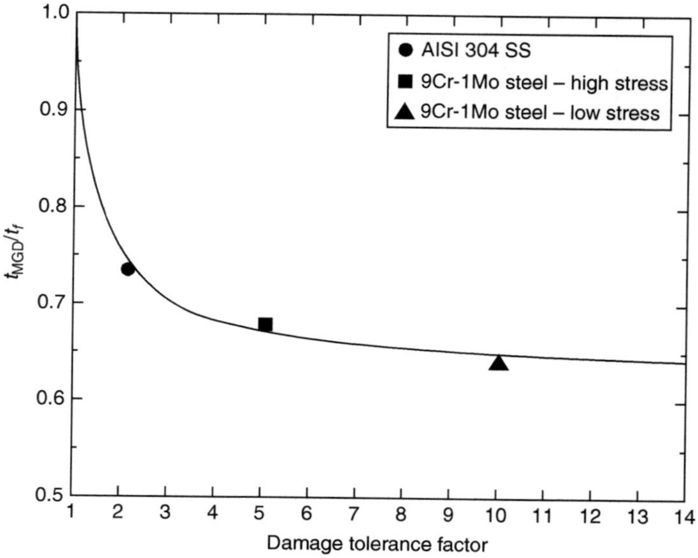

3.24 Validity of Equation [3.60] studied in a 9Cr-1Mo steel and AISI 304 stainless steel. The solid line corresponds to the predictions of Equation [3.60].[98]

3.7.1 Effect of impurities

In Section 3.1 we identified stress, temperature and grain size/microstructure as the three important factors which determine the extent of creep deformation that a material experiences. However there are examples where two materials with similar compositions, grain sizes and second phase distributions might creep at vastly different rates under a given stress and temperature. Such anomalous behavior has been observed in titanium alloys by Mishra *et al.*[99] who found that alloys with nominally similar compositions crept at significantly different rates. It was found that the presence of trace elements such as Fe and Ni degrade the creep properties of the titanium alloy. Even though these elements are present only in the order of ppm, they influence the diffusion rates to an extent as to bring about significant changes in creep rates. The activation energy for diffusion in the higher Fe/Ni containing alloys was found to be smaller and *vice versa*. Mishra *et al.* suggest that the Fe/Ni appear to dissolve interstitially and form foreign atom-vacancy pairs which play a significant role in accelerating the diffusion kinetics of the titanium.

3.7.2 Diffusion creep in ionic solids or ceramics

In the section on diffusion creep mechanisms we discussed the importance of grain size, temperature and stress in determining the rate controlling mechanism. Coble creep is dominant at very fine grain sizes whereas N–H creep is rate controlling at larger grain sizes. Also at relatively low homologous temperatures Coble creep is rate controlling and N–H creep takes precedence at high homologous temperatures. This approach to understanding diffusion creep is quite valid for metals and alloys. Similar phenomena in ceramics become complex due to ambipolar diffusion and stoichiometry. The diffusion flux of both cations and anions constituting the ceramic must be considered to estimate the net diffusion rate. In monovalent materials the vacancies in diffusion creep regime can get transported along the grain boundaries or the lattice and the total strain rate of deformation is given by the sum of N–H and Coble creep mechanisms. But in a ceramic of the type A_pB_q, where A is the cation and B the anion, both the anions and cations participate in the diffusion process and might adopt different transport paths. In this case the total strain rate of deformation in the Coble creep model is given by

$$\dot{\varepsilon} \propto D_{\text{composite}}, D_{\text{composite}} = \frac{(1/p)\left[D_L^+ + 1.2\left(\delta^+ D_{gb}^+/d\right)\right]}{1+(q/p)\left(D_L^+ + 1.2\left(\delta^+ D_{gb}^+/d\right)/D_L^- + 1.2\left(\delta^- D_{gb}^-/d\right)\right)}. \quad [3.61]$$

The transport path of the anions and cations was originally considered by Gordon[100] who suggested that the total transport of vacancies from the horizontal to the vertical boundaries should be in the appropriate stoichiometric ratio. This leads to the prediction that creep would be controlled by the diffusion of the slower moving species along the faster diffusion path. In this scenario, it is possible for the cations and anions to be transported predominantly along different paths as depicted in Fig. 3.25a.

However the transport paths suggested by Gordon might lead to the development of local non-stoichiometry[101] which has not been observed in ceramics. Hence Chokshi[102] suggested that it would be appropriate to constrain diffusion fluxes along each path to be in the appropriate stoichiometric ratio, as depicted schematically in Fig. 3.25b. In this scenario, it is necessary to find the slower moving species along each path, and the rate controlling process is then determined from the faster diffusion path. The difference in transport paths suggested by Gordon[100] and Chokshi[102] has implications for the transitions in diffusion creep mechanisms. Plots of strain rate against T_m/T, for a fixed grain size are shown in Fig 3.26a and 3.26b. The symbols C and N, in these figures, represent Coble and N–H creep, and the superscripts + and – represent cation and anion, respectively. Figure 3.26a, corresponding to transport paths suggested by Gordon, indicates that there will be transitions with an increase in temperature from diffusion creep controlled by cation grain boundary diffusion ($C+$) to cation lattice diffusion (N^+) to

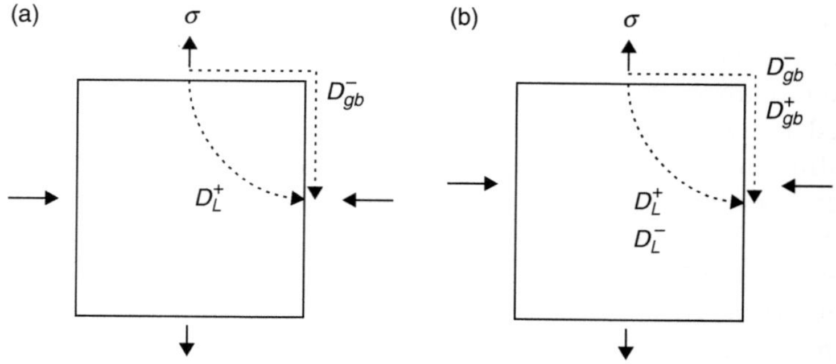

3.25 (a) Transport paths in ceramics as suggested by Gordon where the total flux from horizontal to vertical boundaries is in the appropriate stoichiometric ratio. (b) Transport paths in ceramics as suggested by Chokshi where total flux along each transport path is in the appropriate stoichiometric ratio.

anion grain boundary diffusion (C^-) to anion lattice diffusion (N^-). Figure 3.26b, corresponding to transport paths proposed by Chokshi, indicates that over the same temperature range, there will only be a single transition from Coble creep controlled by cation grain boundary diffusion to N–H creep controlled by anion lattice diffusion.

3.7.3 Presence of a second phase and effect on creep behavior

The constitutive relationships identified in the previous sections are applicable for a wide variety of metals and alloys. However the strain rates of deformation might assume values different from model predictions even while the parametric dependencies remain the same. This was discussed in case study (3.7.2) and was attributed to the presence of impurities. On the other hand, there are cases where the strain rates of deformation as well as parametric dependencies can turn out to be different. Such instances are encountered while dealing with two- or multi-phase alloys that exhibit precipitation or dispersion strengthening. Strain compatibility issues as well as differences in deformation rates of individual phases contribute to discrepancies in experimental observations and traditional creep model predictions. To this end, analytical models have been proposed to understand the creep behavior of multi-phase alloys.[103–105] Here we present a case where the second phase is rigid and is added to enhance the overall strength of the alloy. For particle strengthened alloys, stress exponents higher than those predicted by established creep models[106] and/or anomalous variation of stress exponent with stress are observed.[107] This is rationalized by the introduction of a friction or resisting stress also known as back stress

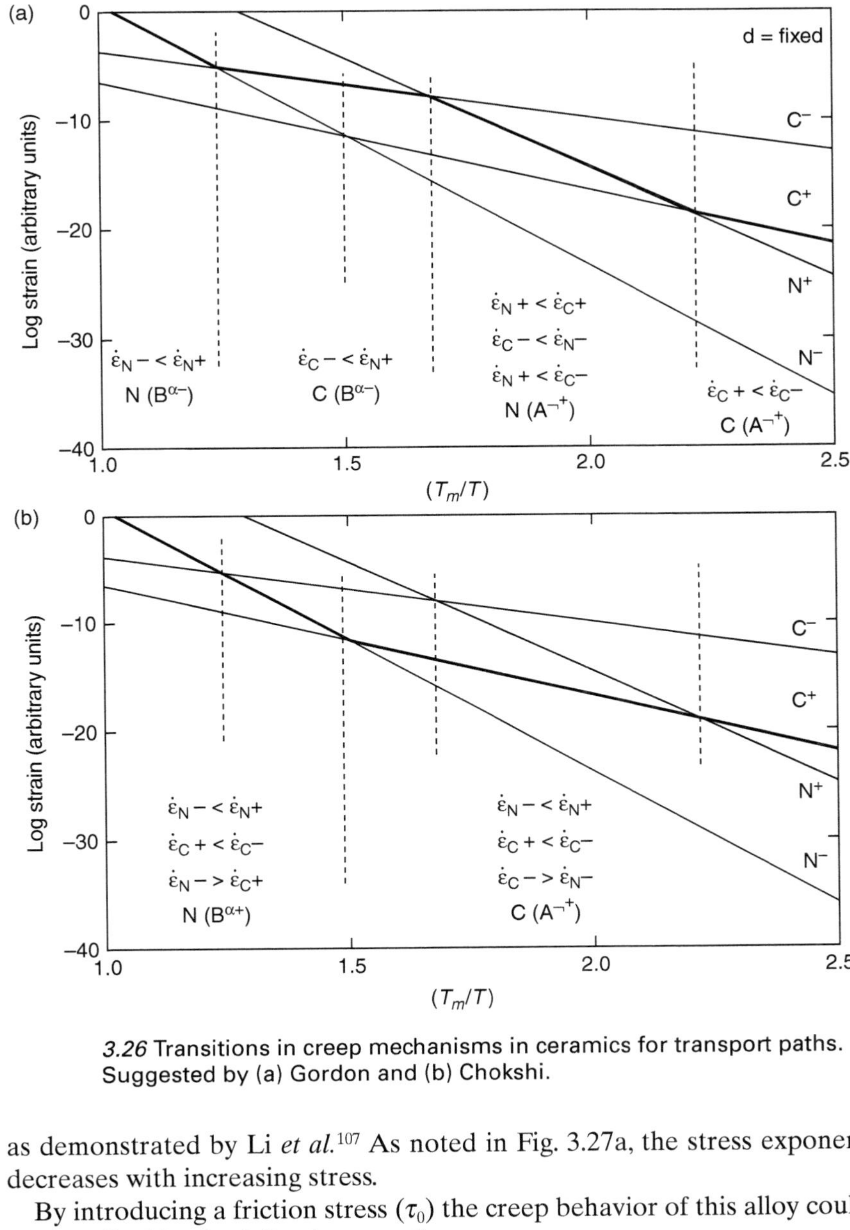

3.26 Transitions in creep mechanisms in ceramics for transport paths. Suggested by (a) Gordon and (b) Chokshi.

as demonstrated by Li *et al.*[107] As noted in Fig. 3.27a, the stress exponent decreases with increasing stress.

By introducing a friction stress (τ_0) the creep behavior of this alloy could be described by the following equation:

$$\dot{\gamma} = A\left(\frac{\tau - \tau_0}{G}\right)^n \exp\left(\frac{-Q_c}{RT}\right) \qquad [3.62]$$

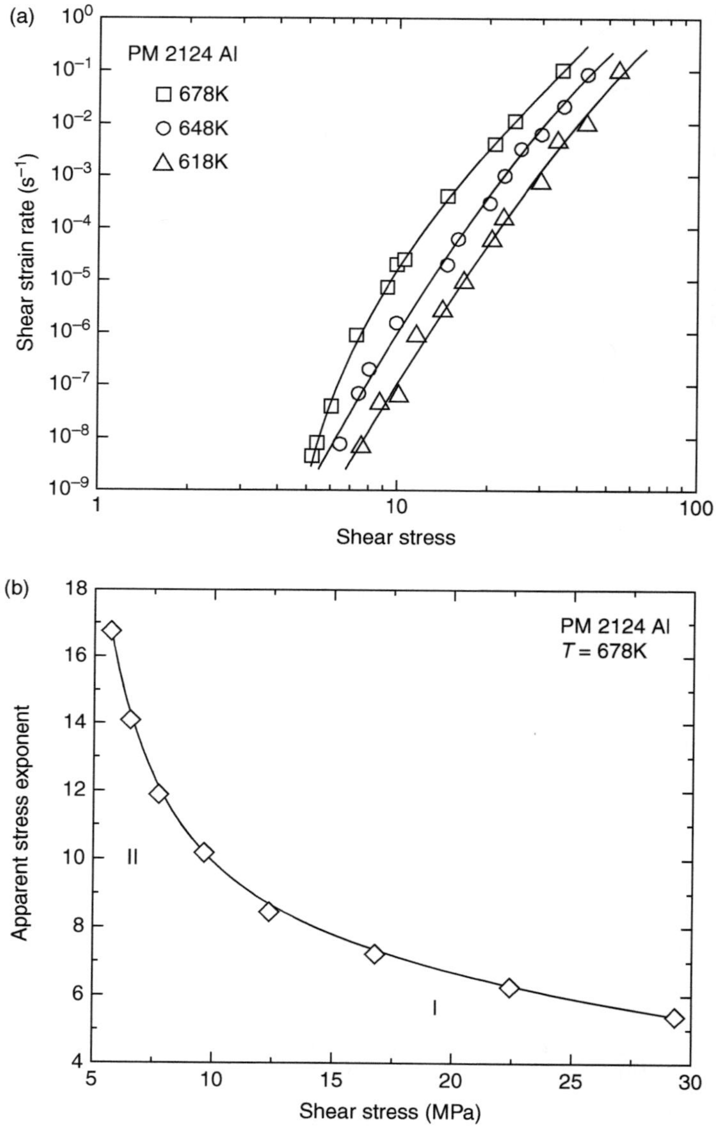

3.27 (a) Strain rate vs stress in PM 2124 Al as a function of temperature and (b) determination of threshold stress through back-extrapolation.[107]

where $\dot{\gamma}$ is the strain rate of deformation, τ is the applied stress, τ_0 is the threshold stress and the rest of the terms are as described previously. Following the introduction of the threshold stress, the creep strain rates, $\dot{\gamma}^{1/n}$, are plotted on a linear scale against the applied stress. Here n is chosen as that value where a linear correlation between $\dot{\gamma}^{1/n}$ and τ is obtained. Extrapolation of the straight line to the x-axis provides a value for the threshold stress (Fig. 3.27b). The threshold stress is attributed to the presence of the second phase which acts as an obstacle to dislocation motion. Such threshold stress based models have been considered in 60 s in analyzing the creep behavior of precipitation and dispersion strengthened alloys.[108]

3.7.4 Effect of a multi-axial state of stress

The final case study deals with understanding the effect of a multi-axial state of stress on creep life predictions. All models discussed inadvertently predict the deformation rates under a uniaxial state of stress. However in real life situations, the engineering structures experience a multi-axial state of stress as well as variations of loading with time. Here we present an example of how a multi-axial state of stress would affect diffusion creep. For a more detailed account of the effects of multi-axial state of stress and loading history, we refer the reader to Penny and Marriott.[82]

Raj[109] proposed relationships to understand the strain rates of deformation during diffusion creep under a multi-axial state of stress. Since strain rate in diffusion creep has a linear correlation with applied stress, Raj suggested that under a multi-axial state of stress the strain rates of deformation can be represented in terms of principal stresses and strain rates,

$$\dot{\varepsilon}_1 = k_1\left(\sigma_1 - \frac{\sigma_2+\sigma_3}{2}\right), \dot{\varepsilon}_2 = k_2\left(\sigma_2 - \frac{\sigma_1+\sigma_3}{2}\right), \dot{\varepsilon}_3 = k_3\left(\sigma_3 - \frac{\sigma_1+\sigma_2}{2}\right). \quad [3.63]$$

These relationships are borrowed from equations of elasticity and Poisson's ratio has been assumed to be 0.5 for constancy of volume. Here k_1, k_2 and k_3 are creep constants that correspond to the appropriate diffusion creep mechanism. These strain rates of deformation will be identical for an isotropic or equiaxed microstructure but would be different for non-isotropic grain configuration. Also the diffusional creep should be independent of the hydrostatic state of stress which has an effect only on the diffusivity term.

Materials such as Zr and other hexagonal close packed metals and alloys exhibit distinct crystallographic textures leading to anisotropic mechanical and creep properties. Zr-alloys are commonly used as thin-walled tubes to clad radioactive fuel such as UO_2 and tube reduction processes render them highly textured during the manufacturing of cladding tubes. Prediction of dimensional changes in-reactor of the cladding along both the diametral and axial directions requires consideration of multi-axial loading particularly with axial load superimposed with internal/external pressure. A modified

Hill formulation was shown to be convenient wherein the generalized stress, σ_g, is defined as the uniaxial stress along the tube axial direction,[110]

$$\sigma_g^2 = \frac{R(\sigma_r - \sigma_\theta)^2 + RP(\sigma_\theta - \sigma_z)^2 + P(\sigma_z - \sigma_r)^2}{P(R+1)} \qquad [3.64]$$

The parameters, R and P, are the mechanical anisotropy parameters. Using the Prandtl-Reuss energy balance in conjunction with the above yield criterion, the strain increments or strain rates along the three orthogonal directions are related to the respective stresses

$$\begin{bmatrix} \dot{\varepsilon}_r \\ \dot{\varepsilon}_\theta \\ \dot{\varepsilon}_z \end{bmatrix} = \frac{\dot{\varepsilon}_g}{P(R+1)\sigma_g} \begin{bmatrix} (R+P) & -R & -P \\ -R & R(P+1) & -RP \\ -P & -RP & P(R+1) \end{bmatrix} \begin{bmatrix} \sigma_r \\ \sigma_\theta \\ \sigma_z \end{bmatrix}, \qquad [3.65]$$

where $\dot{\varepsilon}_g$ is the generalized strain rate corresponding to the generalized stress σ_g. These equations define the yield and flow loci, and the corresponding creep locus is referred to at a constant energy dissipation-rate, $\dot{W} (= \sigma_{ij}\dot{\varepsilon}_{ij} = \sigma_g \dot{\varepsilon}_g)$.[111] It is clear from the above equations that the mechanical anisotropy parameters, R and P, are the transverse contractile strain (-rate) ratios in uniaxial tests. The contractile strain ratios (CSRs), R and P, define the resistance to wall thinning of an anisotropic material and thus control the formability which is of importance to the material manufacturers, in tube reduction processes, sheet drawing and forming, etc.[112] The mechanical anisotropy parameters are directly related to the preferred orientations of the grains. Thus obtained creep loci for cold-worked stress-relieved annealed and following complete recrystallization are shown in Fig. 3.28.[113] Crystallite orientation distribution function (CODF) creep model predictions[113] are shown as solid lines along with the experimental results.

These types of creep studies are not commonly found albeit materials in real engineering structures experience such complex multi-axial loadings. Additional complexity is encountered in attempting to predict transients in creep under such multi-axial loading.

3.8 Creep of zirconium alloys used for LWR cladding

Materials used in the reactor undergo irradiation-assisted creep as well as thermal creep (which predominates if stress and temperature are high enough). The in-pile creep deformation of a material is the net contribution by both of these processes and it is difficult to distinguished between them. Thermal creep rate of unirradiated material is different from that of

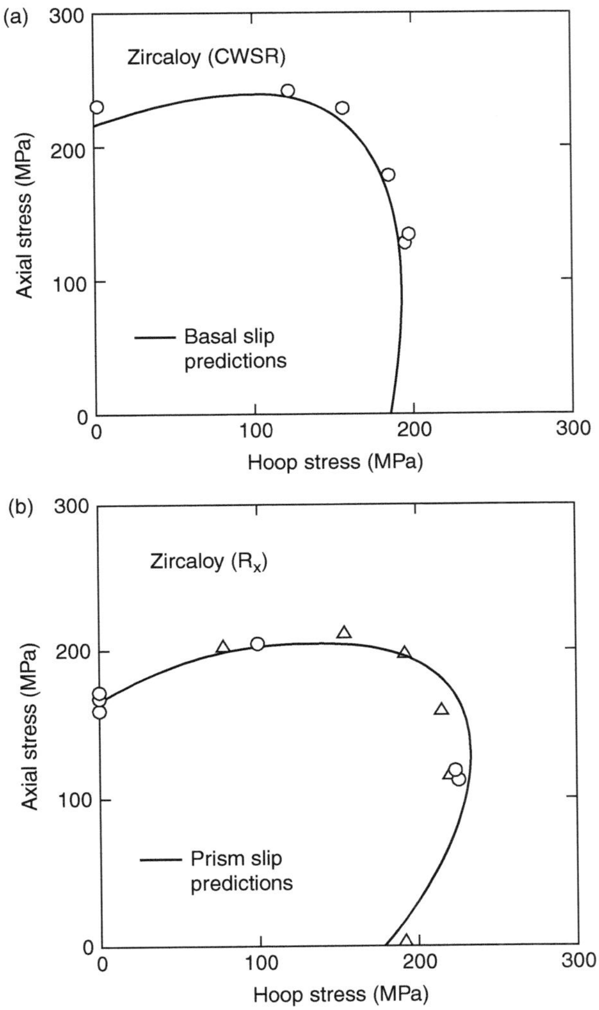

3.28 Creep loci at constant dissipation energy in (a) cold-worked stress-relieved annealed and (b) recrystallized Zircaloy tubing.

irradiated material and both are different from that for a material undergoing irradiation.

Zirconium base alloys, with slightly differing chemical compositions, are used for various components inside a reactor. The clad tubes in BWR and PWR are made of Zircaloy-2 or Zircaloy-4 for most of the operating reactors while new alloys are being proposed for the forthcoming reactors which have to withstand higher burnups (Table 3.2). The Zr-Nb alloy was introduced for spacer grids in place of stainless steel (from 1987 in the WWER-

Table 3.2 Typical composition of some zirconium base alloys

Alloy	Nominal chemical composition (wt.%)							Component	Type of reactor
	Sn	Fe	Cr	Ni	O	Others	Zr		
Zircaloy-2 (UNS grade R60802)	1.5	0.15	0.1	0.05	0.1		Bal	Fuel clad, channel, calandria tube	BWR
Zircaloy-4 (UNS grade R60804)	1.5	0.2	0.1		0.1		Bal	Guide tube, instrument tube, calandria tube	PWR
ZIRLO	1	0.1			0.1	1 Nb	Bal	Fuel clad, spacers	PWR
M5, E110					0.1	1 Nb	Bal	Fuel clad, spacers	PWR WWER
E635	1.2	0.35			0.1	1 Nb	Bal	Fuel clad	WWER
HANA	0.4	0.2	0.1		0.1	1.5 Nb	Bal	Fuel clad	
Zr-2.5wt.%Nb (UNS Grade R60904)					0.1	2.5 Nb	Bal	Pressure tube	
EXCEL alloy	3.5					0.8 Mo 0.8 Nb	Bal		

440 and the mid-1990s in the WWER-1000). With the recent developments in WWER fuels, Zr-1%Nb/Sn/Fe alloys, with higher resistance to irradiation induced growth, creep and corrosion, are being used for guide tubes and for fuel rod cladding with extended residence time (5–6 years).[114]

Fuel cladding is a key barrier in containing fission products and it is essential that this barrier is strong and remains intact over a prolonged period – both in service and during repository storage. Fuel failure occurs when this barrier is degraded and breached. The fuel rod failure rate in LWRs has been significantly reduced since 1987. This achievement, besides design improvements, is due to the introduction of many improved variants of Zr base alloys over the years – the latter ones improved in properties over the earlier ones. The clad tubes in reactors undergo creep extension due to many service conditions. At low burnup, the pellet densifies and the external water pressure causes the clad tube to creep-down. On power ramp, the pellet expands and applies excess strain on the clad. This leads to the pellet touching the clad thus leading to PCI failure or hydride related cracking (which are described in detail in later chapters). The sheath should have good creep rupture properties to withstand this additional strain. A non-symmetric axial growth or creep of the fuel assembly (and guide thimble) can lead to bowing of the assembly. There is another deformation which adds to the creep strain. An analysis performed at Ringhals revealed that the bowing of the rods in this reactor had been due to a large creep deformation caused by excessive compressive forces of the hold down spring on the fuel assemblies and a decrease in lateral stiffness. This problem, though, can be partly overcome by introducing advanced materials with a low growth rate and higher creep resistance (e.g. M5 or ZIRLO) for cladding and guide thimble which improves the dimensional stability of the assemblies albeit irradiation creep remains a matter of concern for these materials.

At the repository the Zircaloy clads of the fuel rods face a challenging environment. The clad temperature – a crucial parameter in influencing the cladding performance in the repository – is estimated to reach a temperature of ~325°C, although the average temperature of the cladding is estimated to be less than 240°C.[115] At this temperature and with a hoop stress of around 100 MPa due to fission gases the clad material can undergo thermal creep. The creep in clad tubes becomes all the more important with dry storage becoming common.[116,117]

3.8.1 Thermal creep of zircaloys

The creep behavior of unirradiated material is taken as a benchmark to postulate its performance in the reactor. Though these out-of-pile tests may not be representative of their in-reactor behavior, they have been successfully used to grade various materials during alloy development programs and to gain basic understanding about the material behavior. It has been recognized that hoop strain in a clad tube is a vital parameter in the breach of fuel clad, and evaluation of their creep and burst behaviors is very important to assess the integrity of the tube.[117] Steady-state creep-rates at relatively high ($>5 \times 10^{-4}E$) followed the same behavior as described earlier exhibiting power-law creep behavior with exponential dependence at higher stresses and were identified as due to dislocation glide-climb creep mechanisms. At low stresses, viscous creep with the characteristic $n = 1$ was indeed reported as expected. Figure 3.29 summarizes the various sets of results in terms of Dorn parameters for Zr-alloys.[118] Bernstein[119] observed that both Zircaloy-2 and pure Zr exhibit a stress exponent value of unity at low stresses which increases to 4.6 and 6, respectively, at higher stresses. The data produced by MacEwen *et al.*[120] also showed that the n increases with stress (for comparable σ/E). On the contrary, data from Ardell and Sherby[121] for α-Zr with comparable purity and in the comparable low stress range, but at slightly higher temperature, showed a stress exponent value of 7.5 and the n value reduced at higher stresses indicating operation of series mechanisms (see

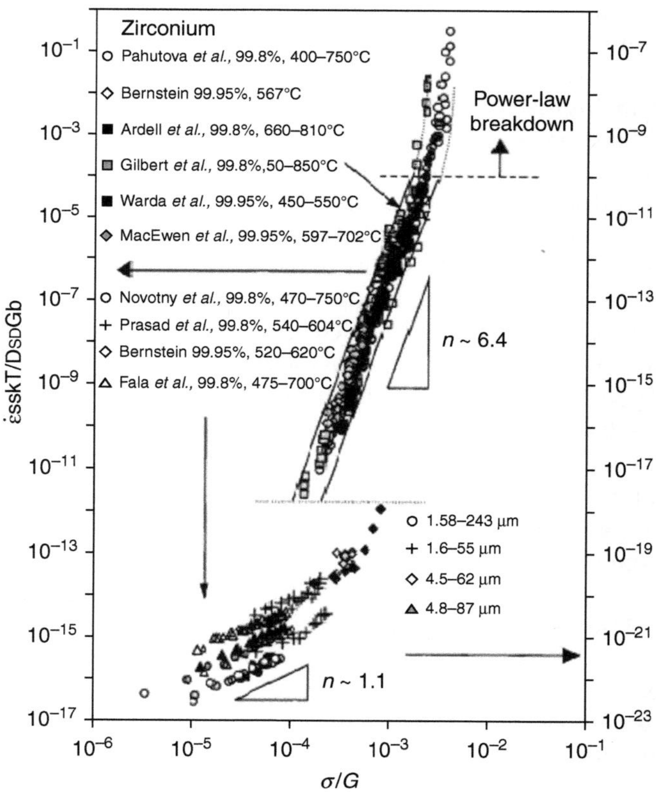

3.29 Steady-state creep of α-Zirconium.[123]

Section 3.4). The reason for these differences is not clear. While the results of Prasad *et al.*[122] indicated a stress exponent value close to 1 for pure zirconium at low stress levels (1–3 MPa) revealing the operation of Coble creep, the mechanism of creep at low stresses (0.2–14 MPa) at intermediate temperatures is ascribed by Ruano *et al.*, to grain boundary sliding than to diffusion mechanism.[123]

3.8.2 Role of alloying elements in creep of Zr-alloys

Although addition of alloying elements is never based solely on the diffusivity criteria, the resulting creep rate of the alloy is the outcome of the diffusivity of the elements added and understanding the diffusion phenomenon in these alloys will help in fine-tuning the concentration of the solute added. For instance, though the strengthening effect of Nb in Zr increases with Nb concentration, the optimum level to obtain a low steady-state creep rate ($<10^{-8}$ /s) as measured from the stress was found to be around 2.5 wt.%.[124] Nb with low diffusion rate and high solubility limit can effectively enhance the creep strength of Zr-2.5Nb alloy.[125] The creep strengthening effect by molybdenum in zirconium is reported to be superior to that of niobium (for comparable alloy fraction) in the temperature range 350–600°C.[126] Further, the modification in the creep mechanism by addition of Nb is clearly seen (Fig. 3.16) to transition from climb-controlled creep as in class-M to viscous glide creep as in alloy class by correlating experimental results on Nb-added Zr-alloy sheet[127] and Zircaloy-4 cladding.[128] Addition of 1 wt.%Nb to Zircaloy-4 reduces the creep rate by about 100 times and a region with $n = 3$ is introduced, absent in Zircaloy-4 and which behaves like pure metal (class M).

3.8.3 Role of hydrogen in creep

Since dry storage of spent fuel (SF) is gaining importance, it is necessary to assure the fuel rod integrity during interim storage for relatively long times. A clad with high burnup is likely to contain large amount of hydrogen (1000 ppm). The initial level of hydrogen is kept very low in order to reduce the in-reactor hydride-related problems and hydrogen pickup during service is controlled by employing new alloys.[129] Short term creep tests in Zircaloy-4 reveal that after a burnup of 64 MWd/kgU, hydrogen did not pose any deleterious effect and the material possessed sufficiently good ductility.[130] But it is interesting to note that hydrogen affects the creep rate in zirconium alloys differently as atomic hydrogen and as hydride.

In Zircaloy-4, the creep rate was reported to depend on the condition of the material – whether in cold-worked stress-relieved (CWSR) or annealed condition; CWSR alloy shows a significant strengthening on addition of hydrogen. The reason for this behavior is attributed to hydrogen influencing the strain hardening rate and static recovery of the material. Biaxial tests in Zircaloy-4 show that the presence of hydrides in the cladding will help to prevent the cold work microstructure from being annealed out of dislocations and thereby lower creep rates are maintained in the spent fuel cladding.[131] The same alloy in annealed condition shows a decrease in creep rate when hydrogen is in solution and an increase when part of the hydrogen is precipitated as hydrides. This behavior is attributed to the reduction in the stacking fault energy of Zr caused by diffusion of hydrogen to the core of the screw dislocations and an increase in their mobility. On the other hand when hydrogen is present in the form of hydrides, it increases the matrix strength and reduces the creep rate. The creep rate of Zircaloy-4 at a temperature of 693K and a stress of 150 MPa, a marginal increase in creep rate is noted for a hydrogen content of 200 wt.ppm. The increase in the creep rate is believed to be brought out by the reduction in the modulus value when hydrogen is added.[132] In a Zr-2.5wt.%Nb alloy, the creep rate at 723K is reported to increase by 2–2.5 times for a hydrogen content of 160 wt.ppm and the stress exponent reduces from 2.41 to 1.59, indicating the change in creep mechanism (Fig. 3.30).[133]

The results above indicate that hydrogen in dissolved state increases the creep rate and this is pertinent to SF which remains in this temperature range (~573K) and has sufficient hydrogen dissolved in it.

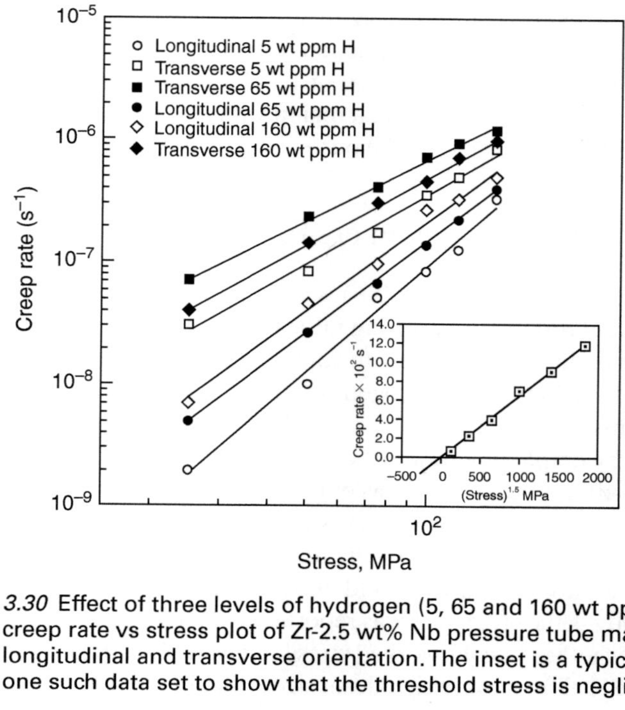

3.30 Effect of three levels of hydrogen (5, 65 and 160 wt ppm) on the creep rate vs stress plot of Zr-2.5 wt% Nb pressure tube material along longitudinal and transverse orientation. The inset is a typical replot of one such data set to show that the threshold stress is negligible.

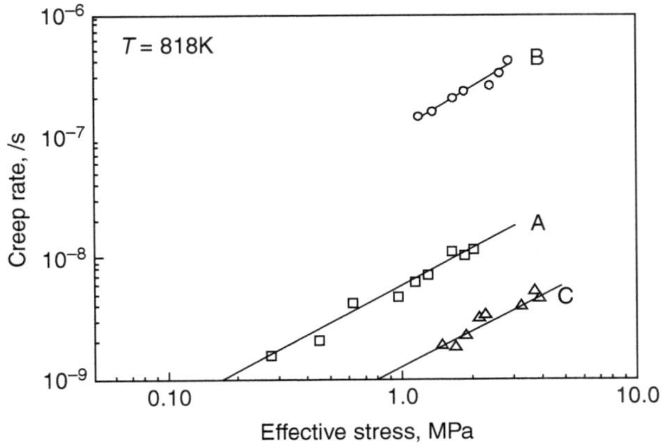

3.31 Abnormal creep in a Zr-2.5wt.%Nb alloy at low stresses.

3.8.4 The effect of thermal treatment and microstructure on creep behavior

An intermediate cooling rate from β phase, in a Zr-2.5wt.%Nb alloy, has resulted in a decrease in creep rate by 100 times over cold worked material and rendered higher anisotropy at 450°C. This increase in creep rate is attributed to segregation of Nb in grains that are favorably oriented for easy slip.[133]

The recent work on the thermal creep of Zr-2.5%Nb alloy by Kishore *et al.*[134] indicates that a microstructure containing a stable phase creeps faster than one with a meta-stable phase and a phase redistribution is established (Fig. 3.31). During creep deformation the stable β phase (with 80 wt.%Nb) dissolves and re-precipitates as β phase (with ~35wt.%Nb), this resulting strain due to phase change adds to the creep strain. Similar phase transformation is reported by Griffiths wherein the Zr-2.5wt.%Nb alloy after 2–14 years of in-reactor service shows that the β phase has a distribution of composition, the Nb concentration varying from 37% to 75%.[135] However, the effect of this phase change on the creep deformation is not well studied.

3.8.5 Irradiation creep

All the load bearing components in the core of the reactor, namely clad tubes, guide tubes (GT), GT assemblies and BWR channels undergo irradiation creep, albeit at different rates. The clad tube is a crucial boundary which has to withstand steep temperature and pressure gradient across its thickness. Steady-state creep dominates the service life of the clad and only in very rare cases the material may enter tertiary creep range. The creep rate of clad material under an irradiation environment is many times higher (depending on the material chemistry and the flux) than that under out-of-pile conditions. Further, the dimensional changes in clad tube (an anisotropic material) happen in a preferential direction which gives rise to other unwanted problems. The irradiation creep is not just the thermal creep imposed with high defect density. In the former the interstitial and vacancy loops that form during irradiation play a major role in the creep mechanism; in the latter the creep rate increases with temperature. The irradiation creep is weakly dependent on irradiation temperature ('athermal') and in-reactor thermal creep controls the deformation above ~400°C.

Two mechanisms are proposed to explain the irradiation creep phenomenon: (a) stress-induced preferential absorption (SIPA), where extra planes of atoms accumulate on crystal planes so as to produce creep strain in the direction of the applied stress and (b) stress-induced preferential nucleation (SIPN), which assumes that nucleation of loops is preferred on planes with a high resolved normal stress. Both of these mechanisms assume that the growth or formation of loops occur at a favorable orientation with respect to applied stress and causes macroscopic strain. Neutron irradiation produces large quantities of point defects – vacancies and self interstitial atoms (SIAs). These defects migrate to different sinks like dislocations and grain boundaries, in a preferential manner due to the anisotropy of the zirconium crystal lattice, in order to reduce the energy of the system. Because of the diffusional anisotropy, interstitial atoms tend to migrate to dislocations lying on prism planes and to grain boundaries oriented parallel to prism planes, while vacancies drift preferentially to dislocations lying on basal planes and to boundaries parallel to basal planes. This gives rise to elongation in one direction and contraction in the other.[136] The creep rate is controlled by dislocation glide and this in turn can be controlled by suitable alloying elements and by choosing an appropriate texture of zirconium matrix.

The total strain measured in an irradiation creep consists of the strain due to thermal creep (ε_{th}), irradiation creep (ε_{irr}) and irradiation growth (ε_g) and is assumed to be additive:

$$\varepsilon = \varepsilon_{\text{th}} + \varepsilon_{\text{irr}} + \varepsilon_g \tag{3.66}$$

The creep rate ($(\dot{\varepsilon})$) is given by the empirical relation

$$\dot{\varepsilon} = f\left(\phi^m \sigma^n e^{-Q/RT}, f, d, \rho, A\right) \tag{3.67}$$

where A is a constant, φ the flux, σ the stress, f the texture parameter, d the grain size, m (~0.4–0.7) and n (~0.8–2) are constants, and the others have their usual meaning.

As the dislocation density in a CWSR material is high and, as glide but not climb is the rate controlling mechanism under reactor conditions, the creep rate of CWSR is higher than that of recrystallized material as shown in the figures. The creep rate (a) increases with increase in the flux, (b) increases with increase in temperature (contribution by thermal creep predominates above 400°C), (c) is higher along the rolling direction than the transverse direction, (d) decreases with fluence (as radiation hardening sets in) and (e) depends upon the type of alloy (Nb and Sn content increases creep resistance). The irradiation creep rates of cold-worked Zr-2.5wt.%Nb alloy are about one-third of those of cold-worked Zircaloy-2[137] at comparable temperature, stress and fast neutron flux while the creep down of HANA (High Performance Alloy for Nuclear Applications) alloy (after a dose of 12 MWd/Kg U) is half that for Zircaloy-4.[138]

3.9 References

1. E. N. Andrade, 'On the viscous flow in metals and allied phenomena', *Proc Roy Soc Lond A*, **84** (1910) 1–12.
2. D. Hanson and M. A. Wheeler, 'The deformation of metals under prolonged loading: Part I: The flow and fracture of aluminum', *J Inst Metals*, **45** (1931) 229–257.
3. G. E. Dieter, *Mechanical Metallurgy*, New York: McGraw-Hill, 1988.
4. J. E. Bird, A. K. Mukherjee and J. E. Dorn, 'Correlations between high-temperature creep behavior and structure', in *Quantitative Relation Between Properties and Microstructure*, D. G. Brandon and A. Rosen (eds) Haifa, Israel (1969) 255–342.
5. J. Cadek, *Creep in Metallic Materials*, Prague, Czechoslovakia: Elsevier, 1988.
6. F. Garofalo, *Fundamentals of Creep and Creep-Rupture in Metals*, New York: McMillan, 1965.
7. W. J. Evans and B. Wilshire, 'The high temperature creep and fracture behavior of 70–30 alpha brass', *Metall Trans*, **1** (1970) 2133–2139.
8. G. A. Webster, A. P. D. Cox and J. E. Dorn, 'A relationship between transient and steady state creep at elevated temperatures', *J Metal Sci*, **3** (1969) 221–225.
9. K. E. Amin, A. K. Mukherjee and J. E. Dorn, 'A universal law for high-temperature diffusion controlled transient creep', *J Mech Phys Solids*, **18** (1970) 413–426.
10. K. L. Murty, 'Biaxial creep of zircaloy cladding', *Trans IIM*, **39** (1986) 357–368.
11. N. Nir, F. H. Huang, E. W. Hart and Che-Yu Li, 'Relationship between anelastic and nonlinear viscoplastic behavior of 316 stainless steel at low homologous temperature', *Metall Mater Trans*, **8** (1977) 583–588.
12. K. L. Murty, 'Creep studies for zircaloy life prediction in water reactors', *J Metals*, Oct. (1999) 32–39.

13. F. H. Norton, *The Creep of Steels at High Temperatures*, New York: McGraw-Hill, 1929.
14. F. A. Mohamed and M. Chauhan, 'Interpretation of the creep behavior of nanocrystalline nickel in terms of dislocation accommodated boundary sliding', *Metall Mater Trans*, **37A** (2006) 3555–3567.
15. S. Gollapudi, K. V. Rajulapati, I. Charit, K. M. Youssef, C. C. Koch, R. O. Scattergood and K. L. Murty, 'Creep in nanocrystalline materials: Role of stress assisted grain growth', *Mater Sci Engg*, **A527** (2010) 5773–5781.
16. O. D. Sherby and P. M. Burke, 'Mechanical behavior of crystalline solids at elevated temperatures', *Prog. Mater Sci*, **13** (1968) 325–390.
17. S. Gollapudi, V. Bhosle, I. Charit and K. L. Murty, 'Newtonian viscous creep in Ti3Al2.5V', *Phil Mag*, **88**:9 (2008) 1357–1367.
18. F. R. N. Nabarro, *Report of Conference on Solids*, London: The Physical Society, 1948.
19. C. Herring, 'Diffusional viscosity of a polycrystalline solid', *J Appl Phys*, **21** (1950) 437–445.
20. R. L. Coble, 'A model for boundary diffusion controlled creep in polycrystalline materials', *J Appl Phys*, **34** (1963) 1679–1682.
21. R. B. Jones, 'Diffusion creep in polycrystalline magnesium', *Nature*, **207** (1965) 70–72.
22. J. Harper and J. E. Dorn, 'Viscous creep in aluminum near its melting temperatures', *Acta Metall*, **5** (1957) 654–665.
23. G. Malakondaiah and P. RamaRao, 'Grain size independent viscous creep in alpha titanium and beta cobalt', *Scripta Metall*, **13** (1979) 1187–1190.
24. G. Malakondaiah and P. Rama Rao, 'Viscous creep of beta-Co', *Mater Sci Eng*, **52** (1982) 207–221.
25. J. Fiala, J. Novotny and J. Cadek, 'Coble and Harper-Dorn creep in iron at homologous temperatures of 0.4–0.54', *Mater Sci Eng*, **60** (1983) 195–206.
26. J. Novotny, J. Fiala and J. Cadek, 'Harper-Dorn creep in alpha-zirconium', *Acta Metall*, **33** (1985) 905–911.
27. T. G. Langdon and P. Yavari, 'An investigation of Harper-Dorn creep – II The flow process', *Acta Metall*, **30** (1982) 881–887.
28. C. R. Barrett, E. C. Muehleisen and W. D. Nix, 'High temperature low stress creep of Al and Al-0.5%Fe', *Mater Sci Eng*, **10** (1972) 33–42.
29. K. L. Murty, 'Viscous creep in Pb-9Sn', *Mater Sci Eng*, **14** (1974) 169–176.
30. P. Kumar, M. E. Kassner and T. G. Langdon, 'Fifty years of Harper-Dorn creep: A viable creep mechanism or a Californian artifact?' *J Mater Sci*, **42** (2007) 409–420.
31. K. Lingamurty, S. Gollapudi and I. Charit, 'Newtonian viscous creep in metals', *Trans IIM*, **63**(2–3) (2010) 85–91.
32. J. R. Spingarn and W. D. Nix, 'A model for creep based on the climb of dislocations at grain boundaries', *Acta Metall*, **27** (1979) 171–177.
33. J. Wadsworth, O. A. Ruano and O. D. Sherby, 'Denuded zones, diffusional creep and grain boundary sliding', *Metall Mater Trans*, **33A** (2002) 219–229.
34. R. L. Squires, R. T. Weiner and M. Phillips, 'Grain boundary denuded zones in a Mg-0.5Zr alloy', *J Nucl Mater*, **8** (1963) 77–80.
35. W. Vickers and P. Greenfield, 'Diffusion creep in magnesium alloys', *J Nucl Mater*, **24** (1967) 249–260.

36. L. E. Raraty, 'A comment on diffusion creep deformation in a magnesium reactor component', *J Nucl Mater*, **20** (1966) 344–348.
37. J. Wolfenstine, O. A. Ruano, J. Wadsworth and O. D. Sherby, 'Refutation of the relationship between denuded zones and diffusional creep', *Scr Metall Mater*, **29** (1993) 515–520.
38. G. W. Greenwood, 'Denuded zones and diffusional creep', *Scripta Metall Mater*, **30** (1994) 1527–1530.
39. J. B. Bilde-Sorensen and D. A. Smith, 'Comment on the refutation of the relationship between denuded zones and diffusional creep', *Scripta Metall Mater*, **30** (1994) 383386.
40. B. Burton and G. L. Reynolds, 'In defense of diffusional creep', *Mater Sci Eng*, **191A** (1995) 135–141.
41. W. Jaeger and H. Gleiter, 'Grain boundaries as vacancy sources in diffusional creep', *Scripta Metall*, **12** (1978) 675–678.
42. K. R. McNee, G. W. Greenwood and H. Jones, 'Observation and interpretation of some microstructural features of low stress creep', *Scripta Mater*, **47** (2002) 619–623.
43. F. A. Mohamed and T. J. Ginter, 'On the nature and origin of Harper-Dorn creep', *Acta Metall*, **30** (1982) 1869–1881.
44. I. M. Lifshitz, 'On the theory of diffusion viscous flow of polycrystalline bodies', *Soviet Phys JETP*, **17** (1963) 909–920.
45. W. A. Rachinger, 'Relative grain translations in the plastic flow of aluminum', *J Inst Met*, **81** (1952–1953), 33–41.
46. M. F. Ashby and R. A. Verall, 'Diffusion accommodated flow and superplasticity', *Acta Metall*, **21** (1973) 149–163.
47. J. R. Spingarn and W. D. Nix, 'Diffusional creep and diffusionally accommodated grain rearrangement', *Acta Metall*, **26** (1978) 1389–1398.
48. A. Ball and M. M. Hutchison, 'Superplasticity in aluminum-zinc eutectoid', *J Met Sci*, **3** (1969) 1–7.
49. T. G. Langdon, 'An estimate of the time to fracture due to triple point cracking', *Phil Mag*, **26** (1970) 945–948.
50. A. K. Mukherjee, 'The rate controlling mechanism in superplasticity', *Mater Sci Eng*, **8A** (1971) 83–89.
51. A. Arieli and A. K. Mukherjee, 'A model for the rate controlling mechanism in superplasticity', *Mater Sci Eng*, **45A** (1980) 61–70.
52. M. E. Kassner and M. T. Perez-Prado, *Fundamentals of Creep in Metals and Alloys*, New York: Elsevier, 2004.
53. F. Ochoa, X. Deng and N. Chawla, 'Effect of cooling rate on creep behavior of a Sn-3.5Ag alloy', *J Electron Mater*, **33** (2004) 1596–1607.
54. J. Weertman, 'Theory of steady state creep based on dislocation climb', *J Appl Phys*, **26** (1955), 1213–1217.
55. J. C. Fisher, 'On the strength of solid solution alloys', *Acta Metall*, **2** (1954) 9–10.
56. R. C. Gifkins, 'Grain boundary sliding and its accommodation during creep and superplasticity', *Metall Trans*, **7A** (1976) 1225–1232.
57. S. Gollapudi, I. Charit and K. L. Murty, 'Creep mechanisms in Ti-3Al-2.5V tubing deformed under closed-end internal gas pressurization', *Acta Mater*, **56** (2008) 2406–2419.

58. A. H. Cottrell and M. A. Jaswon, 'Distribution of solute atoms around a slow dislocation', *Proc Roy Soc London*, **199A** (1949) 104–114.
59. H. Suzuki, *Sci. Rep. Res. Inst. Tohoku University*, 4A (1957) 455.
60. K. L. Murty, 'Transitions in deformation mechanisms in class-A alloys – Historical perspective and recent applications to microelectronic solder and nuclear core materials', in *Proceedings of Seventh International Conference on Creep and Fracture of Engineering Materials and Structures*: 'Mechanisms of High Temperature Deformation' in honor of Prof. Oleg D. Sherby, UC-I, J. C. Earthman and F. A. Mohamed (eds), TMS (1997) 69–78.
61. K. L. Murty and J. Ravi, 'Transitions in Creep Mechanisms in a Zr-1Nb-1Sn-0.2Fe Sheet', in proceedings of the TMS Johannes Weertman Symposium, R. J. Arsenault *et al.* (ed.), TMS (1996) 203–209.
62. K. L. Murty, M. Gold and A. L. Ruoff, 'High temperature creep mechanisms in a-iron and other metals', *J Appl Phys*, **41** (1970) 4917–4928.
63. J. Weertman and J. R. Weertman, 'High temperature creep of rock and mantle viscosity', *Ann Rev Earth Planet Sci*, **3** (1975) 293–315.
64. C. R. Barrett and W. D. Nix, 'A model for steady state creep based on the motion of jogged screw dislocations', *Acta Metall*, **12** (1965) 1247–1258.
65. G. B. Viswanathan, V. K. Vasudevan and M. J. Mills, 'Modification of the jogged screw model for creep of γTiAl', *Acta Mater*, **47** (1999) 1399–1411.
66. G. B. Viswanathan, S. Karthikeyan, R. W. Hayes and M. J. Mills, 'Creep behaviour of Ti-6Al-2Sn-4Zr-2Mo: II. Mechanisms of deformation', *Acta Mater*, **50** (2002) 4965–4980.
67. J. H. Moon, P. E. Cantonwine, K. R. Anderson, S. Karthikeyan and M. J. Mills, 'Characterization of modeling of creep mechanisms in Zircaloy-4', *J Nucl Mater*, **353** (2006) 177–189.
68. L. I. Ivanov and V. A. Yanushkevich, 'The mechanism of fatigue creep of body centered cubic metals', *Fiz Metal Metal*, **17** (1964) 112–117.
69. K. Linga Murty, A. G. Paradkar and S. Nangalia, 'Effect of stress ratio on biaxial creep and dislocation microstructures in recrystallized Ti3Al2.5V tuning', in *Deformation, Processing and Properties of Structural Materials*, E. Taleff, C. K. Syn and D. Leuser (eds), TMS (2000) 169–179.
70. G. M. Pharr, 'Some observations on the relation between dislocation substructure and power law breakdown in creep', *Scripta Metall*, **15** (1981) 713–717.
71. J. P. Poirier, 'Is power law creep diffusion controlled?' *Acta Metall*, **26** (1978) 629–637.
72. S. V. Raj and G. M. Pharr, 'Creep substructure formation in sodium chloride single crystals in the power law and exponential creep regimes', *Mater Sci Eng A*, **122** (1989) 233–242.
73. I. Charit and K. L. Murty, 'Creep behavior of niobium-modified zirconium alloys', *J. Nucl Mater*, **374** (2008) 354–363.
74. K. L. Murty, 'Transitional creep mechanisms in Al-5Mg at high stresses', *Scripta Metall*, **7** (1973) 899–903.
75. P. Yavari and T. G. Langdon, 'An examination of the breakdown in creep by viscous glide in solid solution alloys at high stress levels', *Acta Metall*, **30** (1982) 2181–2189.
76. M. F. Ashby, 'A first report on deformation mechanism maps', *Acta Metall*, **20** (1972) 887–897.

77. M. F. Ashby, 'A first report on sintering diagrams', *Acta Metall*, **22** (1974) 275–289.
78. S. C. Lim and M. F. Ashby, 'Overview no. 55 wear mechanism maps', *Acta Metall*, **35** (1987) 1–24.
79. M. F. Ashby, C. Gandhi and D. M. R. Taplin, 'Overview 3 Fracture mechanism maps and their construction for f.c.c metals and alloys', *Acta Metall*, **27** (1979) 699–729.
80. F. A. Mohamed and T. G. Langdon, 'Deformation mechanism maps for solid solution alloys', *Scripta Metall*, **9** (1975) 137–140.
81. G. R. Leverant and D. N. Duhl, 'The effect of stress and temperature on the extent of primary creep in directionally solidified nickel based superalloys', *Metall Trans*, **2** (1971) 907–908.
82. R. K. Penny and D. L. Marriott, *Design for Creep*, Chapman and Hall, 1995.
83. F. R. Larson and J. Miller, 'A time temperature dependent relationship for rupture and creep stresses', *Trans ASME*, **74** (1952) 765–775.
84. J. E. Dorn and L. A. Shepherd, 'What we need to know about creep', *ASTM Special Tech Publ*, West Conshohocken, PA, **165** (1954).
85. S. S. Manson and A. M. Haferd, 'A linear time temperature relation for extrapolation of creep and stress rupture data', *NACA TN 2890* (1953).
86. G. Murry, 'Extrapolation of the results of creep tests by means of parametric formulae', *ASME/ASTM/I Mech E Proceedings Joint Conference on Creep*, New York/London (1963).
87. N. J. Grant and A. G. Bucklin, 'On the extrapolation of short time stress rupture data', in *Deformation and Fracture at Elevated Temperature*, N. J. Grant and A. W. Mullendore (eds), MIT Press, Boston (1965) 35–66.
88. J. Glen, 'A new approach to the problem of creep', *J Iron Steel Inst*, **189** (1958) 333–343.
89. A. Mendelsohn and S. S. Manson, 'Optimization of time temperature parameters for creep and stress rupture with application to date from German cooperative long time creep program', *NACA TN d/1975* (1963).
90. F. C. Monkman and N. J. Grant, 'An empirical relationship between rupture life and minimum creep rate in creep-rupture tests', in *Deformation and Fracture at Elevated Temperature*, N. J. Grant and A. W. Mullendore (eds), MIT Press, Boston (1950) 91–103.
91. G. Srikant, B. Marple, I. Charit and K. L. Murty, 'Characterization of stress rupture behavior of cp-Ti via burst testing', *Mat Sci Eng*, **A463** (2007) 203–207.
92. B. Wilshire, 'Observations, theories and predictions of high temperature creep behavior', *Metall Mater Trans*, **33A** (2002) 241–248.
93. B. Wilshire and T. D. Lieu, 'Deformation and damage processes during creep of Incoloy MA957', *Mater Sci Eng*, **386A** (2004) 81–90.
94. I. Le May, *Principles of Mechanical Metallurgy*, Elsevier, New York (1981).
95. L. M. Kachanov, *Introduction to Continuum Damage Mechanics*, Martinus Nijhoff, Dordrecht, the Netherlands (1958).
96. Y. N. Rabotnov, *Creep Problems in Structural Members*, North-Holland, Amsterdam, the Netherlands (1969).
97. M. F. Ashby and B. F. Dyson, *Advances in Fracture Research*, Valluri *et al.* (eds), Pergamon Press, Oxford (1984).

98. C. Phaniraj, B. K. Choudhary, K. B. S. Rao and B. Raj, 'Relationship between time to reach Monkman-Grant ductility and rupture life', *Scripta Mater*, **48** (2003) 1313–1318.
99. H. Mishra, D. V. V. Satyanarayana, T. K. Nandy and P. K. Sagar, 'Effect of trace impurities on the creep behavior of a near alpha titanium alloy', *Scripta Mater*, **59** (2008) 591–594.
100. R. S. Gordon, 'Mass transport in diffusional creep of ionic solids', *J Am Ceram Soc*, **56** (1973) 147–152.
101. A. Dominguez-Rodriguez and J. Castaing, 'Diffusion fluxes and creep of polycrystalline compounds: application to alumina', *Scripta Metall Mater*, **29** (1993) 1207–1211.
102. A. H. Chokshi, 'Diffusion creep in oxide ceramics', *J Eur Ceram Soc*, **22** (2002) 2469–2478.
103. I. D. Choi, D. K. Matlock and D. L. Olsen, 'Creep behavior of nickel-copper laminate composites with controlled composition gradient', *Metall Trans*, **21A** (1990) 2513–2520.
104. K. S. Ravichandran and V. Seetharaman, 'Prediction of steady state creep behavior of two phase composites', *Acta Metall Mater*, **41** (1993) 3351–3361.
105. I. W. Chen, 'Diffusional creep of two phase materials', *Acta Metall*, **30** (1982) 1655–1664.
106. Z. Y. Ma, R. S. Mishra, S. C. Tjong, 'High temperature creep behavior of TiC particulate reinforced Ti-6Al-4V alloy composite', *Acta Mater*, **50** (2002) 4293–4302.
107. Y. Li, S. R. Nutt and F. A. Mohamed, 'An investigation of creep and substructure formation in 2124 Al', *Acta Mater*, **45** (1997) 2607–2620.
108. G. S. Ansell, 'The mechanism of dispersion strengthening: A review', in *Oxide Dispersion Strengthening Conference*, V. Lenel (ed.), Gordon and Breach, New York (1968) 61–141.
109. Rishi Raj, 'Equations for diffusional creep under multiaxial stress states', *Scripta Metall*, **15** (1981) 273–274.
110. K. L. Murty, 'Deformation mechanisms and transients in creep of zircaloys: Applications to nuclear technology', *Trans IIM*, **53**(3) (2000) 107–120.
111. H. Stehle, E. Steinberg and E. Tenckhoff, 'Mechanical anisotropy and microstructure of zircaloy canning tubes', in *Zirconium in the Nuclear Industry*, ASTM STP 633 (1977) 486–507.
112. K. L. Murty and I. Charit, 'Texture development and anisotropic deformation of zircaloys', *Prog Nucl Energy*, **48** (2006) 325–359.
113. B. L. Adams and K. L. Murty, 'Biaxial creep of textured zircaloy: Part I – phenomenological and experimental descriptions', *Mater Sci Eng*, **70** (1985) 169–180.
114. Iaea Nuclear Energy Series No. NF-T-2.1, 'Review of Fuel Failures in Water Cooled Reactors', International Atomic Energy Agency, Vienna, 2010.
115. 'Hydride-Related Degradation of SNF Cladding under Repository Conditions', ANL-EBS-MD-000011 REV 00 ICN 01 December 2000.
116. K. L. Murty, 'Internal pressurization creep of thin-walled tubing of zr-alloys for dry storage feasibility of nuclear spent fuel', *J Met*, September (2000) 34–38.

117. Review of Fuel Failures in Water Cooled Reactors, No. NF-T-2.1, STI/PUB/1445 ISBN 978–92–0–102610–1 Printed by the IAEA in Austria June 2010.
118. T. Hayes, M. E. Kassner and R. S. Rosen, 'Steady state creep of Zr at temp upto 850°C', *Metall Mater Trans*, **33A** (2002) 337.
119. I. M Bernstein, 'Diffusion creep in zirconium and certain zirconium alloys', *Trans AIME*, **239** (1967) 1518–1522.
120. S. R. MacEwen, R. G. Fleck, E. T. C. Ho and O. T. Woo, 'Deformation of α-zirconium in the vicinity of 0.5Tm', *Metall Trans A*, **12** (1981) 1751–59.
121. A. J. Ardell and O. D. Sherby, 'The steady state creep of polycrystalline alpha zirconium at elevated temperatures', *Trans AIME*, **239** (1967) 1547–1556.
122. N. Prasad, G. Malakondaiah and T. Rama Rao, 'Low stress creep behavior of zirconium', *Trans Indian Inst Metals*, **42** (1989) S165–S174.
123. O. A. Ruano, J. Wadswoth and O. D. Sherby, 'Low temperature creep of a-Zr at intermediate temperatures', *Mater Sci Eng*, **84** (1986) L1–L6.
124. M. Pahutova, J. Cadek and V. Cerny, 'Steady state creep rate of Zr-Nb alloys in a temperature interval 350–550°C', *J Nucl Mater*, **61** (1976) 285–296.
125. H. G. Kim, Y. H. Kim, B. K. Choi and Y. H. Jeong, 'Effect of alloying elements (Cu, Fe, and Nb) on the creep properties of Zr alloys', *J Nucl Mater*, **359** (2006) 268–273.
126. M. Pahutova and J. Cadek, 'Effect of molybdenum on some basic creep characteristics of alpha zirconium in a temperature interval of 350° to 600°C', *Mater Sci Eng*, **20** (1975) 277–285.
127. K. L. Murty and J. Ravi, Wiratmo, 'Transitions in creep mechanisms and creep anisotropy in Zr-1Nb- 1Sn-0.2Fe sheet', *Nucl Eng Design*, **156** (1995) 359–371.
128. Y. Zhou, B. Devarajan and K. L. Murty, 'Short-term rupture studies of Zircaloy-4 and Nb-modified Zircaloy-4 tubing using closed-end internal pressurization', *Nucl Eng Design*, **228** (2004) 3–13.
129. J. R. Theaker, R. Choubey, G. D. Maon, S. A. Aldridge, L., Davis, R. A. Graham and C. E. Coleman, in: E. R. Bradley and G. P. Sabol (eds), *Zirconium in Nuclear Industry 10th International Symposium*, ASTM-STP-1245 (1994) 221.
130. W. Goll, H. Spilker and E. H. Toscano, 'Short-time creep and rupture tests on high burnup fuel rod cladding', *J Nucl Mater*, **289** (2001) 247–253.
131. N. Rupa, M. Clavel, P. Bouffloux, C. Domain and A. Legris, 'Impact of hydrogen on plasticity and creep of unirradiated zircaloy-4 cladding tubes', in: G. D. Moan and P. Rudling (eds), *Zirconium in Nuclear Industry 13th International Symposium*, ASTM STP 1423, ASTM International, West Conshohocken, PA (2000) 399–405.
132. D. Setoyama and S. Yamanaka, 'Indentation creep study of zirconium hydrogen solid solution', *J Alloys Compounds*, **379** (2004) 193.
133. R. Kishore, 'Effect of hydrogen on the creep behavior of Zr-2.5wt%Nb alloy at 723K', *J Nucl Mater*, **385** (2009) 591–594.
134. R. Kishore, S. Banerjee and P. Rama Rao, 'First report on observation of abnormal creep in Zr-2.5%wt%Nb alloy at low stress', *J Mater Sci*, **44** (2009) 2247–2256.
135. M. Griffiths, J. F. Mecke, J. E. Winegar, in E. R. Bradley and G. P. Sabol (eds), *Zirconium in Nuclear Industry: 11th International Symposium*, September 11–14, 1995, ASTM STP 1295, West Conshohocken, 580.

136. R. Adamson, F. Garzarolli and C. Patterson, 'In-reactor creep of zirconium alloys', R. Adamson (ed.), September 2009, Advanced Nuclear Technology International, Krongjutarvägen 2C, SE-730 50, Skultuna, Sweden.
137. P. A. Ross-Ross and C. E. L. Hunt, 'The in-reactor creep of cold-worked Zircaloy-4 and Zirconium-2.5wt.% niobium pressure tubes', *J Nucl Mater*, **26** (1968) 2–17.
138. Y. H. Jeong, S-Y. Park, M-H. Lee, B-K. Choi, J-H. Baek, J-Y. Park, J-H. Kim and H-G. Kim, 'Out-of-pile and in-pile performance of advanced zirconium alloys (HANA) for high burn-up fuel', *J Nucl Sci Technol*, **43**(9) (2006) 977–983.

Part II

Materials ageing and degradation in particular light water reactor (LWR) components

4

Properties of zirconium alloys and their applications in light water reactors (LWRs)

R. B. ADAMSON, Zircology Plus, USA and
P. RUDLING, ANT International, Sweden

DOI: 10.1533/9780857097453.2.151

Abstract: This chapter highlights the various uses and properties of zirconium alloy cladding and structural components used in nuclear power light water reactors. Specific attributes including dimensional stability, corrosion resistance, irradiation effects and mechanical properties are discussed in detail.

Key words: zirconium alloys, nuclear reactors, dimensional stability, radiation effects, mechanical properties, corrosion.

4.1 Introduction

Zirconium alloys are used as the prime structural material in light water reactors (LWRs). As such, they have to meet several requirements: low neutron absorption cross section; corrosion resistance in 280–350°C water; resistance to radiation in both mechanical behaviours and dimensional stability; reasonable strength, ductility and fabricability; affordable cost; and availability in large quantities.

Unalloyed zirconium was used as the structural material in the prototype core for nuclear submarines in 1953 (Rickover, 1975 in Adamson, 2010). However, variability in corrosion resistance, strength and cost issues prompted development of a stronger, more corrosion-resistant alloy named Zircaloy-2. This alloy was used in the first nuclear powered submarine, Nautilus 1954, and in the first commercial electricity-generating reactor, Shippingport 1957. Today, a variety of zirconium alloys (see below for details) are used in all LWRs throughout the world.

This chapter covers the following topics relevant to the uniqueness of zirconium alloys: Section 4.2 on fuel assembly design; Sections 4.3–4.6 on material and performance issues; Section 4.7 covers future trends in materials; and Section 4.8 provides sources of further information.

Zirconium and hafnium (used as a neutron absorber) are unique among materials used in LWRs in that they have the hexagonal close packed (HCP)

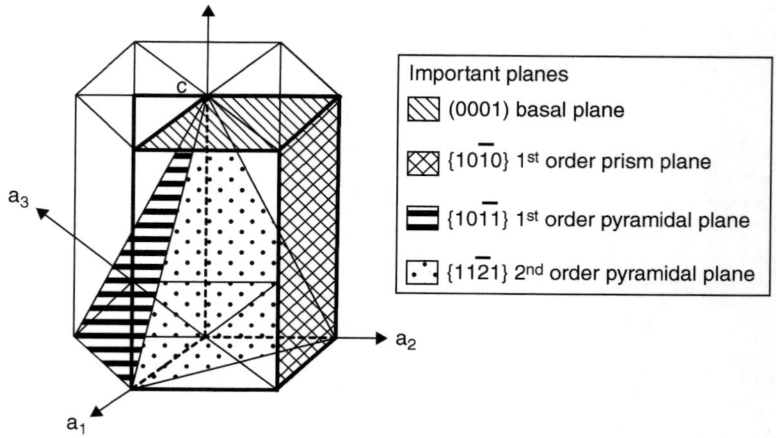

4.1 HCP crystallographic cell for zirconium alloys, showing the two major sets of plane: prism planes and, normal to them, the basal planes (see Schemel, 1989). Reprinted, with permission, from (Tenckhoff, 2005), copyright ASTM International, 100 Barr Harbor Drive, West Conshohocken, PA 19428.

crystallographic structure. This structure results in significant preferred orientation during fabrication, which results in significant anisotropy of all properties. Deformation in unirradiated zirconium alloys occurs mainly on prism planes, as illustrated in Fig. 4.1. During irradiation, dislocation loops of different types form on prism and basal planes (Section 4.3). Orientation distributions of the HCP cell are described by 'texture' parameters, which give the volumetric average orientations of the basal and prism planes in a particular component. However, this chapter will not explore crystallography and fabrication in any detail. Excellent reviews are given by Tenckhoff (1988); Tenckhoff (2005); Nikulina *et al.* (2006/2007); Rudling *et al.* (2007); and Strasser and Rudling (2004).

4.2 Fuel assembly designs

There are essentially, four different types of commercial light water cooled reactors, whose main characteristics are provided in Table 4.1 (Cox *et al.*, 2006).

4.2.1 Pressurized water reactors (PWRs) and boiling water reactors (BWRs)

There is a wide variety of fuel assembly (FA) types for BWRs and PWRs. The fuel rod array for BWRs was initially 7 × 7 but there has been a trend over the years to increase the number of FA rods and today most designs

Table 4.1 Design parameters in water cooled reactors

Parameter	Western type PWR	*VVER* (440/1000) MW	*BWR*	*RBMK*
1. Coolant	Pressurized H_2O	Pressurized H_2O	Boiling H_2O	Boiling H_2O
2. Fuel assembly materials (pressure tube materials)	Zr-4, ZIRLO, DUPLEX, M5, *MDA, NDA,* Inconel, SS	E110, E635	Zry-2, Zry-4, Inconel, *SS*	E110, E635 (Zr-2.5Nb)
3. Average power rating, (kW/l)	80–125	83–108	40–57	5
4. Fast neutron flux, average, n/cm^2·s	6–9E13	5–E13	4–7E13	1–2E13
5. Temperature, °C				
Average coolant inlet	279–294	267–290	272–278	270
Average coolant outlet	313–329	298–320	280–300	284
Max cladding *O.D.*	320–350	335–352	285–305	290
Steam mass content, %			7–14	14
6. System pressure, bar	155–158	125–165	70	67
7. Coolant flow, m/s	3–6*	3.5–6	2–5*	3.7
8. Coolant chemistry				
Oxygen, ppb	<0.05	<0.1	200–400	<20
Hydrogen (D_2),	2–4	30–60	–	–
ppm	25–50	0–1400	–	–
cc/kg	0–2200	0.05–0.6	–	–
Boron (as boric	0.5–3.5			–
acid), ppm	–			
Li (as LiOH), ppm				
K (as KOH), ppm		5–20		
NH_3, ppm		6–30		
NaOH, ppm		0.03–0.35		

* Variation from lower to upper part of the core and from plant to plant.
Source: A.N.T. International (2011) and Cox *et al.* (2006).

are either of 9 × 9 or 10 × 10 square configuration design (Cox *et al.*, 2006). The driving force for this trend was to reduce the linear heat generation rate (LHGR), which resulted in a number of fuel performance benefits such as lower fission gas release (FGR) and increased pellet clad interaction (PCI) margins. However, to increase utility competitiveness, the LHGRs of 9 × 9 and 10 × 10 FA have successively been increased, and peak LHGRs are today almost comparable to those of the older 7 × 7 and 8 × 8 designs.

Also for PWRs there has been a trend to greater subdivision of fuel rods, for example from the Westinghouse 15 × 15 to 17 × 17 design (Cox *et al.*, 2006). However, since PWRs do not have the same flexibility with core internals and control rods as BWRs, to accomplish this requires modification of the reactor internals. There is, however, one exception, namely DC

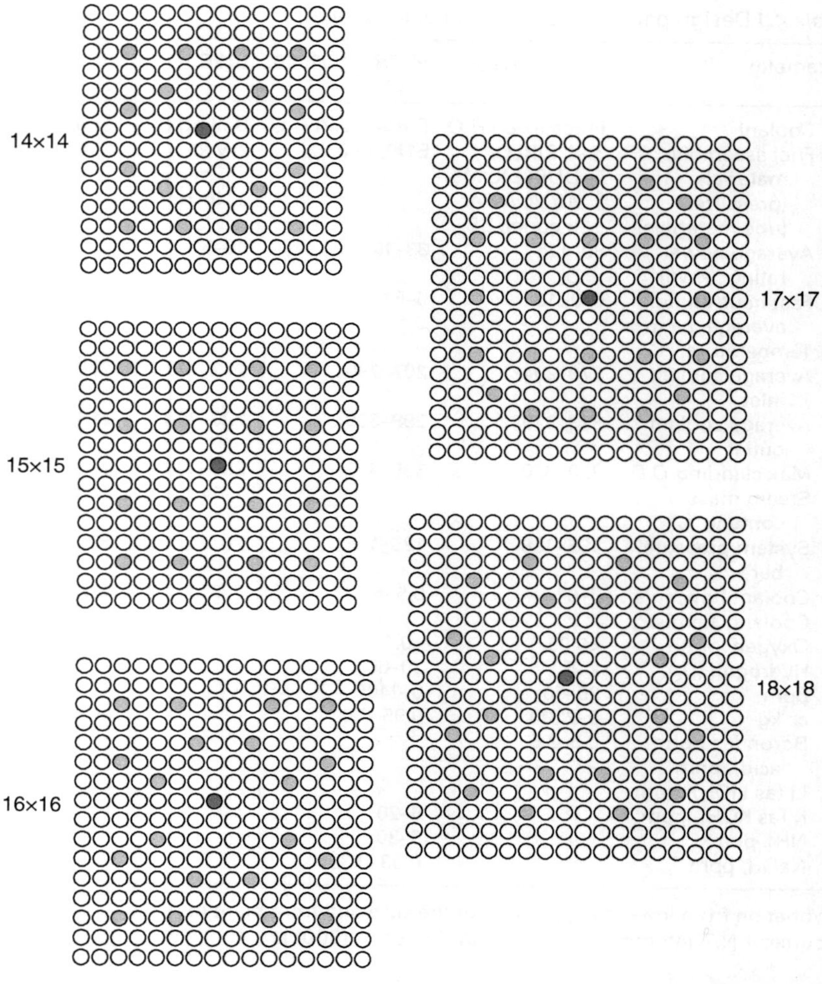

4.2 Layouts of different PWR FA design. Rods marked with light grey colour are GTs into which the control rod cluster is inserted. The position marked by a dark grey filled circle is the IT position (Cox *et al.*, 2006). (Source: A.N.T. International, 2011.)

Cook 1, which is switching to 17 × 17 through changing the reactor internals. Figure 4.2 shows the current PWR fuel rod array designs.

In most PWRs, the assemblies are positioned in the core by bottom and top fittings, and the lateral clearances are restricted by the assembly-to-assembly contacts at the spacer-grid levels (Cox *et al.*, 2006). Furthermore, the control rods consist of rod cluster control assemblies (RCCAs), the poison part of which moves into guide thimbles (GTs). These guide thimbles are an integral part of the assembly structure.

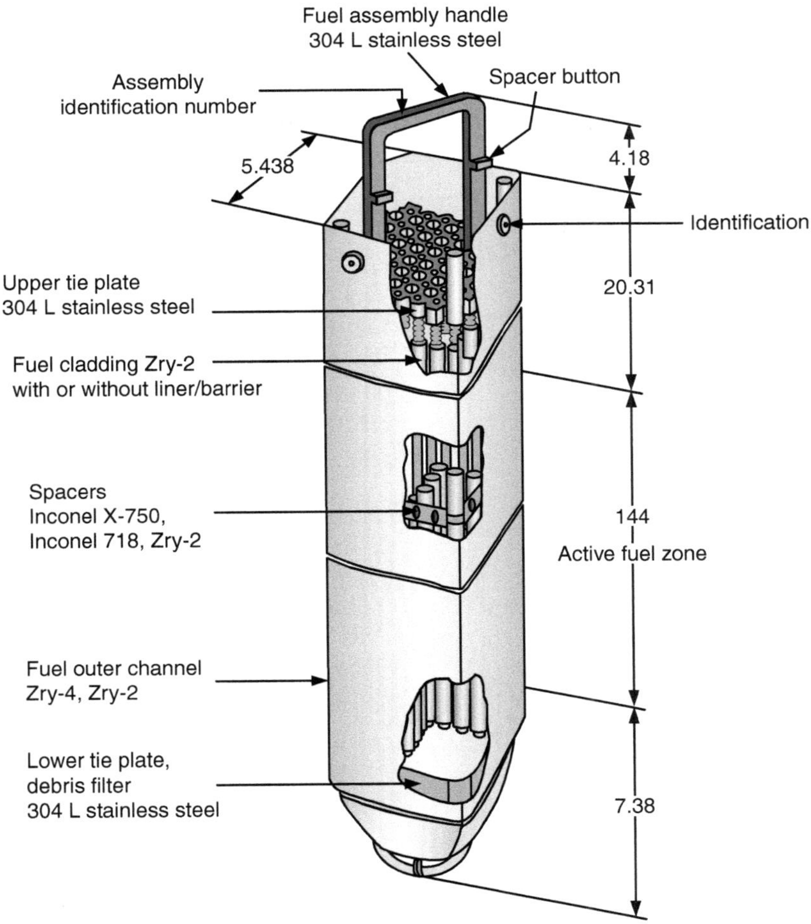

4.3 Typical BWR FA in inches (Cox *et al.*, 2006).

In all BWRs, the assemblies are enclosed in 'fuel channels' surrounding the assemblies and between which the blades of the control rods moves.

Irrespective of the many possible different shapes, sizes and configurations, the common FA design requirements are (Cox *et al.*, 2006):

- Maintain proper positioning of the fuel rods under normal operating conditions and in design basis accidents (DBAs) (e.g. seismic effects, LOCA, RIA).
- Permit handling capability before and after irradiation.

Figures 4.3 and 4.4 show a typical BWR and PWR FA, respectively. Also, the different FA components are shown and the material selections for these

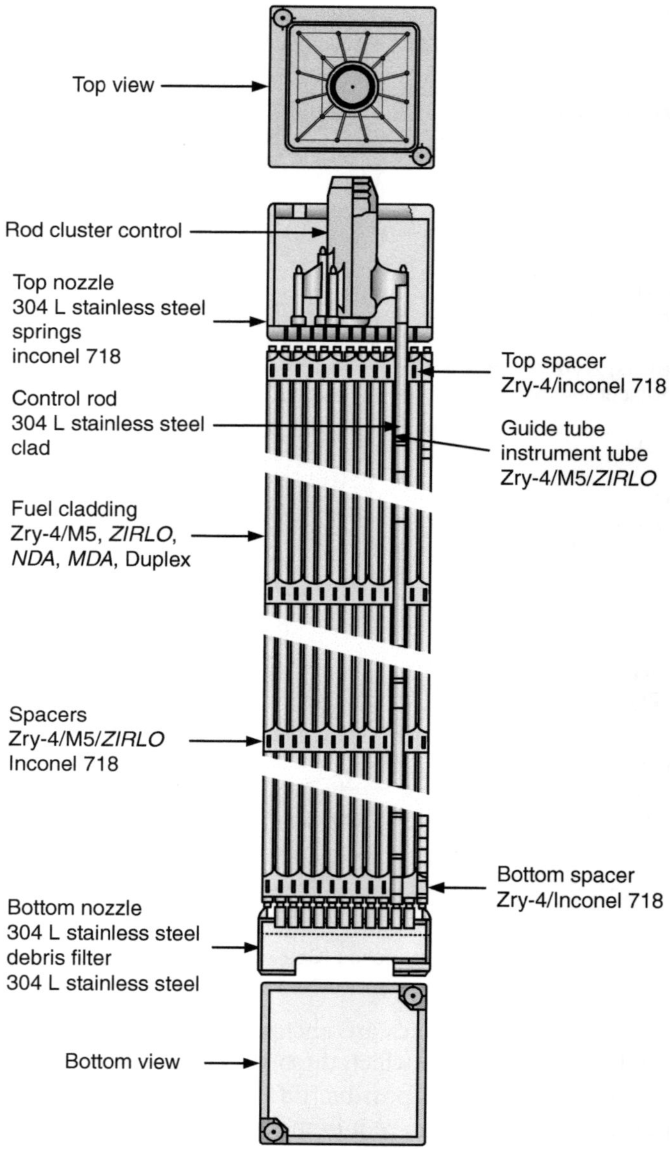

4.4 Typical PWR FA (Cox *et al.*, 2006).

components are provided. The selection of the different structural materials is based on their nuclear and mechanical properties as well as their cost, in order to ensure acceptable performance during normal operation and accidents.

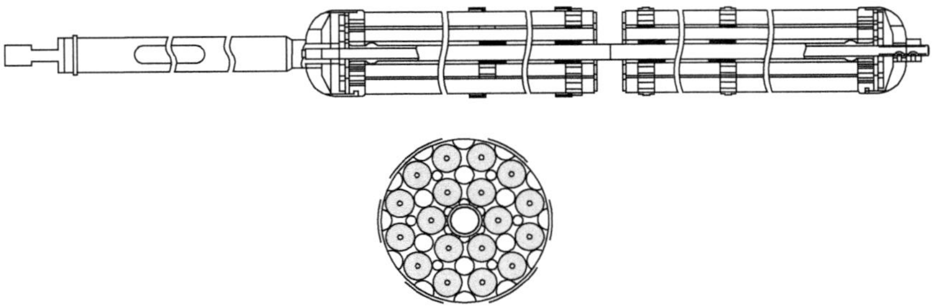

4.5 Draft of the RBMK-1500 fuel assembly. (Cox *et al.*, 2006).

4.2.2 RBMKs

RBMKs are basically vertical channel-type boiling water reactors (Cox *et al.*, 2004). They use E110 alloy (Zr-1%Nb) fuel cladding in the fully recrystallized conditions. The fuel assembly consists of two fuel bundles of 18 rods supported on a central stainless steel rod, see Fig. 4.5, with 11 stainless steel grids on each bundle. The overall length of the assembly is 10.01 m (Kupalov-Yaropolk *et al.*, 1998).

4.2.3 VVERs

E110 alloy (Zr-1%Nb) has been the universal fuel cladding in the Voda Voda Energo Reactor (VVER) design, which is a pressurized water reactor (Cox *et al.*, 2004). In the early VVER-440 (MWe) reactors the fuel assemblies were hexagonal arrays of 126 E110 clad fuel rods, with 10 or 11 stainless steel grid spacers, enclosed in an E125 hexagonal wrapper (which performs the same function as a BWR channel) (Smirnov *et al.*, 1994) see Fig. 4.6a. In the VVER-1000 type reactors each assembly contains 312 fuel rods, a central E125 alloy support tube and 18 stainless steel guide tubes for control and shut-down rods. Except for the first VVER-1000 (Novo-Voronezh-5) such VVER-1000 fuel assemblies do not have 'wrappers'.

4.2.4 Material used in fuel assemblies

The materials used for the FA components are Zr alloys, Inconel (precipitation hardened Inconel X-750, Inconel 718 and solution treated Inconel 625) and stainless steel (SS 304L) (Cox *et al.*, 2006). A low cobalt content is desired in the stainless steel and nickel base alloys to keep the activity transport by the coolant and the radiological exposure of the workers (man-rem) low. Since these alloys have high thermal neutron capture cross sections

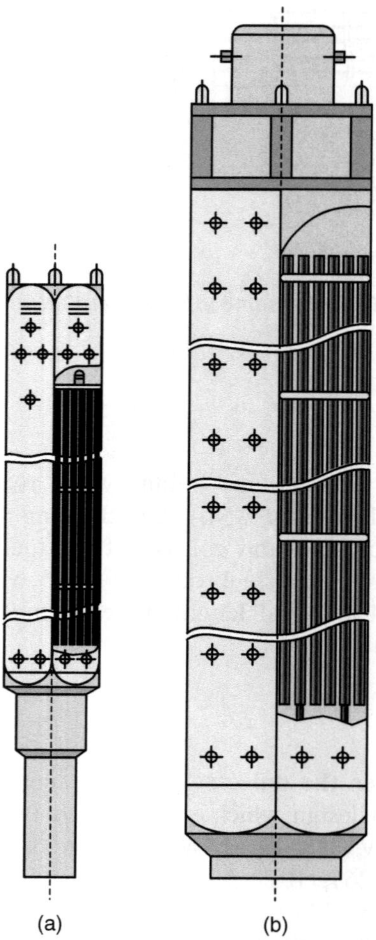

4.6 (a) VVER-440 operating FA; (b) VVER-1000 FA (Cox *et al.*, 2006).

that lead to reactivity losses, most of the FA components within the core are made of low cross section Zr alloys. These Zr-Sn based alloys are heat treated to reach optimum mechanical and corrosion resistance properties. Spring materials need to be made of materials with low stress relaxation rates, such as Inconel X-750 or Inconel 718. These Ni base alloys are generally heat treated to reach an optimum precipitation hardening. Table 4.2 presents an overview of alloys used in LWR and their typical compositions.

Table 4.3 provides data for different Zr alloys used by different fuel vendors (Cox *et al.*, 2006). It is noteworthy that there are so many different Zr-alloys for PWR applications. Originally, Zircaloy-4 (Zr-4) was used in PWRs, but increased corrosion rates at extended burnups resulted in the

Table 4.2 Chemical compositions of various stainless steels and Ni base alloys

Material	Fe (wt%)	Ni (wt%)	Cr (wt%)	Mn (wt%)	Si (wt%)	Mo (wt%)	Ti (wt%)	Nb (wt%)	Al (wt%)
AISI 304	Bal.	10	19	≤2	≤0.75				
DIN 1.4541	Bal.	11	18	≤2	≤0.75		0.4		
Inconel X-750	7	Bal.	15	≤1	≤1		2.6	1	0.7
Inconel 718	17	Bal.	19	0.5	0.75	3	0.7	5	0.6
Inconel 625	2.5	Bal.	22	0.3	0.1	8.8	0.3	3.9*	0.2

*(Nb + Ta) = 3.9 wt%.
Source: A.N.T. International (2011) and Cox *et al.* (2006).

need to develop more corrosion-resistant alloys. However for BWRs, whilst the material originally selected, namely Zircaloy-2 (Zry-2), appeared to have adequate corrosion performance, recent burnup experience has shown that alloys with better corrosion and hydriding resistance are needed. Improved versions of Zry-2 and advanced alloys are being developed. Initially pure Zr-sponge liner was used as a PCI remedy for BWR applications. It was later found that the pure Zr sponge material results in a tendency for secondary degradation of failed fuel and therefore all fuel vendors added some alloying elements to increase the resistance towards secondary degradation of failed rods. The most potent alloying element to obtain this increased resistance is Fe by improving corrosion resistance of the liner material. However, Fe also has a tendency to decrease PCI performance. In RBMK and VVER reactors, E110 has always been used as fuel cladding material while E125 is being used for some of the structural components in the fuel assemblies.

4.3 Effects of irradiation on zirconium alloys

We proceed with sections describing fundamental metallurgical properties and phenonema which ultimately affect core component behavior.

4.3.1 Basic irradiation damage

In structural materials like Zircaloy, the overwhelming majority of defects are caused by neutrons, and the most important type of defect is the dislocation loop. Two types of loops predominate: <a> and <c> loops. The <a> loop lies on a prism plane and has a Burgers vector in the <a> direction of the HCP lattice. Table 4.4 lists some important characteristics. Both vacancy and interstitial loops exist, but more than half have vacancy character. They are very small (100 nm 'black spots') and are difficult to analyze even with the transmission electron microscopy (TEM) (see Fig. 4.7).

Table 4.3 Commercial Zr base materials currently used for zirconium alloy fuel components in PWRs, BWRs, VVERs and RBMKs (Cox *et al.*, 2006)

Alloy	Sn %	Nb %	Fe %	Cr %	Ni %	O %	Fuel vendor
BWRs							
Zircaloy-2 (SRA[a]/(RXA[b])	1.2–1.7	–	0.07–0.2	0.05–0.15	0.03–0.08	0.1–0.14	All fuel vendors
Zr-Liner[b]							
Sponge	–	–	0.015–0.06	–	–	0.05–0.1	Only used in Japan and Russia
ZrSn	0.25	–	0.03–0.06	–	–	0.05–0.1	W
ZrFe	–	–	0.4	–	–	0.05–0.1	AREVA
ZrFe	–	–	0.10	–	–	0.05–0.1	GNF[c]
PWRs							
Zircaloy-4 (SRA)	1.2–1.7	–	0.18–0.24	0.07–0.13	–	0.1–0.14	
ZIRLO (SRA)	1	1	0.1	–	–	0.12	W
Optimized ZIRLO (SRA/pRXA[d])	0.7	1	0.1	–	–	0.12	W
M5 (RXA)	–	0.8–1.2	0.015–0.06	–	–	0.09–0.12	AREVA
HPA-4[e] (SRA/RXA)	0.6	–	Fe+V	–	–	0.12	AREVA
NDA[f] (SRA)	1	0.1	0.3	0.2		0.12	NFI[g]
MDA[h] (SRA)	0.8	0.5	0.2	0.1		0.12	MHI[i]
VVER, RBMK							
E110 (RXA)	–	0.9–1.1	0.014	<0.003	0.0035	0.05–0.07	Fuel cladding
Alloy E125 (SRA)	–	2.5	–	–	–	0.06	Structural components

[a] Stress relieved annealed.
[b] Recrystallized annealed.
[c] Global nuclear fuel.
[d] Partially recrystallized condition.
[e] High performance alloy.
[f] New developed alloy.
[g] Nuclear fuel industries.
[h] Mitsubishi developed alloy.
[i] Mitsubishi heavy industries.
Source: A.N.T. International (2011).

Table 4.4 Radiation damage: <a> loops in Zircaloy

Nature	Vacancy(+), interstitial
Size	8–20 nm (80–100 Å)
Density	8×10^{14} m^{-2}
Saturation fluence	1×10^{25} n/m^2 (E>1 MeV)
Thermal stability	To about 400°C (673K)
Effect	Strength, ductility, dimensional stability

Source: A.N.T. International (2011).

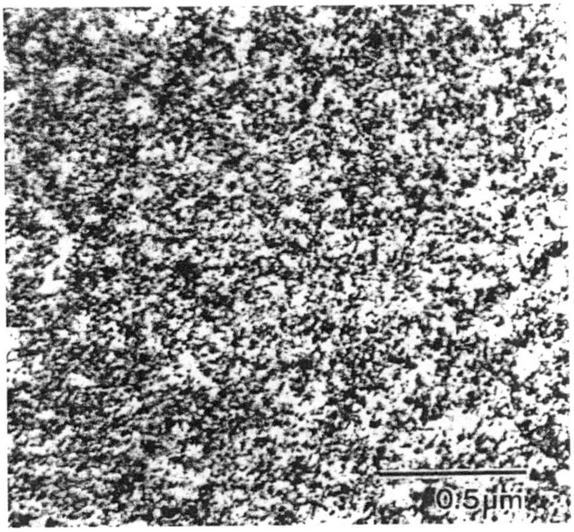

4.7 <a> type dislocation loops in neutron irradiated Zircaloy-2 (after post-irradiation annealing at 723K for 1 h). (Source: Adamson, 2000.)

These <a> loops form early in the irradiation and the number density reaches a saturation value at a fuel burnup below 5 GWd/MT (1×10^{21} n/cm^2, $E > 1$ MeV). The size of the loops increases with irradiation temperature, and the loops become unstable (start to disappear) at about 673K (400°C). As will be discussed later they have a strong effect on mechanical properties and dimensional stability.

The <c> type of loop lies on the basal plane and has its Burgers vector, or at least a strong component of it, in the c-direction of the HCP cell. As indicated in Table 4.5, and unlike the <a> loop, it is strictly a vacancy-type loop, is relatively large (100 nm) and does not form until considerable irradiation effects have occurred. In Zircaloy, <c> loops are first observed by TEM at a burnup of around 15 GWd/MT (~3×10^{25} n/m^2, $E > 1$ MeV) and increase in density for the remainder of the fuel lifetime. They are thermally stable

Table 4.5 Radiation damage: <c> loops in Zircaloy

Nature	Vacancy
Size	>100 nm (1000 Å)
Density	0.5 × 1018 m^{-2} (for Fig. 4.8)
Incubation fluence	3 × 1025 n/m^2 (*E*>1 MeV)
Thermal	Stable to >560°C (833K) Form at >200°C (475K)
Effect	growth, creep?

Source: A.N.T. International (2011).

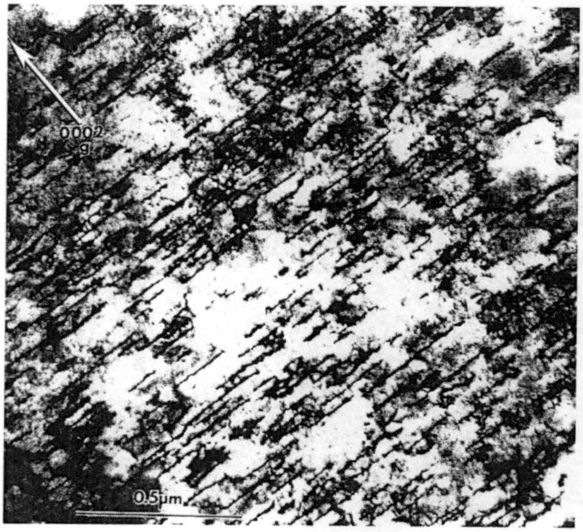

4.8 <c> type dislocations in Zircaloy-4 after a fluence of 12 × 1025 n/m^2 at 561K. (Source: Adamson, 2000.)

to high temperature (>833K). It is thought that <c> loops strongly influence irradiation growth and creep behaviour and probably do not affect mechanical properties. Figure 4.8 shows TEM images of a high density of <c> loops in highly irradiated Zircaloy. Such <c> loops, unlike <a> loops, do not appear to form in all zirconium alloys, particularly in those having additions of Nb, or Nb and Fe (Shishov *et al.*, 2002), until high fluences are experienced.

As outlined in Tables 4.4 and 4.5, the formation kinetics of <a>- and <c>-type loops differ. The density of <a> type dislocation builds up quickly and saturates at a fluence less than 1×10^{25} n/m^2, $E > 1$ MeV, as illustrated in Fig. 4.9. It appears that a fluence-incubation period exists before <c> type loops begin to form at about 3×10^{25} n/m^2, $E > 1$ MeV for typical reactor temperatures, as illustrated in Fig. 4.10.

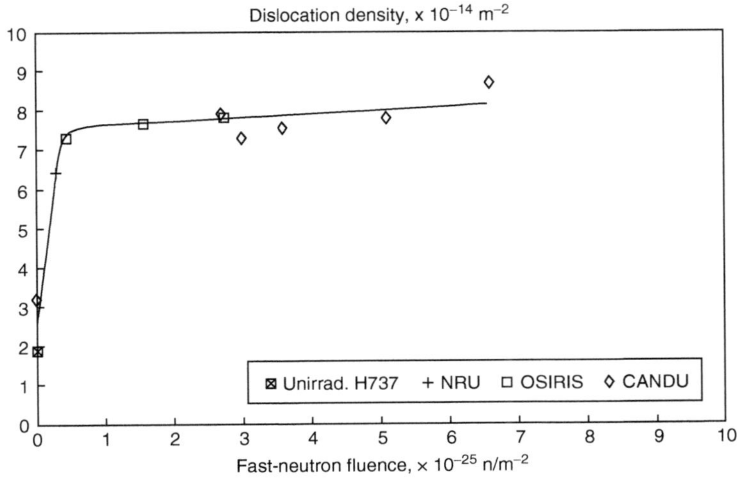

4.9 Variation of <a> type dislocation loops as a function of fluence in various reactors at 250–290°C (523–563K). (Source: Reprinted, with permission, from Davies *et al.* (1994), copyright ASTM International, 100 Barr Harbor Drive, West Conshohocken, PA 19428.)

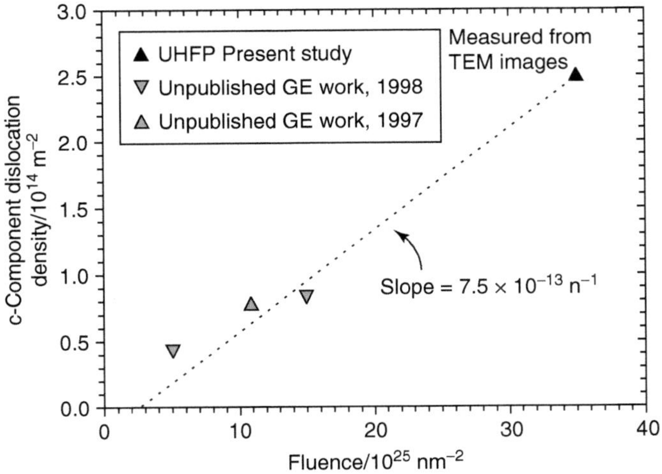

4.10 Variation of <c> type dislocation density as a function of fluence for Zircaloy-2 irradiated at 290°C (563K). (Source: Reprinted, with permission, from Mahmood *et al.* (2000), copyright ASTM International, 100 Barr Harbor Drive, West Conshohocken, PA 19428.)

For a straightforward review of the relationship between irradiation-induced microstructure and Zircaloy properties, see Adamson (2000). More technical details and references can be found there.

4.3.2 Effects of irradiation on precipitates

Corrosion resistance in zirconium alloys is intimately related to the presence of second phase particles (SPPs) formed in the zirconium matrix by deliberate additions of alloying elements. The precipitates are usually incoherent crystalline intermetallic compounds, meaning that their physical structure is unrelated to the Zr matrix in which they are imbedded. In as-fabricated Zircaloy-4 the most common SPP is $Zr(Fe,Cr)_2$, while in Zircaloy-2 they are $Zr(Fe,Cr)_2$ and $Zr_2(Fe,Ni)$. For the ZrNb type alloys the most common is βNb (which is not an intermetallic) and for the ZrSnNbFe alloy types are $Zr(Nb,Fe)_2$ and βNb. Table 4.6 gives a more complete description, also indicating some neutron irradiation effects.

At normal *LWR* temperatures (270–370°C, 543–643K) the SPPs change under irradiation in a combination of two ways – amorphization and dissolution.

Amorphization means that the original SPP crystalline structure is converted to an amorphous structure. Amorphization is a complex process, described in some detail by Griffiths *et al.* (1987); Yang (1989); Motta (1997); Bajaj *et al.* (2002); and Taylor *et al.* (1999). It occurs when an intermetallic compound accumulates enough irradiation-induced defects to cause it to thermodynamically favour an amorphous rather than a crystalline structure. The rate of amorphization depends on the relative rates of damage creation and damage annealing in the SPP; therefore important parameters are neutron flux, irradiation temperature and SPP chemistry. A critical temperature exists above which the annealing processes are fast enough to prevent the damage accumulation of defects needed for transformation. For typical reactor irradiations amorphization of both $Zr(FeCr)_2$ and $Zr_2(Fe,Ni)$ occurs readily at temperatures near 100°C (373K) (although Fe is not related from the SPPs into the Zr matrix, as discussed later). At typical (LWR) temperatures (300°C, 573K) and neutron flux, $Zr(Fe,Cr)_2$ becomes amorphous but $Zr_2(Fe,Ni)$ does not. Above about 330°C (603K) neither SPP becomes amorphous.

The amorphization process begins at the outside surface of the SPP and works its way inward with increasing fluence. This is illustrated in Fig. 4.11 (Etoh & Shimada, 1993) where the SPP on the left has an amorphous rim (dark area) and the one on the right, at higher fluence, is fully amorphous. There appears to be an incubation period prior to amorphization initiation, with the incubation fluence decreasing with temperature in the range 270–330°C (543–603K).

Table 4.6 SPPs at normal LWR temperatures

Reference	Material	As- fabricated	Moderate burnup		High burnup	
			<330°C	>330°C	<330°C	>330°C
Adamson, 2000; Griffiths *et al.*, 1996	Zircaloy-2 or Zircaloy-4	$Zr(Fe,Cr)_2$	PA PD	X PD	A D	X D
Garzarolli, *et al.*, 1996(a); Adamson & Rudling, 2002		$Zr_2(Fe,Si)$	X PD	X PD	X D	X D
Garzarolli *et al.*, 2002(a)		Zr_3Fe	X S	X	X PD	X
Griffiths *et al.*, 1996; Etoh & Shimada, 1993	Zircaloy-2	$Zr_2(Fe,Ni)$	X PD	X PD	X D	X D
	Zr-1Nb					
Shishov *et al.*, 1996	E110	βNb		X		
Bossis *et al.*, 2002 Gilbon *et al.*, 2000	M5	βNb		X IEβNb		
Doriot *et al.*, 2004/2005	M5 (RX)	βNb $Zr(Fe,Nb)_2$				X, PD, IE X, PD
Shishov *et al.*, 2007	E110(RX)(0.1 Fe)	βNb $Zr(Fe,Nb)_2$				X, PD, IE X,D
	E110(RX) (0.01 Fe)	βNb				X, PD ,IE
	Zr-2.5Nb					
Urbanic & Griffiths, 2000	(27% CW)	ω(in β phase)	X,PD IE βNb (in α phase)			
Averin *et al.*, 2000; Shishov *et al.*, 1996	E125 (RX,PRX)	βNb	X,PD IE βNb	X,PD IE βNb		
	ZrSnNbFe					
Nystrand & Bergquist, 1997; Comstock *et al.*, 1996	ZIRconium Low Oxidation (ZIRLO) (Stress Relieved (SR))	ZrFeNb βNb	X,PD, βNb X			
Averin *et al.*, 2000; Shishov *et al.*, 1996	E635 (RX,PRX)	$Zr(Nb,Fe)_2$ -	PA,PD	X,PD IE βNb	A,D	X,D IE βNb
Nikulina *et al.*, 1996		$(Zr,Nb)_2Fe$	X,PD, IEZr3–4Fe	X	X	X, IEZr3–4Fe
Shishov *et al.*, 2002	E635,RX(0.15Fe)	$Zr(Nb,Fe)_2$ βNb		X,PD,βNb		D, IEβNb
Shishov *et al.*, 2007	E635(RX)	$Zr(Fe,Nb)_2$				D, IE

Notes: All are crystalline (X) as-fabricated.
A – amorphous; D – dissolved; P – partially…; X – crystalline; S – stable; IE – irradiation-enhanced precipitation
Source: A.N.T. International (2011).

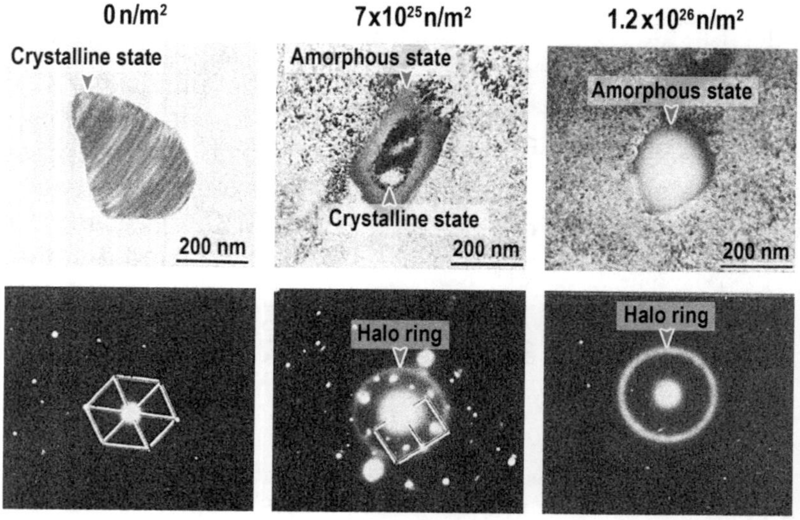

4.11 The fluence dependence of the amorphous transformation of Zr(Fe,Cr)2 precipitate in recrystallized annealed (RXA) Zircaloy-2, neutron irradiated at 288°C (561K). Diffraction patterns indicate stages of the transformation (Etoh & Shimada, 1993).

Amorphization rate increases as temperature decreases, as neutron flux increases and as SPP size decreases. Literature evaluation therefore needs to consider reactor and material conditions of specific interest.

The fluence required to produce complete amorphization depends on neutron flux, temperature and SPP size, but for typical $Zr(Fe,Cr)_2$ SPPs of initial size near 0.1 μm and the entire SPP is amorphous by the end of bundle life burnups <50 MWd/KgU (1×10^{22} n/cm^2, $E > 1$ MeV). Interestingly, under well controlled conditions of flux and temperature, the amorphization rate of $Zr(Fe,Cr)_2$ in Zircaloys can be used to estimate the neutron fluence (Motta & Lemaignan, 1992; Taylor *et al.*, 1999; Bajaj *et al.*, 2002).

For the Zr-Nb type alloys neither the βNb nor $Zr(Nb,Fe)_2$ SPPs become amorphous for irradiation temperature >330°C (603K). However, at 60°C (333K) $Zr(Nb,Fe)_2$ does become amorphous at high fluences.

SPP amorphization in itself does not appear to affect material behaviour; however, dissolution of both amorphous and crystalline SPPs does influence corrosion, growth and mechanical properties, to be discussed later. At typical LWR operating temperatures, SPP dissolution occurs relentlessly until the SPP essentially disappears.

As SPPs dissolve, the zirconium matrix becomes enriched (well beyond the normal solubility limit) in the dissolving element. For instance in Zircaloy-2, Fe leaves both $Zr(Fe,Cr)_2$ and $Zr_2(Fe,Ni)$ SPPs as schematically illustrated in Fig. 4.12 (Mahmood *et al.*, 2000). This process is given in more detail by Takagawa *et al.* (2004) and in Fig. 4.13. Here it is seen

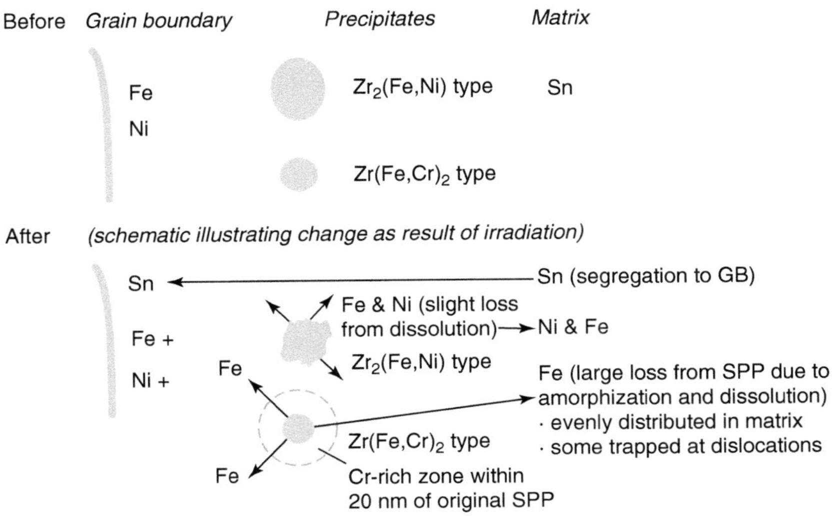

4.12 Schematic illustrating SPP dissolution and solute redistribution for small SPP Zircaloy-2 irradiated near 300°C. (Source: Reprinted, with permission, from Mahmood *et al.* (2000), copyright ASTM International, 100 Barr Harbor Drive, West Conshohocken, PA 19428.)

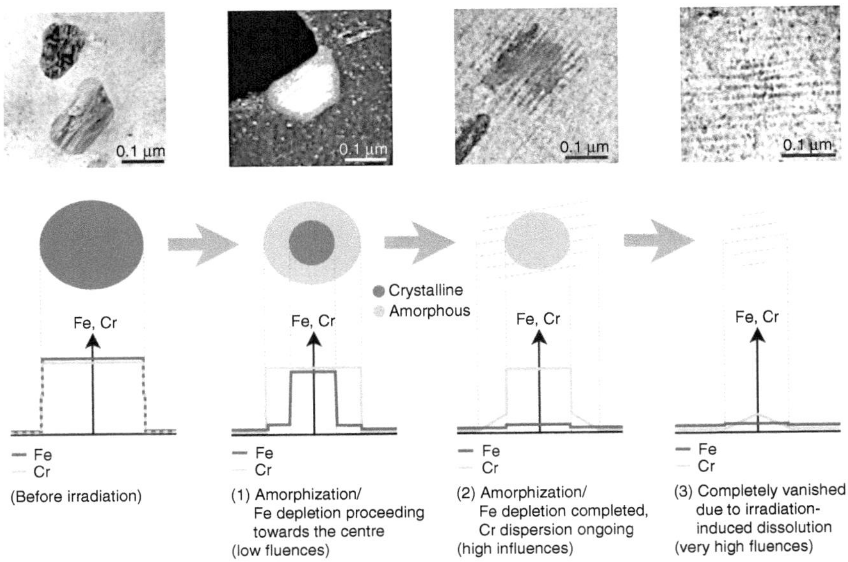

4.13 Evolution of a Zr-Fe-Cr particle under BWR irradiation. Upper figures: TEM micrographs; middle diagrams: schematic illustration of amorphization; lower figures: schematic illustrations of the chemical compositions. (Source: Reprinted, with permission, from Takagawa *et al.* (2004), copyright ASTM International, 100 Barr Harbor Drive, West Conshohocken, PA 19428.)

that Fe rapidly diffuses from the amorphous rim into the matrix, while Cr diffusion is sluggish. At high fluence (~1×10^{22} n/m^2, $E > 1$ MeV) complete amorphization and Fe-depletion has occurred, while the Cr level is still high. Only at very high fluence (~1.5×10^{22} n/m^2, $E > 1$ MeV) is the Cr dispersed into the matrix, and the SPP essentially disappears.

The rate of dissolution depends on the SPP size (higher rate for smaller sizes), and the extent of dissolution depends on size and fluence. It has been demonstrated in a BWR that small (<.04 μm) SPPs can completely dissolve at low to moderate burnups, (Huang *et al.*, 1996). Also in a PWR, but at temperature near 290°C, SPPs with an average size of 0.2 μm were >80% dissolved at moderate burnup (1×10^{26} n/m^2, $E > 1$ MeV) (Garzarolli *et al.*, 2002).

Modelling of the dissolution process gives insight into the alloying concentration of the matrix (Mahmood *et al.*, 1997). Figure 4.14 illustrates the model for release of solute into the matrix for various size SPPs. For the small SPPs (1R, 2R, 3R) all the Fe is released by moderate burnup.

For the channel material with very large (0.6 μm) SPPs only a small amount of Fe would be released even at high burnups. However, modern materials have SPPs with an average size < 0.3 μm.

In another study, experimental measurement of Fe released from Zr $(Fe,Nb)_2$ SPP in an E635 alloy containing 0.35% Fe during irradiation at 330–350°C is shown in Fig. 4.15 (Shishov *et al.*, 2002). (In Fig. 4.15, fluence has been converted from E > 0.1 MeV to E > 1.0 MeV by dividing by 4.) Here it is seen that the Fe has diffused from the SPP to the alpha Zr matrix such that all of the Fe is in the matrix by moderate burnup. Extending to high burnup (2×10^{26} n/m^2) in this case may only increase the probability of re-precipitation of Fe in the matrix. It should be noted that the 'normal' solubility of Fe in unirradiated Zr is <0.02 wt%.

Table 4.6 outlines changes in SPPs to be expected at moderate (50 MWd/kgU) to high (100 MWd/kgU) burnup for various alloys now in use. To illustrate interpretation of the table, consider the as-fabricated crystalline (X), $Zr(Fe,Cr)_2$ SPP for Zircaloy-2 or -4.

For moderate burnup at <330°C, the SPP would become partially amorphous (PA) and partially dissolved (PD), depending on its initial size. At >330°C it would remain crystalline (X) and become PD, the extent of which would depend on initial size. At high burnup for <330°C it would very likely become totally amorphous (A) and could completely dissolve (D), depending on its initial size. At >330°C it would remain crystalline (X), although it would become strongly fragmented as it eventually totally dissolved (D).

For the ZrNb and ZrSnNb alloys the most common SPPs are βNb and the (Laves phase) (L) $Zr(Nb,Fe)_2$. Also observed is the T-phase $(Zr,Nb)_2Fe$. Details of the SPPs present in the Nb-Fe corner of the phase diagram are presented in Fig. 4.16 (Shishov *et al.*, 2007). Also, a simplified diagram is presented in Fig. 4.17 (Garzarolli in Nikulina *et al.*, 2006). Such phase diagrams

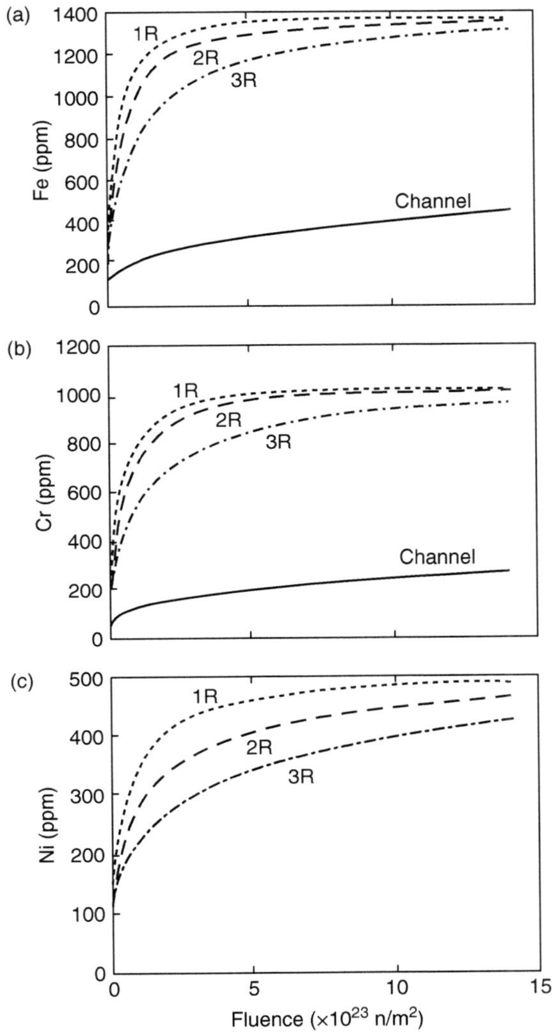

4.14 Modelling predictions for solute release to the matrix as a function of fluence for Zircaloy-2 (SPP size: 1R = 0.026 μm; 2R = 0.042 μm; 3R = 0.056 μm and Zircaloy-4 channel= 0.6 μm) irradiated near 300°C: (a) Fe, (b) Cr and (c) Ni (Mahmood *et al.*, 1997). Copyright 1997 by the American Nuclear Society, La Grange Park, Illinois.

are only approximate, and may vary because the heat treatments used may not achieve equilibrium conditions. Figure 4.16 points out that there is a region, not reported in Fig. 4.17, where only the Laves phase exists. This tends to be for 1%Nb and 0.2–0.4 Fe; for example in the 'normal' E635 alloy Zr-1.2Sn-1.0Nb-0.4Fe.

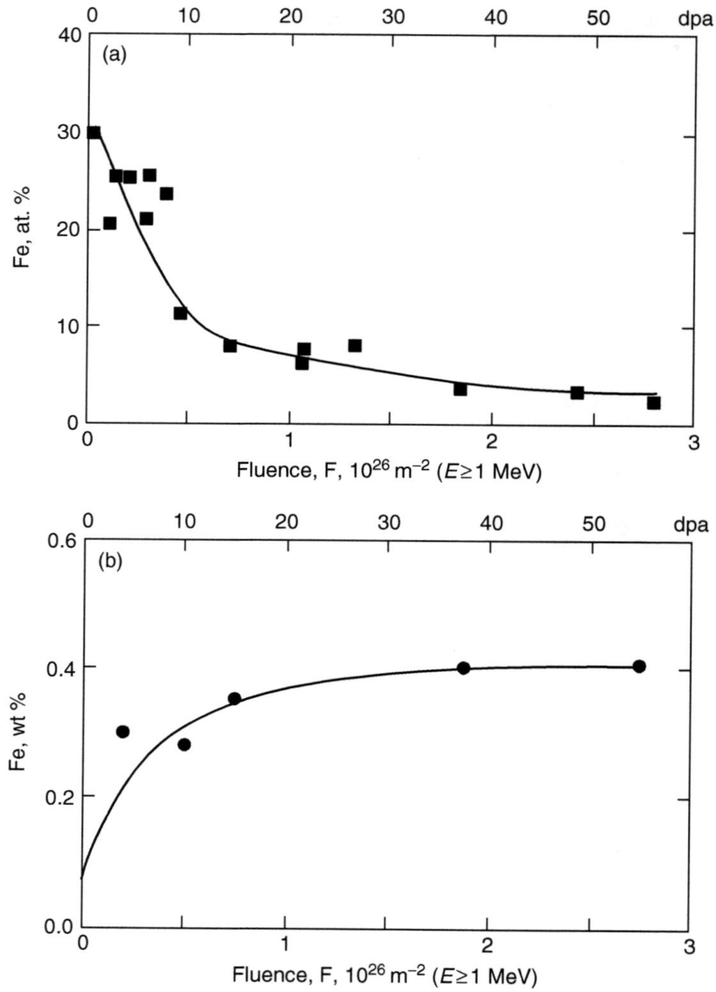

4.15 Fe content as a function of fluence ($E > 0.1$ MeV/4) at 330–350°C in alloy E635, (a) SPP Zr (Nb,Fe)2 and (b) α–Zr solid solution. (Source: Reprinted, with permission, from Shishov *et al.* (2002), copyright ASTM International, 100 Barr Harbor Drive, West Conshohocken, PA 19428.)

None of the SPPs become amorphous at normal reactor temperatures, but at high burnup at 60°C (333K) the Laves phase does become at least partially amorphous. However, all of the SPPs undergo irradiation-induced dissolution. The βNb SPP loses Nb to the matrix, but the excess Nb then re-precipitates as a very fine βNb. The Laves phase $Zr(Fe,Nb)_2$ transforms to a fine βNb SPP, with essentially all Fe ending up in the matrix (see Fig. 4.15). Behaviour of the T-phase $(Zr,Nb)_2Fe$ is more complicated, with Fe diffusing

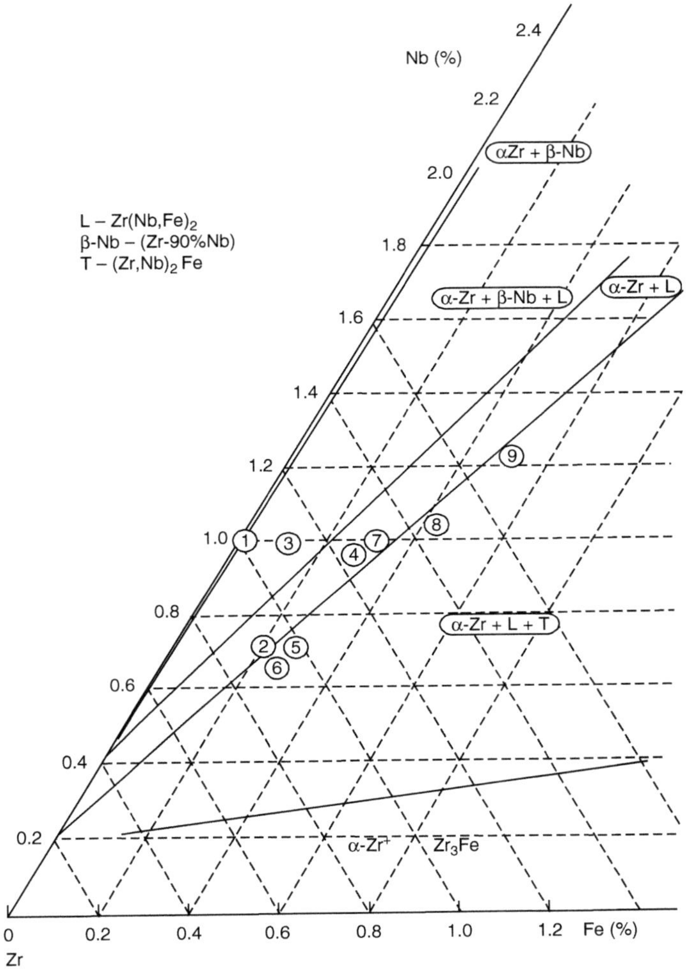

4.16 Zr-Nb-Fe ternary alloy phase diagram, zirconium corner at 580°C (853K), non-equilibrium conditions. (Source: Reprinted, with permission, from Shishov *et al.* (2005), copyright ASTM International, 100 Barr Harbor Drive, West Conshohocken, PA 19428.)

to the matrix, and Nb and Sn concentrating in the outer shell of the SPP. The core remains a T-phase. Details can be obtained in the following references: Shishov *et al.* (2002, 2007) and Shishov (2011) and Doriot *et al.* (2004).

4.3.3 Effects of post-irradiation annealing

Irradiation temperature does have an effect on microstructure – for instance higher irradiation temperature results in larger <a> loops, <c> loops do not

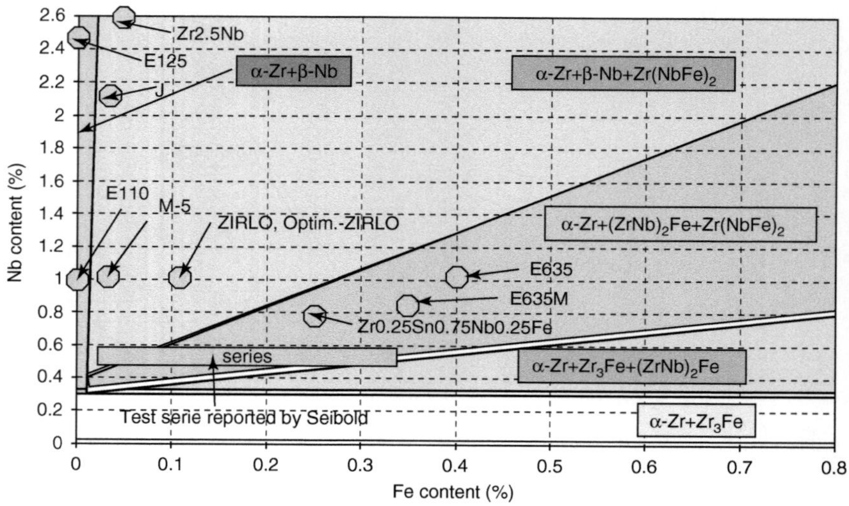

4.17 Zr-Nb-Fe ternary alloy phase diagram constructed from information in Toffolon *et al.* (2002); Shishov *et al.* (2005); Nikulina *et al.* (2006) – in Rudling *et al.* (2007).

form at 77°C (350K), and $Zr_2(Fe,Ni)$ SPPs do not become amorphous above an irradiation temperature of about 100°C (453K) (see e.g. Griffiths *et al.*, 1996). In addition, post-irradiation temperatures cause effects that give insight to the microstructure stability.

Damage in the form of <a> loops appears to be stable in post-irradiation annealing conditions to about 400°C (673K). Figure 4.18 (Adamson & Bell, 1986) shows that 1 hour at 400°C is a threshold condition for damage in size and density of <a> loops. Above that temperature, or quite likely longer times at that temperature, results in a marked increase in loop size and decrease in loop density. A temperature of 550°C (823K) for 1 h is sufficient to reduce the loop density to zero. This is accompanied by a dramatic decrease in hardness, as discussed below. Complementary data (Cheng *et al.*, 1994) indicate no changes in <a> loops after 200 days at 316°C (588K).

On the other hand, <c> component dislocations are quite resistant to change over the whole temperature range where <a> loops disappear. Yang (1989) and Kruger (1990) have shown that 1 h at 560°C (833K) or 575°C (848K) causes little or no change in <c> loop density or size. One hour at 675°C results in a 50% reduction in <c> loop density, while 1 h at 750°C (1023K) results in removal of all loops.

Figure 4.18 indicates hardness decreases in concert with changes in the <a> loop size and density. This is an indication that <c> loops do not have influence on the hardness. A summary is given by Adamson (2006). An additional study (Ribis *et al.*, 2007), confirms the results of Adamson and Bell

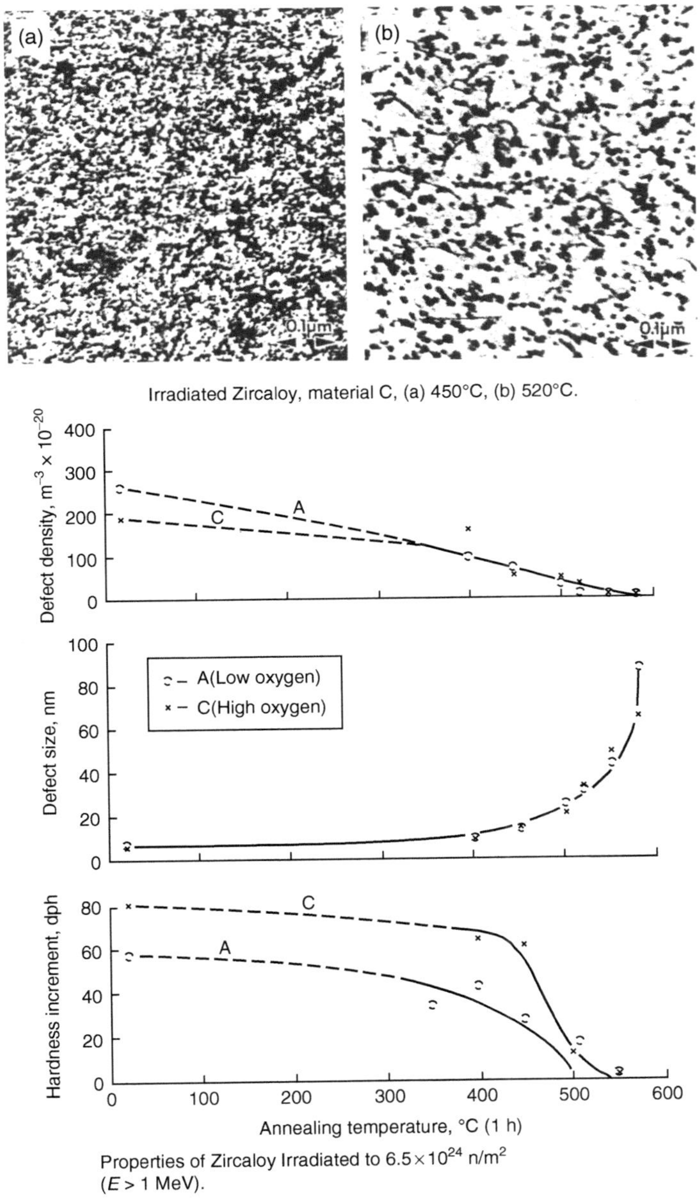

4.18 Post-irradiation microstructure (<a> loop density and size) and hardness of Zircaloy-2 irradiated to a fluence of 6.5×10^{24} n/m^2 ($E > 1$ MeV). Upper: TEM after annealing at indicated temperatures. Lower: density, size and hardness as functions of annealing temperature (Adamson & Bell, 1986).

(1986), and add modelling equations for the recovery process. Bourdiliau *et al.* (2010) go a step further and show that there is a direct relation between recovery of hardness and recovery of ultimate tensile stress (UTS) for both SRA Zircaloy-4 and Zr1Nb. However the recovery for Zr1Nb is more sluggish than for Zircaloy-4, as shown in Fig. 4.19. Zr1Nb does not fully recover the irradiation-induced hardening, primarily due the effects of the thermally stable, irradiation-induced phase which forms in that alloy.

Post-irradiation annealing also has effects on irradiation-affected SPPs. The observed phenomena give important insights into, for instance,

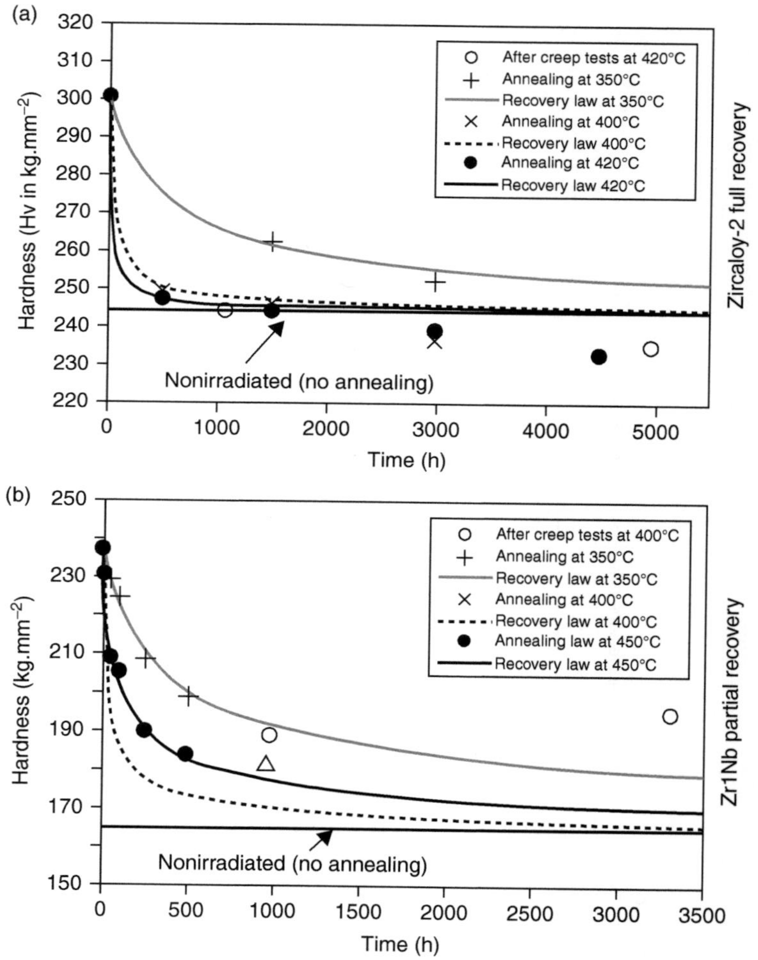

4.19 Comparison of the hardness recovery of (a) SRA Zircaloy-4 and (b) RXA Zr1Nb. (Source: Reprinted, with permission, from Bourdiliau *et al.* (2010), copyright ASTM International, 100 Barr Harbor Drive, West Conshohocken, PA 19428.)

corrosion mechanisms. For Zircaloy Yang (1989), Kruger (1990) and Cheng *et al.* (1994) report that post-irradiation annealing causes SPPs to recrystallize, to regain Fe and Ni, and to form under specific conditions of time and temperature. Minimal effects are observed for 316°C (589K) for 30 days, but for 200 days significant amounts of Fe diffuse back to the precipitates. At 400°C (673K) Fe diffuses back to precipitates in less than 10 days, and Fe-rich precipitates form at grain boundaries. At higher temperatures >560°C (833K) amorphized SPPs recrystallize, Fe and Cr diffuse back to SPPs, and re-precipitation occurs in the matrix and grain boundaries. Recent studies by Vizcaino *et al.* (2010) tend to confirm the earlier results.

4.4 Mechanical properties of zirconium alloys

By 'mechanical properties' we essentially mean strength and ductility. Strength is expressed in terms of hardness, tensile strength, burst strength, fatigue strength, etc. Ductility is likewise expressed in terms of strain-to-failure or strain-to-some limit for the various loading conditions. Fracture toughness is a combination of strength and ductility which describes the stress required to propagate a specific crack geometry under specific loading conditions. In this section we discuss various mechanical properties as affected by reactor neutron irradiation. In addition, we describe mechanisms and parameters which are related to mechanical properties and which affect reactor component behaviour. This section deals primarily with properties which can be determined by out-of-reactor (or post-irradiation) testing. For instance, tensile properties (strength and ductility) of interest for in-reactor performance are mainly dependent on fluence and independent of flux. However, if the rate at which strain is applied becomes very low ($<10^{-6}$ s^{-1}), the mechanism of deformation changes, and flux becomes an important variable (Azzarto *et al.*, 1969). That phenomenon is dealt with more as creep, in the section on dimensional stability.

As described in Section 4.3.1, neutron irradiation dramatically alters the microstructure of zirconium alloys. Of importance for mechanical properties are creation of <a> dislocation loops, and to a lesser extent, dissolution of precipitates (SPPs). Irradiation increases strength and hardness, and decreases ductility. The effect on fatigue life (or strength) is less clear and depends on testing technique, but generally appears to be small, with some reduction of fatigue life in the low cycle region (Wisner *et al.*, 1994). Fracture toughness is clearly reduced by irradiation in Zr-2.5Nb (Davies *et al.*, 1994), with concurrent effects of trace elements like chlorine (Coleman & Theaker, 2004), and there are indications of a smaller reduction in Zircaloys, for example (Bertsch *et al.*, 2010). The combination of irradiation and hydride effects is important; for uniformly distributed hydrides the observed reduction in ductility is mainly an irradiation, rather than a hydride effect; however concentrations of hydrides (rims, blisters) or low test temperature can overwhelm irradiation effects. Details are given in the following sub-sections.

4.4.1 Strength and ductility

As outlined in Section 4.3.1, irradiation produces damage in the form of small dislocation loops (<a> component loops) which harden the material. The result is an increase in strength and decrease in ductility.

At reactor start-up, the tensile properties are the unirradiated properties reported by the fuel supplier. Mechanical properties begin changing immediately upon startup, and by an exposure of 5 MWd/KgU or a fluence of about 1×10^{25} n/m^2 ($E > 1$ MeV) an increase in strength and decrease in ductility reach fluence-saturated values. Figure 4.20 illustrates this point for Zircaloy-4 irradiated and tested at 315°C (588K) (after Morize *et al.*, 1987). Note also that the UTSs of cold worked stress relieved (CWSR) and recrystallized (RX) materials become similar at low exposures. This is a general trend which depends on the balance of hardening by pre-existing dislocations (cold work) and irradiation-produced defects.

Fuel cladding requires sufficient strength to prevent inward plastic deformation of the cladding at beginning-of-service conditions. PWR strength must be higher than for BWRs due to the higher water pressure needed to suppress boiling; therefore, PWR Zircaloy cladding has traditionally been in the cold work stress relieved annealed (SRA) condition. The discussion above points out that the difference in strength between SRA and RXA materials is short-lived under reactor conditions.

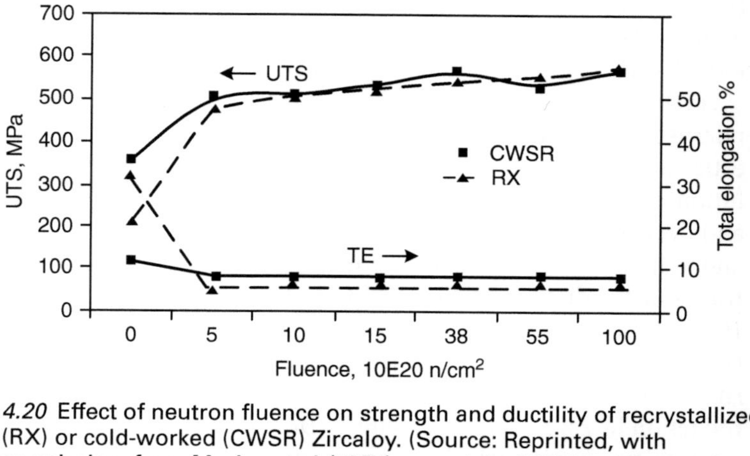

4.20 Effect of neutron fluence on strength and ductility of recrystallized (RX) or cold-worked (CWSR) Zircaloy. (Source: Reprinted, with permission, from Morize *et al.* (1987), copyright ASTM International, 100 Barr Harbor Drive, West Conshohocken, PA 19428.)

4.4.2 Deformation

In order for significant strain to occur, dislocations must overcome the obstacles to their motion – the irradiation-induced <a> loops. At low stresses this may happen by the process of dislocation climb, which means irradiation-induced point defects (PDs) diffuse to the dislocation and allow it to move around the obstacle. This is an important creep process, to be covered further in Section 4.6. At high stress or high strain rates, as in a power excursion, or at all practical strain rates out-of-reactor, the dislocations can actually interact with the <a> loop defects and remove them from the microstructure. In effect this creates a localized soft area, where additional deformation tends to concentrate: this process is called dislocation channelling. The physical process is illustrated in Fig. 4.21. The long straight

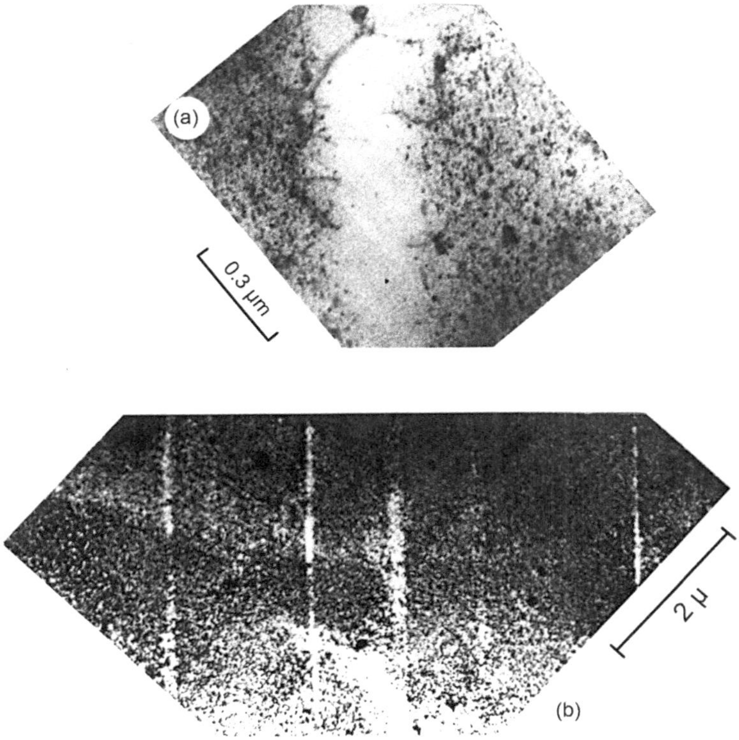

4.21 Dislocation channels in zirconium alloys: (a) zirconium, showing a channel with no radiation damage; (b) Zircaloy-4, showing channels along traces of prism planes $(10\bar{1}0)$. (Source: Reprinted, with permission, from Adamson *et al.* (1986) and Cheadle *et al.* (1974), copyright ASTM International, 100 Barr Harbor Drive, West Conshohocken, PA 19428.)

white bands are dislocation channels in which irradiation damage (black areas and black spots) is removed. In zirconium alloys the channels are in the order of 0.01–0.30 μm wide depending on fluence and irradiation temperature, and each channel can accommodate large (50–300%) local strains. The channels intersect the surface to cause large protrusions or slip steps there (Adamson, 1968; Sharp, 1972).

In Zircaloy, the dislocation channels tend to form in a very a localized area called a deformation band. For a simple uniaxial tensile test specimen the sequence of formation is illustrated in Fig. 4.22. At point (A) the deformation band begins to form and is fully formed at (B). At point (B) a second deformation band forms perpendicular to the first, and the specimen fractures at point (D). Because virtually all the strain forms in the deformation band, there is little or no deformation in the rest of the specimen gauge length. A plot of measured strain along the length of a typical specimen is given in Fig. 4.23. Since little plastic strain occurred outside deformation bands, the true gauge length of the specimen is much shorter than the nominal specimen gauge length. Therefore, specimen geometry greatly influences reported strain values. The effect of test specimen geometry on failure strain is illustrated in Fig. 4.24 where the conventional value of uniform elongation (UE) is plotted against gauge length for different specimen geometries of

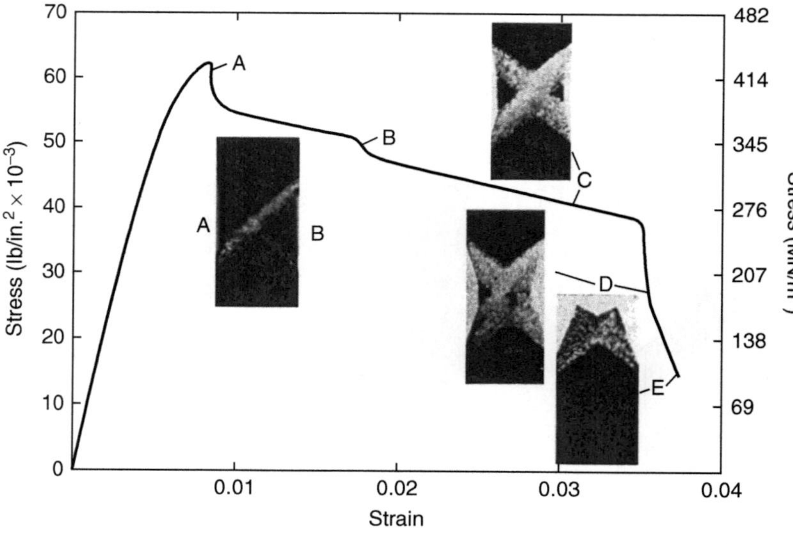

4.22 Engineering stress-strain curve for Zircaloy-2 sheet that had been irradiated at 280°C to a neutron fluence of 5 × 1020 n/cm^2 and subsequently tested at 300°C. (Source: Reprinted, with permission, from Bement *et al.* (1965), copyright ASTM International, 100 Barr Harbor Drive, West Conshohocken, PA 19428.)

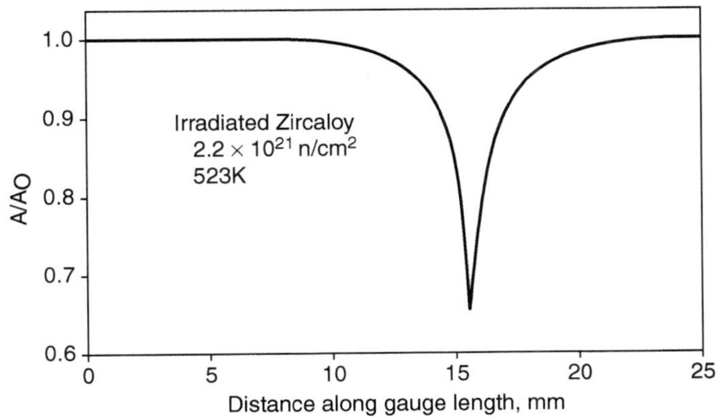

4.23 Deformation expressed as ratio of cross sectional area to original area measured along the specimen gauge length (Williams *et al.*, 1974).

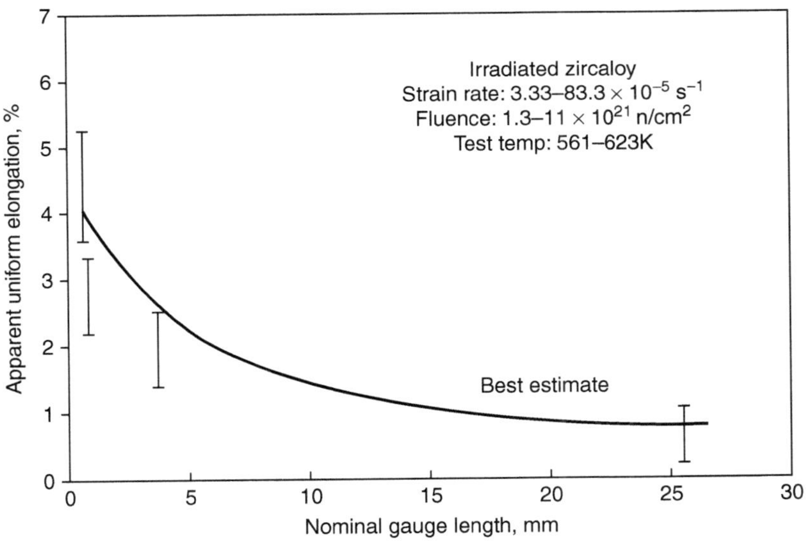

4.24 Effect of specimen gauge length on uniform elongation. (Source: Reprinted, with permission, from Adamson *et al.* (1986), copyright ASTM International, 100 Barr Harbor Drive, West Conshohocken, PA 19428.)

basically the same material. These data show that strain values developed for use as failure criteria or strain limits are not real material properties, but are strongly influenced by the specimen design used to obtain the data.

For burst tests of irradiated materials at 350°C (623K) (Onimus *et al.*, 2004) careful laser imaging measurements indicate that tubing deformed homogeneously throughout the gauge section until strain concentrates in the burst region. However, this strain (i.e. the number of channels) is small compared to the burst strain.

A main reason that anisotropic deformation is decreased relative to unirradiated material (Mahmood *et al.*, 2000) is that the high stresses needed to reach the yield point activate alternate slip systems in irradiated Zircaloy. The primary slip plane in unirradiated Zircaloy is the prism plane, the so-called $\langle 11\bar{2}0\rangle(10\bar{1}0)$ system. As the applied stress becomes high, both the pyramidal and basal planes can become active. Observations of prism plane dislocation channels have been well documented (Adamson *et al.*, 1986; Bell, 1974; Adamson & Bell, 1986; Bourdiliau *et al.*, 2010), but observation of pyramidal and basal channels have also been reported (Bell, 1974; Fregonese *et al.*, 2000; Regnard *et al.*, 2001; Onimus *et al.*, 2004, 2005; Bourdiliau *et al.*, 2010). In fact the CEA group show with considerable data and justification that, for 350°C (623K) testing temperature, basal slip predominates, but that may yet prove to be a function of irradiation and testing temperature, testing mode and impurity level (Bourdiliau *et al.*, 2010). Dislocation channelling phenomena themselves and details about which channelling planes predominate are important when modelling crack propagation and material response to actual in-reactor loading patterns. Onimus *et al.* (2005) have made good progress in modelling the phenomena for the CEA conditions. The data is summarized in a ZIRAT 15 Annual Report (Adamson *et al.*, 2010).

Specimen design plays a dual role, influencing ductility through both geometry and stress state. The type of plane stress specimens shown in Fig. 4.22 result in a 'classical' deformation band formation. The dislocation channels can freely extend from surface to surface. In the plane strain specimens of Fig. 4.25, the channels run into specimen regions where the stress is significantly lower before a free surface is reached, therefore preventing formation of a well-developed deformation band. The latter case, constrained plane strain, more realistically represents deformation in most reactor component situations.

Hardness of zirconium and Zircaloy

Knoop microhardness in a hot cell was used to determine hardness of zirconium and Zircaloy as a function of fluence, purity and irradiation temperature (Tucker & Adamson, 1984). The ranking of hardness was the same in both unirradiated and irradiated materials. Hardness saturated with fluence, as it also does for tensile properties. Figure 4.26 gives some of the data. The general trends are the same for all zirconium alloys.

The effects of post-irradiation annealing on mechanical properties and radiation damage are given in an earlier section. In general both

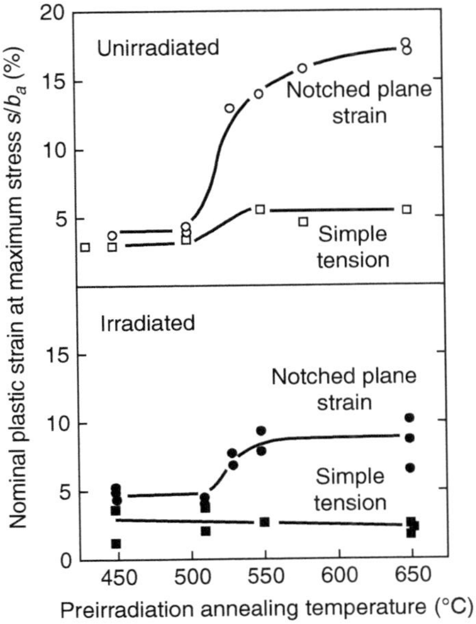

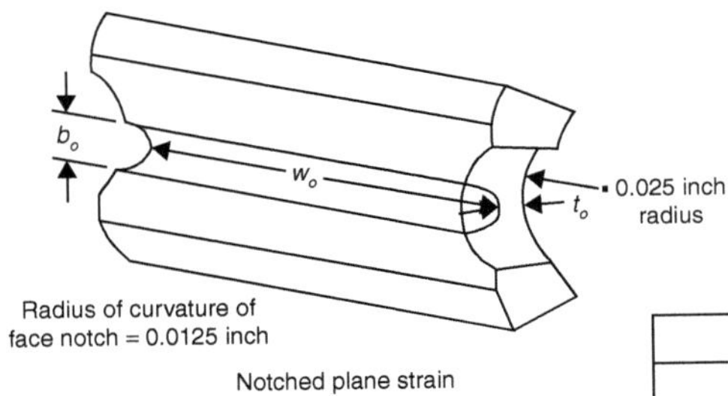

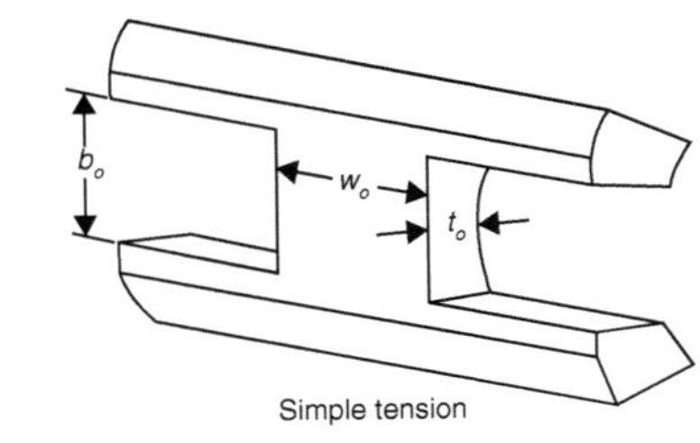

Nominal dimensions (inches)			
	b_o	w_o	t_o
Notched plane strain	.026	.250	.013
Simple tension	.050	.050	.026

4.25 Zircaloy-2 ductility as a function of irradiation and pre-irradiation annealing temperature for the simple tension plane stress and notched plane strain specimens shown. (Source: Reprinted, with permission, from Tomalin (1977), copyright ASTM International, 100 Barr Harbor Drive, West Conshohocken, PA 19428.)

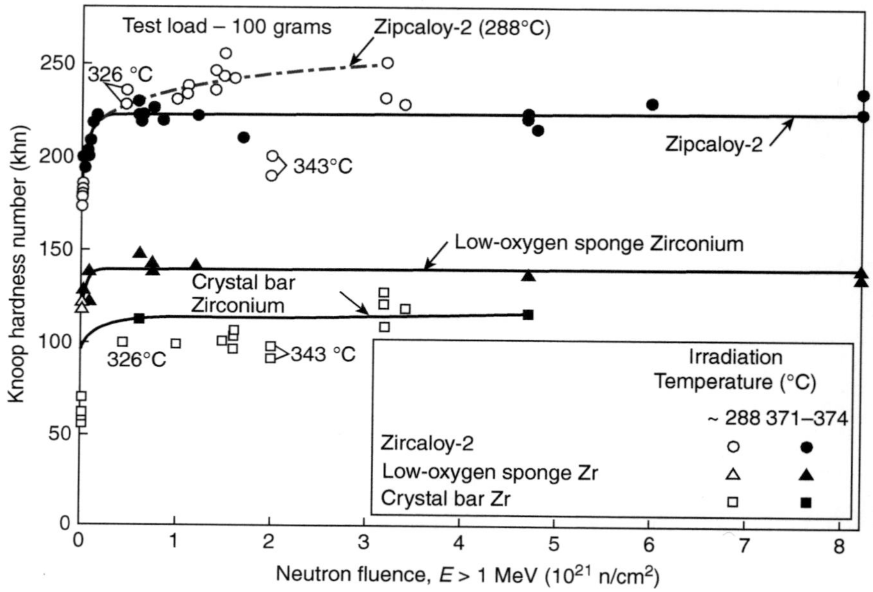

4.26 Knoop microhardness vs fast neutron fluence for zirconium and Zircaloy-2 (Tucker & Adamson, 1984).

characteristics anneal out at temperatures above about 400°C, although the rate of annealing is more sluggish for Nb-containing alloys.

4.4.3 Effects of hydrides on ductility

A brief summary of hydride effects is given here to provide background for pellet-cladding mechanical interaction (PCMI) type failures. All zirconium alloy reactor components absorb hydrogen during reactor service through the corrosion reaction between zirconium and water. Basics of these phenomena are given in ZIRAT Special Topical Reports (Cox & Rudling, 2000; Adamson *et al.*, 2006; Strasser *et al.*, 2008). Hydrides tend to embrittle zirconium alloys and therefore their effects are important for in-reactor normal service, for ex-reactor handling operations and for accident and transient scenarios such as LOCA and RIA. It is thought that individual hydrides themselves are actually brittle at all normal reactor temperatures (Simpson & Cann, 1979; Shi & Puls, 1999); and it is clear that high concentrations of hydrides (5000–16000 ppm) are very brittle, as in hydride blisters or rims.

Under normal conditions, hydride platelets form in the circumferential direction in fuel cladding illustrated in Fig. 4.27a, but under some circumstances such as during long term storage or during power transients they

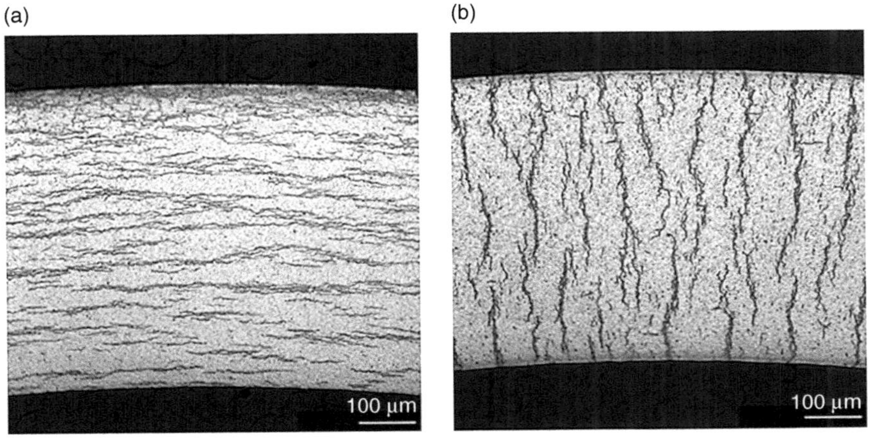

4.27 Hydride orientation in Zircaloy-4 (SRA) cladding: (a) circumferential, (b) radial (Chu *et al.*, 2005).

can form in the radial direction (Fig. 4.27b). Because in high power rods a temperature gradient encourages hydrogen to diffuse to the colder outer clad surface, rims of hydrides can form, illustrated in Fig. 4.28a.

Hydrides effects are listed here, giving appropriate figures and references.

- The effect of hydrides is strongly dependent on testing temperature. Material at 300°C (573K) (reactor operating temperature regime) retains much more ductility than at 20°C. Figures 4.28b and 4.29 indicate the ductile-to-brittle transition for unirradiated material is less than 200°C, for circumferentially oriented hydrides. Figures 4.30 and 4.31 indicate that at 332°C the primary reduction in ductility comes from the irradiation effect, while at room temperature the effect on ductility of irradiation and hydrides is additive for uniformly distributed hydrides below about 1000 ppm. It is apparent that below 100°C ductility is very low.
- The distribution of hydrides is important. Dense layers of hydrides (for instance at fuel cladding surfaces) retain little ductility at any temperature, and are susceptible to crack formation. Whether or not the crack will be arrested by the relatively ductile zirconium matrix depends on the layer thickness, as shown in Fig. 4.32.
- The strength of irradiated or unirradiated Zircaloy is insensitive to hydrogen content. See Fig. 4.33.
- Existence of radial hydrides can substantially reduce ductility, particularly at room temperature. Figure 4.34 shows the failure strains for the range of hydride orientations given in Fig. 4.35. When radial hydrides exist as in Fig. 4.35c failure strain is low. Figure 4.36 indicates that a high percentage of radial hydrides reduces the failure strain at room

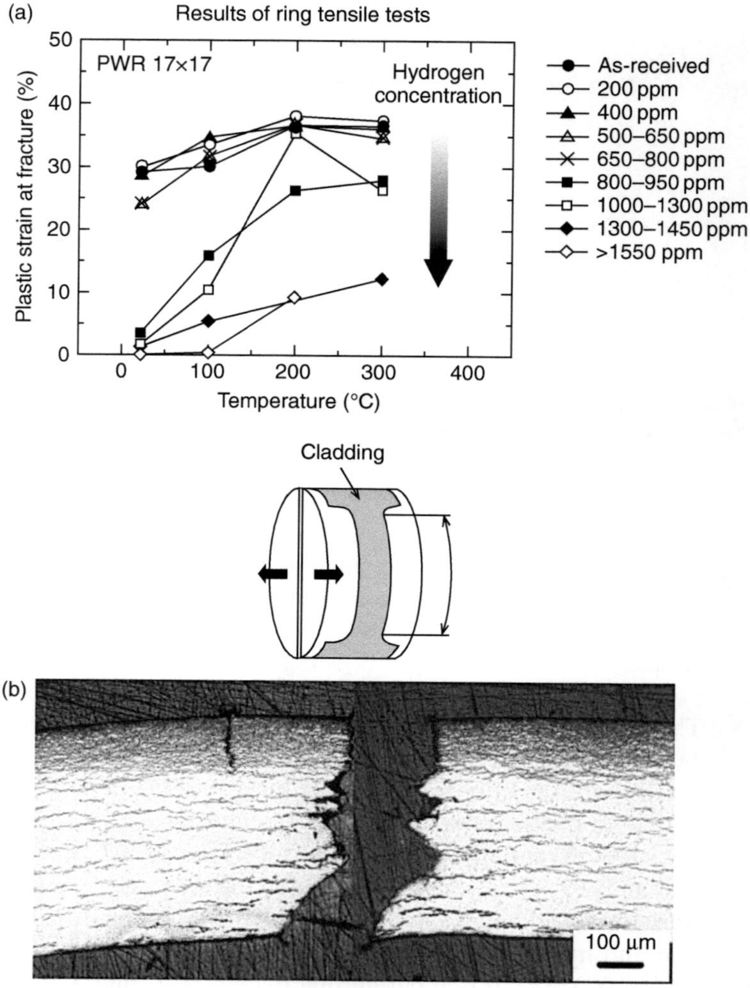

4.28 (a) Hydride rim and associated cracks in cladding failed in a room temperature burst test (Nagase & Fuketa, 2005). (b) Temperature effect on cladding ductility (Fuketa *et al.*, 2003).

temperature but not at 300°C (573K). All specimens are unirradiated and are tested with applied stress normal to the hydride platelet. For similar materials having the applied stress parallel to the hydride platelet, no hydride effect is seen (Yagnik *et al.*, 2004).

Hydride orientation

Careful metallographic examinations show hydrides to be short, thin platelets that have precipitated along a variety of crystallographic planes, very

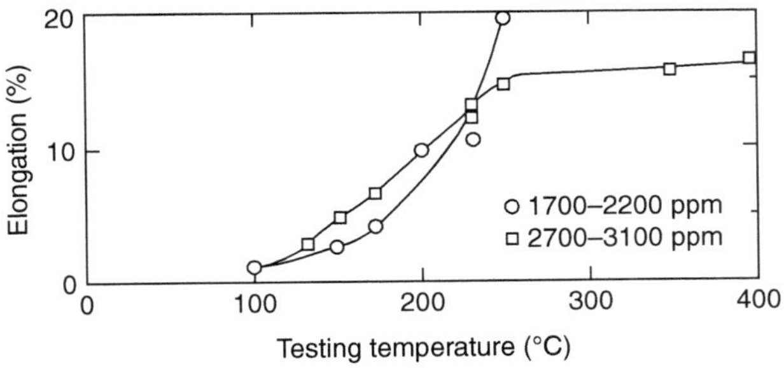

4.29 Elongation (%) as a function of the testing temperature for the specimens hydrided at 700°C (Bai, 1996).

commonly on grain boundaries or intragranularly. For pure zirconium the most common habit plane is near $[10\bar{1}0]$ and for Zircaloy or alloys it is $[10\bar{1}7]$, which is ~15° from the basal plane. Intragranular precipitation is less common and is more likely to occur in materials with large grain size, at intermetallic particles, at dislocations or as a result of very rapid cooling rates. Factors that determine the orientation of the precipitating hydrides in addition to grain size include stress, texture and cold work. As a result, the fabrication process will have a strong effect on the hydride orientation. Good reviews of the various hydride precipitates are given by Ells (1968), Cox and Rudling (2000) and Coleman (2003). Figure 4.37 illustrates hydrides in 3-dimensions after the metallographic etching process, which exaggerates the actual length of the hydrides.

The orientation of the hydride platelets that form during normal reactor operation, preferentially near the cooler cladding OD, usually have axial-circumferential orientation in tubes (respectively axial-tangential orientation in strips) and they remain so during wet storage of the spent fuel. The hydrides can become oriented in the radial (through thickness) direction if they are precipitated during operation under high tensile stresses, for example from fuel swelling, or precipitated from solid solution by cooling the alloy from a higher temperature under a tensile hoop stress. Reorientation could occur during reactor operation during cool-down or power cycling, although it is generally unlikely. However, it can occur during dry storage if the internally pressurized cladding is at a high temperature, holds sufficient hydrogen in solution and is then cooled under a sufficiently high hoop stress. The hydrides in solution will precipitate in the radial orientation, while the hydrides that did not dissolve will remain in their original circumferential orientation.

This is most likely to occur during rapid cool-down from high temperatures when a cask drying or evacuation procedure is applied rather than during storage.

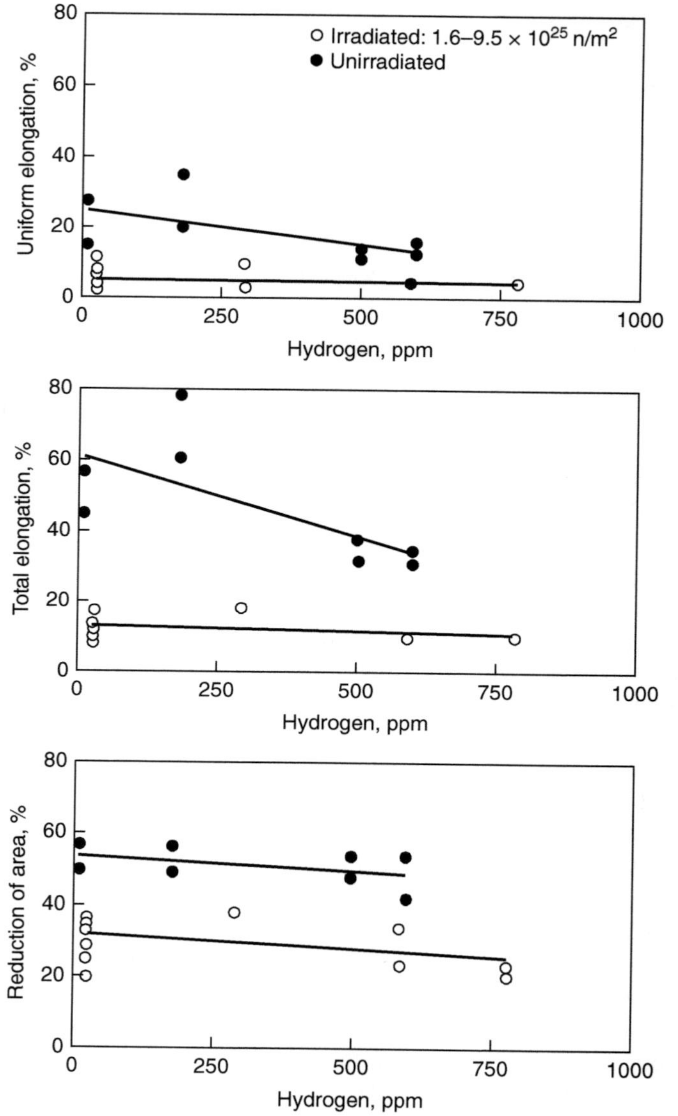

4.30 Uniform (top) and total elongation (middle) and reduction of area (bottom) as a function of hydrogen for unirradiated and irradiated Zircaloy-2 tested at 332°C (605K) (circumferential hydrides) (Wisner & Adamson, 1998).

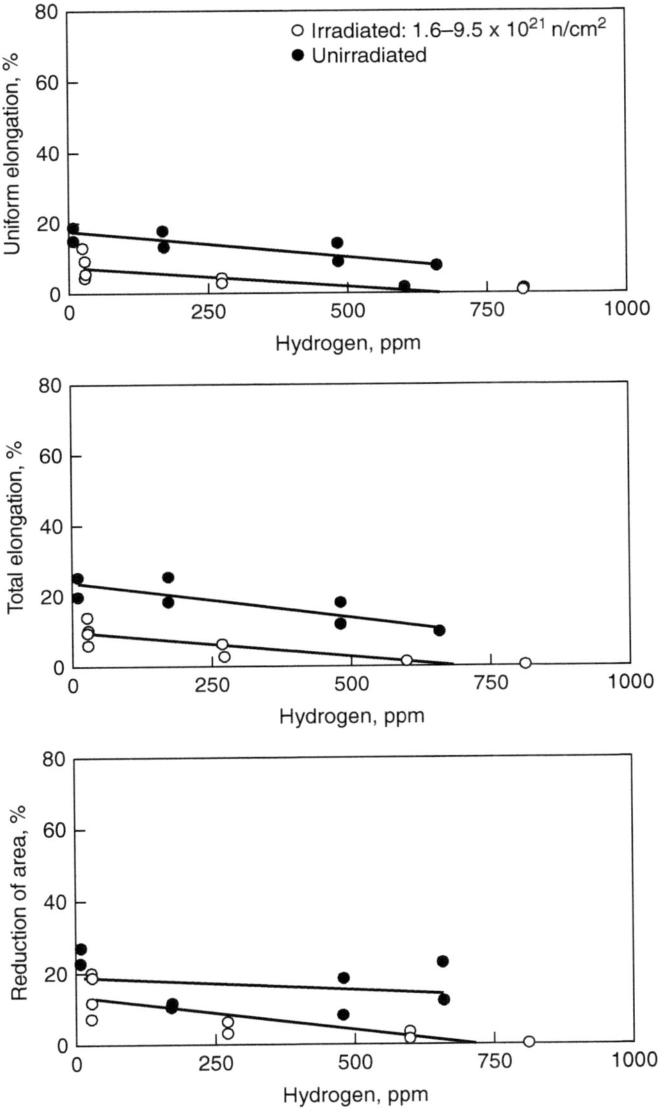

4.31 Uniform (top) and total elongation (middle) and reduction of area (bottom) as a function of hydrogen for unirradiated and irradiated Zircaloy-2 tested at 22°C (295K) (circumferential hydrides) (Wisner & Adamson, 1998).

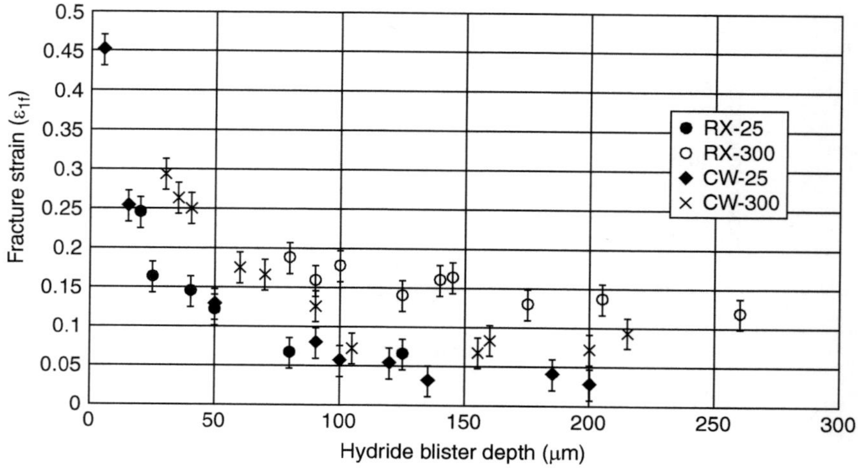

4.32 Local fracture strain versus hydride blister thickness for both cold worked stress relieved (CWSR) and recrystallized (RX) Zircaloy-4 sheet tested at either 25°C or 300°C. All data are for 3 mm blisters (Pierron *et al.*, 2003).

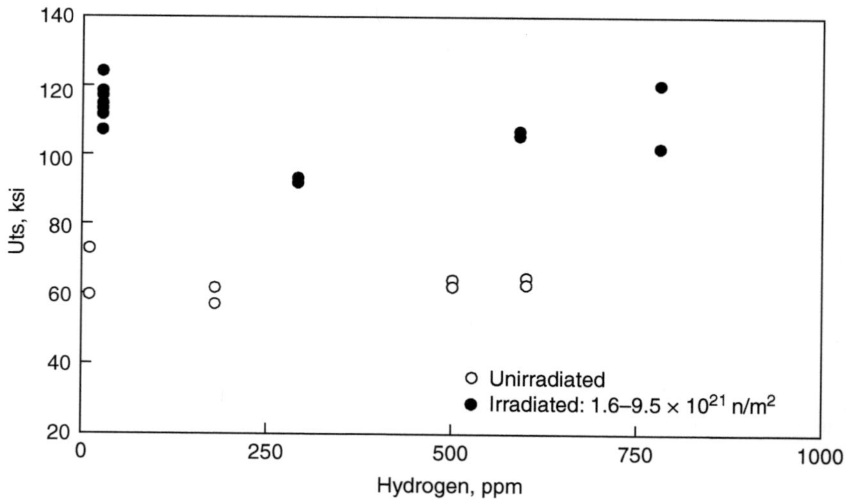

4.33 Strength as a function of hydrogen content for irradiated and unirradiated Zircaloy-2 tested at 332°C (605K) (Wisner & Adamson, 1998).

The factors that affect hydride reorientation in irradiated cladding are:

- hoop stress, or tensile and compressive stresses;
- maximum temperature;
- cool-down rate and final temperature;

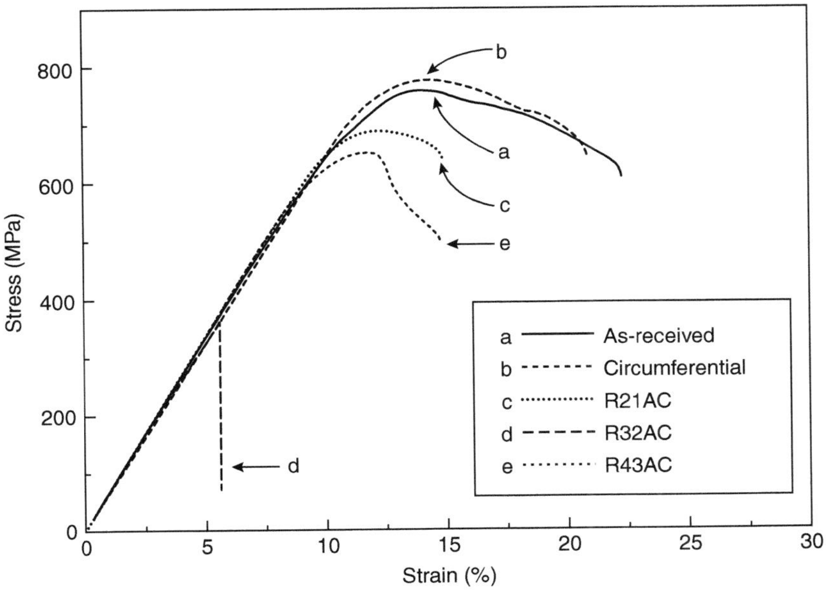

4.34 Stress-strain response of hydrided Zircaloy-4 tubes stressed in circumferential direction at room temperature. Specimen R32AC has the most radial hydrides (Hong and Lee, 2005).

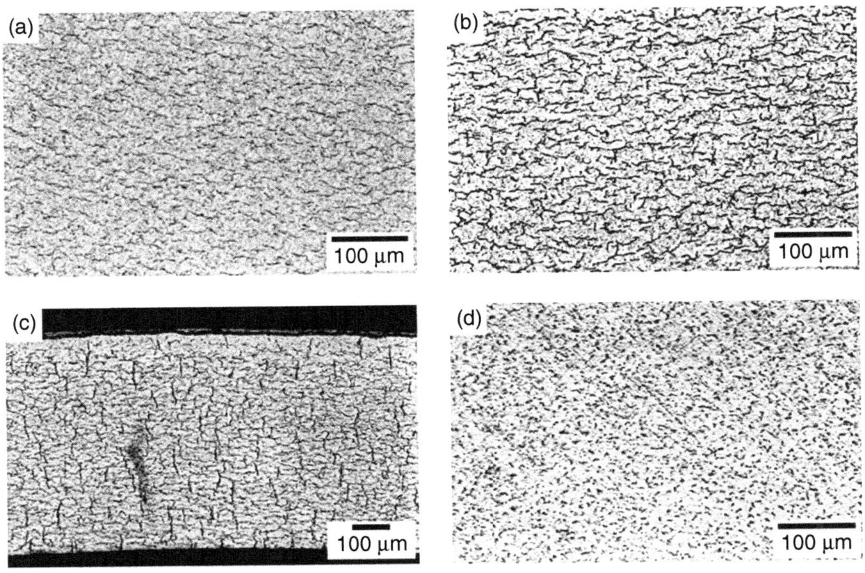

4.35 Hydride distribution in the radial-circumferential plane of SRA Zircaloy-4 (a) as received, (b) R21AC, (c) R32AC and (d) R43AC (Hong & Lee, 2005).

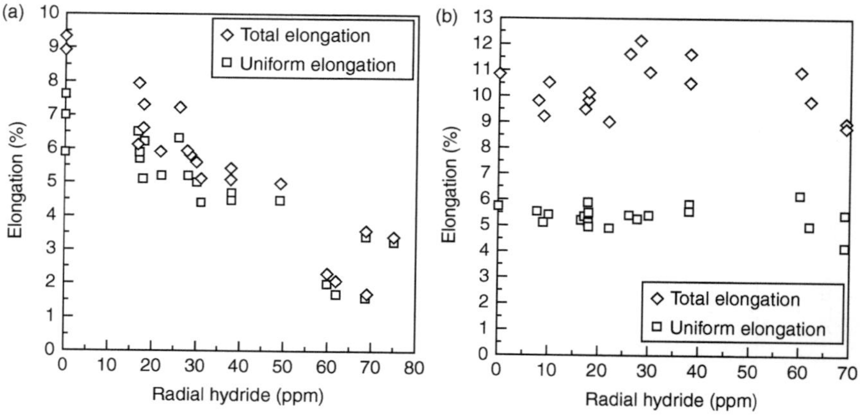

4.36 Effect of radial hydrides on elongation of Zircaloy-4 cladding specimens with ~200 ppm hydrogen. Tested in circumferential direction at (a) room temperature and (b) 300°C (573K) (Yagnik *et al.*, 2004).

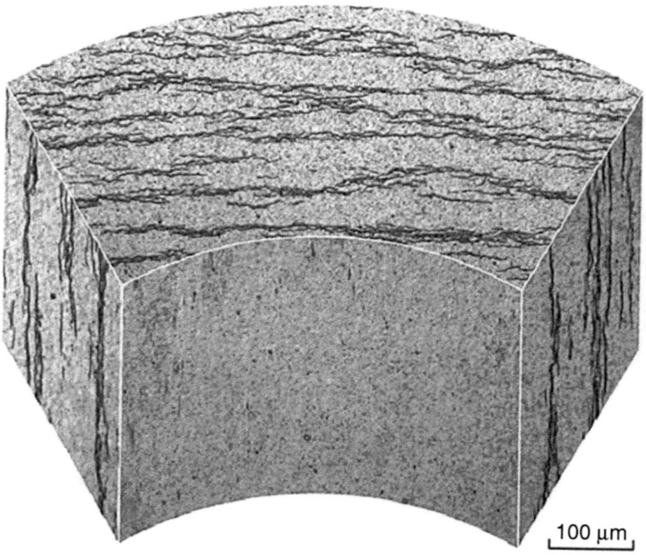

4.37 Typical hydride orientation in a cold worked and stress relieved Zircaloy-4 cladding (~230 ppm hydrogen) (Chu *et al.*, 2008).

- solubility of H in the specific alloy at its specific burnup that will determine the amount of H in solution at the maximum temperature and the amount of circumferential hydrides;
- microstructure features such as grain size, amount of cold work and dislocation structures;

- texture;
- time.

The radial hydrides visible in metallographic cross sections can be present in a wide variety of sizes and distributions as well as fractions of the total hydrides present.

Radial hydrides in zirconium alloy cladding are undesirable because they reduce the critical stress intensity required to propagate a radial crack through the wall of the cladding during handling or transportation, as shown by mechanical property data in the previous section. This is illustrated in Fig. 4.38, where it is seen that cracks propagate along radial hydrides, but are blunted in the circumferential hydride region. For this reason considerable attention and effort is expended to define the conditions for radial hydride formation and evaluate their effect on mechanical properties and the performance of the fuel, particularly during hypothetical accidents. One of the objectives of the dry storage regulations in the United States is to limit the conditions that could result in hydride reorientation and affect fuel reconfiguration during handling and transport.

Since one of the preferred hydride sites is the grain boundary, RXA material with equi-axed grains is more susceptible to radial hydride formation compared to SRA material with grains elongated in the axial direction. This is illustrated in Fig. 4.40 for SRA Zircaloy-4 and RXA Zircaloy-2.

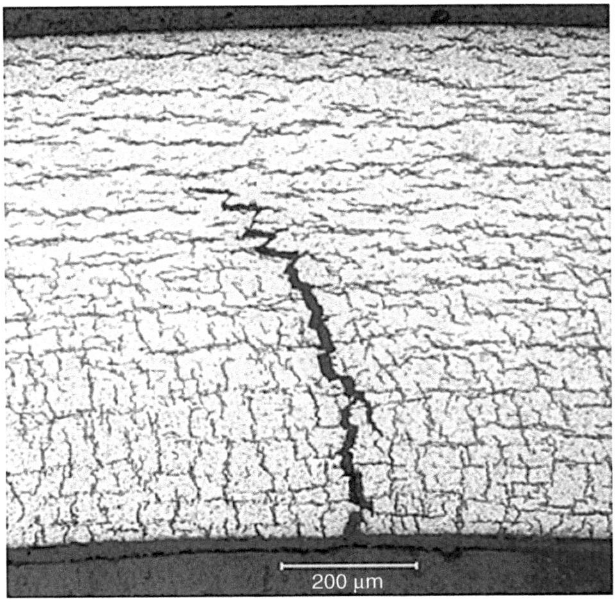

4.38 Cracks propagating due to a hoop stress (along horizontal direction in figure) (Daum *et al.*, 2005).

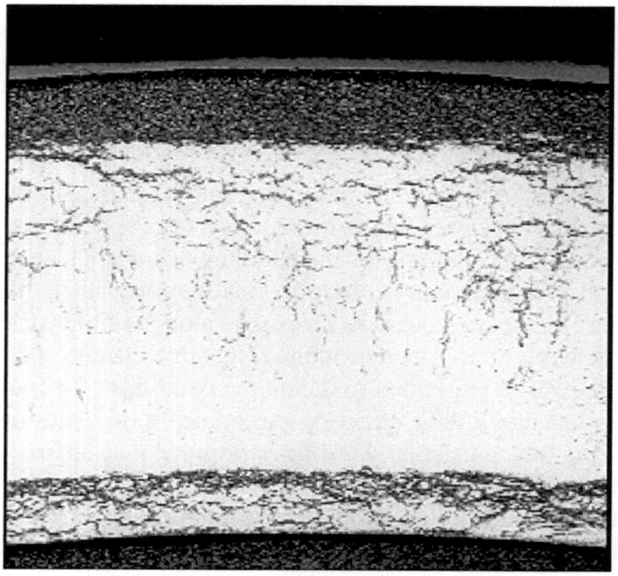

4.39 Dense hydride rim on the outer side, BWR liner fuel rods with low Fe and Si Zry-2 cladding exposed at corner position to high heat fluxes to 53.5 MWd/kgU with an average hydrogen content of 1600 ppm (Miyashita *et al.*, 2007). Copyright 2007 by the American Nuclear Society, La Grange Park, Illinois.

Hydride distribution in fuel rods having very high heat flux can be quite complicated, as illustrated for an extreme case in Fig. 4.39 for a Zircaloy-2 BWR cladding with a liner of zirconium. A dense hydride rim is seen at the outer surface (a condition more common in PWR rods than BWR ones), mixed radial and circumferential in the outer interior, a zone denuded of hydrides on the inner interior and substantial hydriding of the inner zirconium liner. High burnup performance of both BWR and PWR rods may be affected by such hydride distribution. One evaluation is given by Garzarolli *et al.* (2010).

4.5 Corrosion of zirconium alloys

We proceed here with descriptions of corrosion phenomena, which often limit the lifetime of core components.

4.5.1 Types of corrosion and comparison between PWRs and BWRs

Corrosion of zirconium alloys used in the core of nuclear power plants (and the accompanying absorption of hydrogen in the zirconium metal matrix)

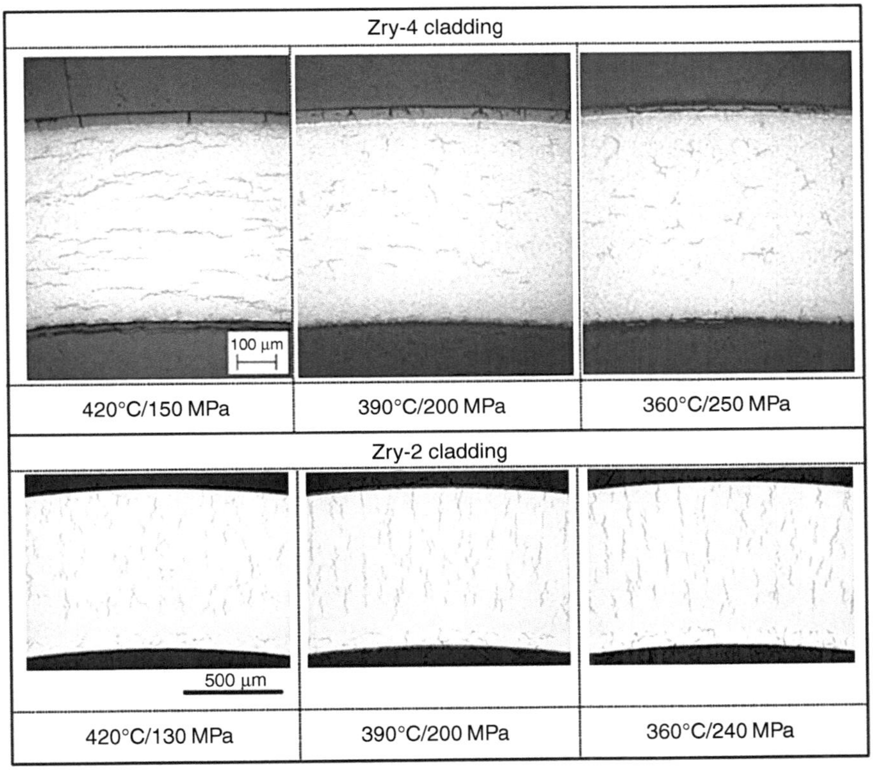

4.40 Comparison of hydride precipitation between CWSRA Zircaloy-4 and RXA Zircaloy-2 cladding (Ito *et al.*, 2004). Copyright 2004 by the American Nuclear Society, La Grange Park, Illinois.

is of prime interest when considering performance of the core components and therefore the performance of the entire reactor. For instance, for PWRs a practical corrosion limit exists (about 100 μm oxide thickness which is associated with a critical amount of hydrogen absorption) that has driven a material change away from Zircaloy-4.

The technical literature and many conferences are full of papers dealing with corrosion issues. Of particular interest is the series 'Zirconium in the Nuclear Industry: International Symposium', ASTM STP, American Society of Testing and Materials, which provides most relevant details. Also reviews of the topic are given in ZIRAT reports: Adamson *et al.* (2002), Cox *et al.* (2004), Adamson *et al.* (2007); and 'Waterside corrosion of zirconium alloys in nuclear power plants', International Atomic Energy Agency (IAEA)-TECDOC-996, January 1998. The most recent open literature review of mechanisms is by Cox (2005).

Corrosion of zirconium alloys is an electrochemically-driven process affected by the microstructure and microchemistry of the alloy surface, the nature of the oxide layer that forms, the temperature at the metal/oxide interface, the chemistry and thermohydraulics of the corrodent water, the effects of irradiation and the effects of time. Table 4.7 gives information on the various types of commercial power reactor systems currently being used throughout the world. In comparing BWRs with PWRs, with corrosion mechanisms in mind, the main features are:

- BWR coolant boils; PWR coolant does not. This has an important effect at the oxide/water interface.
- PWR coolant contains a high concentration of hydrogen; BWR coolant does not. Complementarily, BWR coolant contains a high concentration of oxygen, PWR coolant does not. This has an important effect on corrosion processes.
- PWR components generally operate at higher temperatures than BWR components. Corrosion processes are temperature dependent.
- Both reactor types employ chemical additions to the coolant which may affect corrosion and buildup of deposits on fuel rods.

It also should be noted that BWR zirconium alloys continue to be primarily Zircaloy-2 or slight variants of Zircaloy-2. PWR zirconium alloys no longer tend to be Zircaloy-4, for reasons of insufficient corrosion resistance (and hydriding resistance) at high burnup, but have moved toward zirconium alloys with Nb additions.

The type of oxides which form during corrosion in reactor water can be classified into several categories. The two most basic are uniform and nodular corrosion. The 'uniform' category has an extension – 'patch' or accelerated uniform. The fourth category is 'shadow corrosion', which can look like thick uniform corrosion but has some characteristics of nodular corrosion. The fifth category is crud-related corrosion, which is a temperature driven process induced by poor heat transfer in crud-impregnated corrosion layers. These categories will be discussed later, but are introduced here. Table 4.8 (Garzarolli in Adamson *et al.*, 2002) gives a useful summary of characteristics of various corrosion types.

Uniform corrosion occurs in both PWRs and BWRs. The oxide itself is uniform in thickness and consists of several different layers. For either in or out of reactor, the initial shape of the corrosion-versus-time curve is as shown in Fig. 4.41 in the pre-transition region. The first transition point occurs at around 2 μm oxide thickness in PWRs. The shape of the post-transition curve in PWRs depends on several variables: initial SPP size, irradiation, amount of cold work, specific alloy, water chemistry, temperature, local thermohydraulics and hydride concentration. For Zircaloy-4 the corrosion

Table 4.7 Design parameters for water cooled reactors (Adamson *et al.*, 2002)

Parameter	Western type PWR	VVER (440/1000) MW	CANDU[b]	BWR	RBMK[c]
1. Coolant	Pressurized H_2O	Pressurized H_2O	Pressurized D_2O	Boiling H_2O	Boiling H_2O
2. Fuel materials (Pressure tube materials)	Zry-4, ZIRLO[e], DUPLEX, M5, Inconel, SS[e]	Zr–alloy E110	Zry-4 (Zr-2.5Nb)	Zr-2, Zry-4, Inconel, SS	Zr–alloy E110, (Zr–2.5Nb)
3. Average power rating, kW/l	80–125	83–108	9–19	40–57	5
4. Fast neutron flux, Average, n/cm².s	$6–9 \times 10^{13}$	$5–7 \times 10^{13}$	$1.5–2 \times 10^{12}$	$4–7 \times 10^{13}$	$1–2 \times 10^{13}$
5. Temperatures, °C					
Average coolant inlet	279–294	267–290	249–257	272–278	270
Average coolant outlet	313–329	298–320	293–305	280–300	284
Max cladding OD	320–350	335–352	310	285–305	290
Steam mass content, %				7–14	14
6. System pressure, bar	155–158	125–165	96	70	67
7. Coolant flow, m/s	3–6[f]	3.5–6	3–5	2–5[f]	3.7
8. Coolant chemistry[g]					
Oxygen, ppb	<0.05	<0.1		200–400	<20
Hydrogen (D_2),	2–4		(3–10)	0.05–0.30	–
ppm	25–50	30–60			
cc/kg					
Boron (as boric acid), ppm	0–2200	0–1400	–	–	–
Li (as LiOH), ppm	0.5–3.5	0.05–0.6	1	–	–
K (as KOH), ppm	–	5–20		–	–
NH3, ppm		6–30			
NaOH, ppm		0.03–0.35			

[a] Voda Voda Energo Reactor (VVER).
[b] Canadian Deuterium Uranium (CANDU).
[c] Reaktor Bolshoi Mozhnosti Kanalov (RBMK).
[d] Zirconium Low Oxidation (ZIRLO).
[e] Stainless Steel.
[f] Variation from lower to upper part of the core and from plant to plant.
[g] Zn in ppb quantities may be added for BWRs and PWRs; Pt and Rh in ppb quantities may be added for BWRs.
Source: A.N.T. International (2011).

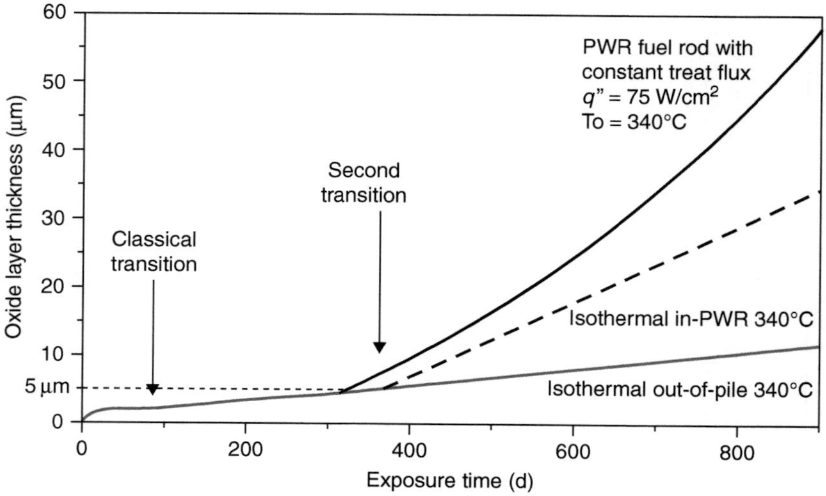

4.41 Schematic of PWR corrosion kinetics (Adamson *et al.*, 2006/2007). (Source: Reprinted, with permission, from Garzarolli *et al.* (1996), copyright ASTM International, 100 Barr Harbor Drive, West Conshohocken, PA 19428.)

thickness at high burnup may reach or exceed 100 μm, while some of the newer alloys have less than half of that. As noted in Table 4.8, only uniform corrosion occurs in PWRs under non-boiling, hydrogenated conditions.

The study of uniform corrosion in the laboratory at temperatures relevant to reactor operation suffers from three problems: (1) irradiation effects are absent, (2) corrosion rates are very low at 280–360°C (553–633K) and (3) very long times (hundreds of days) are required to differentiate between material variables. The most widely used test appears to be 360°C (633K) pressurized water for very long time periods. An example is given in Fig. 4.42 where a series of Zircaloy-type alloys with different Sn contents are compared. Garde *et al.* (1994), show that in-reactor results after high burnup give the same ranking as the autoclave laboratory tests. Also, adding Li to the water is thought to improve comparisons for PWRs (Sabol *et al.*, 1994).

Both Zircaloy-2 and -4 form a protective uniform oxide in typical BWRs and PWRs. In BWRs the various types of oxidation kinetic are shown in Fig. 4.43.

Uniform corrosion continues at a very low rate out to high burnups. During reactor exposure the microstructure of the Zircaloys used in BWRs is continually evolving due to irradiation damage and SPP dissolution. In the range 30–50 MWd/kgU (6–10 × 10^{21} n/cm², $E > 1$ MeV), changes in the microstructure induce an acceleration of uniform corrosion, as described earlier. First, patches of white oxide appear in the otherwise black or grey uniform background, as illustrated in Fig. 4.44. These patches remain very

Table 4.8 Types of corrosion observed for Zircaloy in-reactor and out-of-reactor

Type of corrosion	Comment	Observed	Irradiation effect in oxygenated coolant (BWR)	Irradiation effect in hydrogenated coolant PWR
Uniform	Normal mode	Out-of-pile, in-BWR and in-PWR	5–10× increased from beginning, low flux dependency	2–4× increased after 1st corrosion rate transition, low flux dependency
Nodular	Local breakdown of oxide protectiveness	In-BWR and out-of-pile >500°C, Zry with large SPP	Increases almost linearly with fast flux, little temp. dependency	Not observed
Shadow corrosion	Probably driven by potential differences	Only under irradiation and oxidative coolant conditions (BWR)	Increases probably almost linearly with fast flux	Not observed
Crevice corrosion	Change of environment in small gaps	Out-of-pile, in-BWR and in-PWR	Observed	Observed
Increased corrosion with fine SPP	Reduction of protectivity	Out-of-pile, in-BWR and in-PWR	Increases almost linear with fast flux	Increases almost linearly with fast flux, little temp. dependency
Increased corrosion at high fluences	Reduction of protectivity	In-PWR	?	Increases with increasing fluence above a critical threshold
Increased corrosion at high hydride concentrations	Lower corrosion resistance of hydride	Out-of-pile, in-BWR and in-PWR	Observed	Flux dependent but less temperature dependent

Source: A.N.T. International (2011) and Garzarolli in ZIRAT7 STR 'Corrosion in Zirconium Alloys', Adamson *et al.* (2002).

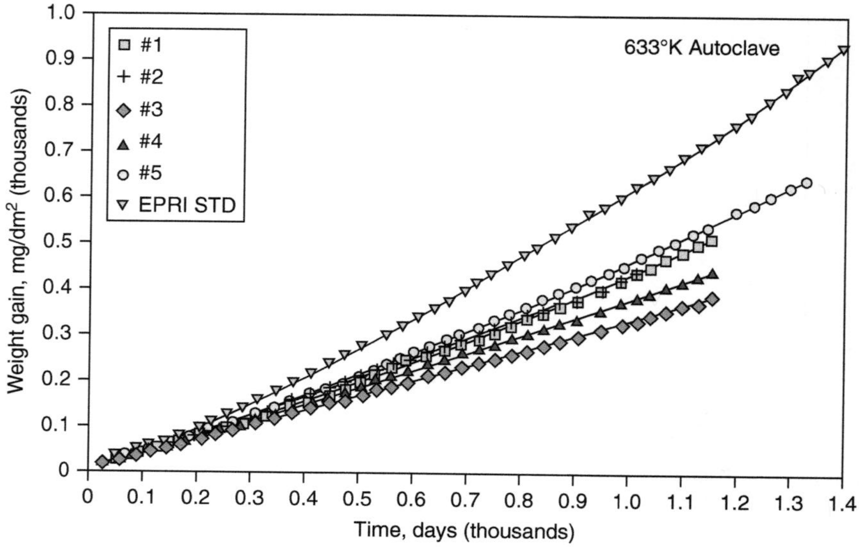

4.42 Corrosion weight gain as a function of laboratory autoclave test exposure time for several Zircaloy-4 tube variants. (Source: Reprinted, with permission, from Garde *et al.* (1994), copyright ASTM International, 100 Barr Harbor Drive, West Conshohocken, PA 19428.)

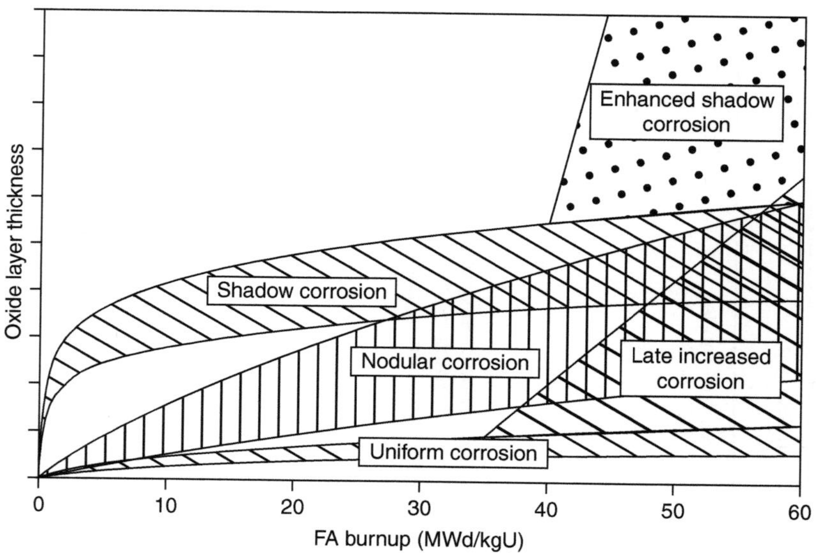

4.43 Characteristics of different types of corrosion observed in BWRs. Oxide layer thickness in arbitrary units versus fuel assembly burnup.

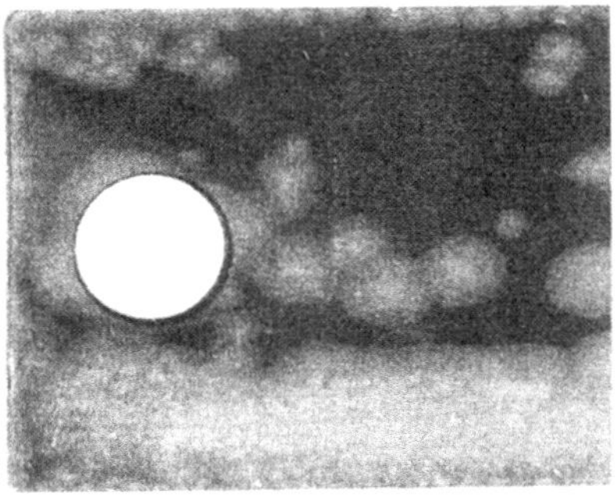

4.44 Patch oxide formation on Zircaloy-2 after an exposure of 8.5 × 10^{21} n/cm², *E* > 1 MeV. The oxide is thinnest at uniform black oxide (5) and patch oxide (4) and is thickest at coalesced patches (3). (Source: Reprinted, with permission, from Huang *et al.* (1996), copyright ASTM International, 100 Barr Harbor Drive, West Conshohocken, PA 19428.)

thin, about the same as the black uniform corrosion film. At some point the patches cover 100% of the surface and oxide thickening occurs at an accelerated rate, labelled 'late increased corrosion' in Fig. 4.43.

Importantly, at the same or somewhat higher fluences, a marked increase in hydrogen pickup fraction occurs. This hydriding is potentially a more serious issue than the corrosion increase and is discussed later.

Figure 4.45 gives optical micrographs and schematics of various types of corrosion. Schematics A and C show normal and increased uniform oxide, and the others illustrate nodular corrosion. Zr-Nb alloys and small-SPP Zircaloys (average SPP size less than about 0.1 μm) in general do not form nodules. In large-SPP Zircaloys, nodules initiate early in life and grow at a decreasing rate with fluence. In some cases, for aggressive water chemistries and susceptible material, nodules can coalesce to cover the entire surface. Nodular oxide thickness does not generally cause performance problems; however in severe cases spalled oxide can be a source of 'grit' in control drive mechanisms. Serious fuel failure problems can be induced by a combination of heavy nodular corrosion and copper-zinc-laden crud, resulting in the crud-induced localized corrosion (CILC) phenomenon discussed later.

Susceptibility of Zircaloys to in-reactor nodular corrosion can be identified by laboratory high temperature steam tests. The most effective testing procedures are variations of the 'two-step test' described by Cheng *et al.* (1987) where Zircaloys are exposed to steam at 410°C/1500 psi (683K/102 bars) for 8 h followed by 510°/1500 psi (783K/102 bars) for 16 h. In such tests, Zr-Nb alloys do not exhibit nodular corrosion.

The fifth type of corrosion, indicated in Fig. 4.43, is so-called shadow corrosion, described in more detail later. Shadow corrosion is induced on all zirconium alloys when they are in close proximity to many non-zirconium

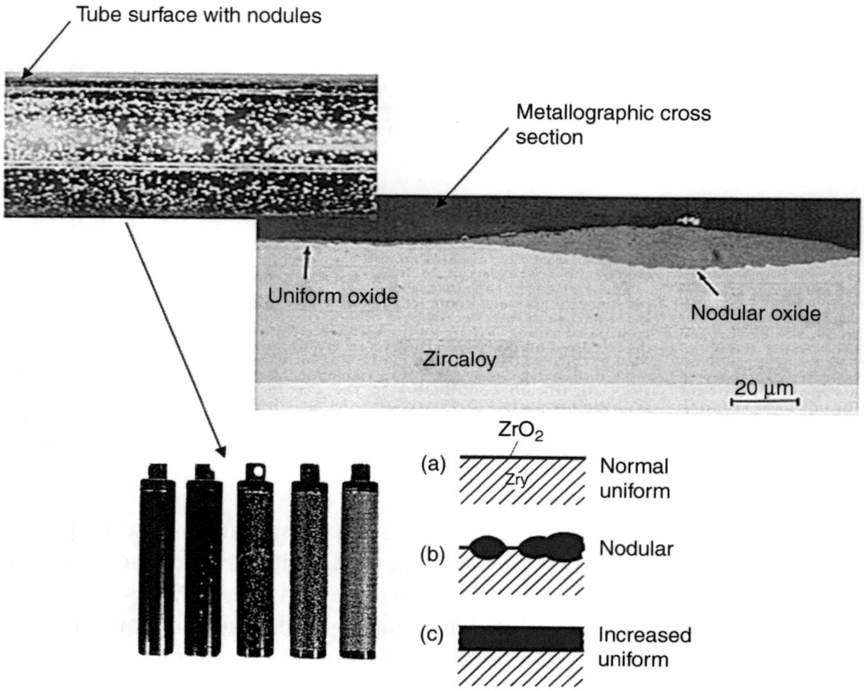

4.45 Corrosion morphology for Zircaloy in BWRs (Adamson *et al.*, 2007).

alloys such as stainless steel or Inconel. The oxide thickness is unusually large and often appears to be particularly dense and uncracked. For example shadow corrosion oxide induced by a stainless steel control blade bundle is shown in Fig. 4.46. Shadow corrosion has 'always' been present in BWRs, but not in PWRs primarily related to the high PWR hydrogen concentration which reduces or eliminates galvanic potentials between dissimilar alloy components. In BWRs shadow corrosion caused no performance issues until recently when at one reactor fuel failures were induced by unusually severe 'enhanced spacer shadow corrosion' (Zwicky *et al.*, 2000). More recently, shadow corrosion has been alleged to be involved in BWR channel bow problems (Mahmood *et al.*, 2010). Both issues are addressed later in this chapter.

Accelerated uniform corrosion

The observation that uniform corrosion of BWR materials may increase at high fluence (burnup) has been introduced. The main factor driving this increase is connected to the initial size distribution of the SPPs and their

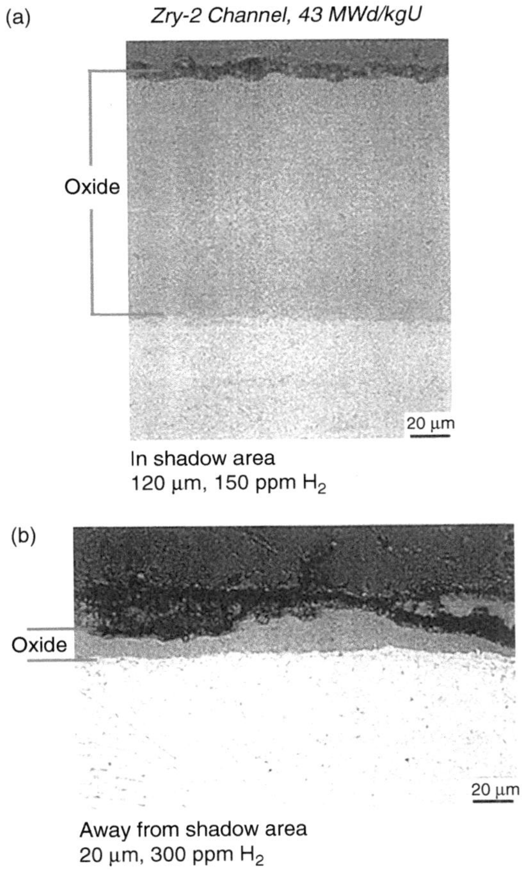

4.46 Zirconium oxides (a) away from and (b) near a stainless steel control blade bundle (Adamson *et al.*, 2000).

dissolution during irradiation. This was noted by Cheng and Adamson (1987) and then by Yang and Adamson (1989), in reference to thick uniform oxide observed in welded regions of Zircaloy-4 having totally dissolved SPPs. A clear correlation between SPP size and increased corrosion at high fluence was given by Garzarolli *et al.* (1994) (Fig. 4.47) where it was shown that 'small SPP sizes' resulted in relatively thick corrosion films after 3 or 4 cycles in-reactor but not after 1 or 2 cycles. Huang *et al.* (1996) showed that when SPPs virtually 'disappeared' (within the resolution of STEM at that time) corrosion increased, as did hydrogen pickup. Similar results were reported by Tägtström *et al.* (2002), Takagawa *et al.* (2004) and Ishimoto *et al.* (2006). It is clear that loss of SPPs affects corrosion performance, and even earlier in fluence, hydrogen pickup.

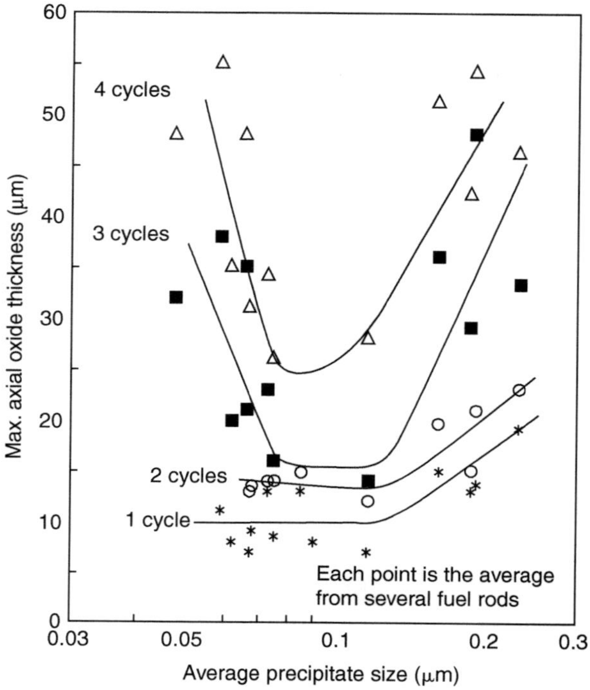

4.47 Effect of SPP size on corrosion of Zircaloy-2 cladding. (Source: Reprinted, with permission, from Garzarolli *et al.* (1994), copyright ASTM International, 100 Barr Harbor Drive, West Conshohocken, PA 19428.)

Zircaloy-4, which does not contain alloying quantities of Ni, is also susceptible to increased corrosion when SPPs dissolve. An example is given in Fig. 4.42 for Zircaloy-4 with a relatively large SPP, about 0.2 μm average size determined by TEM (Garzarolli, *et al.*, 2002). It is seen that when the volume fraction of SPPs gets very low, corrosion increases dramatically.

A compilation of data for Zircaloy-2 and -4 given by Garzarolli in Adamson *et al.* (2006) and Garzarolli *et al.* (2011b) illustrates that Zircaloy-4 generally has higher corrosion than Zircaloy-2 (see Fig. 4.49). However, the hydrogen pickup fraction (HPUF) for Zircaloy-4 appears to stay remarkably low at high burnup, as indicated by the data of Miyashita *et al.* (2006) and the correlation given in Fig. 4.50.

In PWRs nodular corrosion in unlikely to occur due to high hydrogen and low oxygen in the water. Accelerated uniform corrosion does occur for Zircaloy-4 in PWRs as seen in Fig. 4.51 (and Fig. 4.48) and the HPUF is relatively low, as in BWRs (Fig. 4.52). Those figures also show that M5 (which is basically a Zr1Nb alloy with 300–500 ppm Fe) does not undergo accelerated

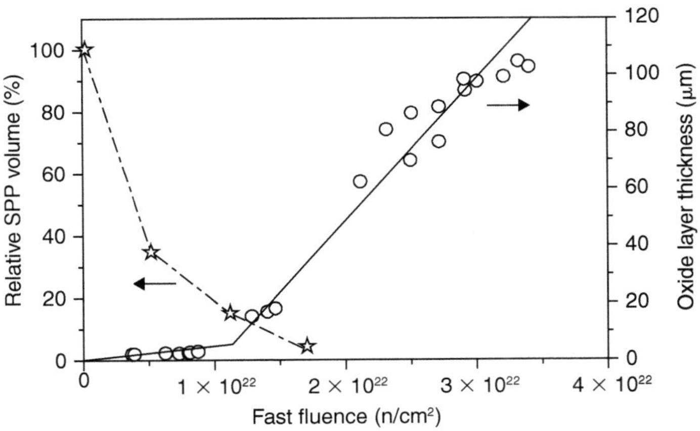

4.48 Influence by irradiation to very high fluences at 290°C (563K) on corrosion and SPP dissolution of Zircaloy-4 with large SPPs. (Source: Reprinted, with permission, from Garzarolli *et al.* (2002), copyright ASTM International, 100 Barr Harbor Drive, West Conshohocken, PA 19428.)

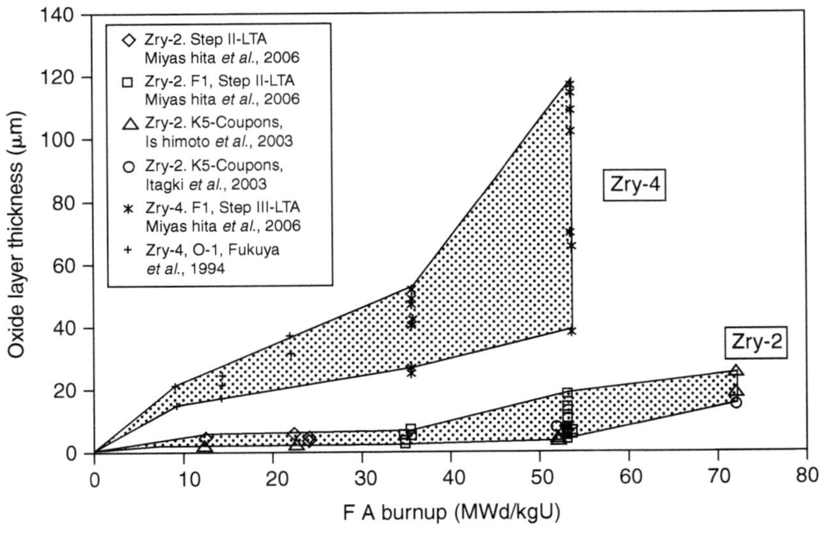

4.49 Oxide thickness of Zircaloy-2 and Zircaloy-4 under isothermal irradiation in different Japanese studies. Particularly important is the data of Miyashita *et al.* (2006); ZIRAT 11 compilation (Adamson *et al.*, 2006).

uniform corrosion out to exposures equivalent to 70 GWd/mt at the PWR irradiation temperature of 315°C (587K). At that temperature, irradiation effects on SPPs are similar in BWRs and PWRs. Two types of SPPs exist for M5 – βNb and the Laves phase $Zr(Fe,Nb)_2$. At high fluence all the Fe will

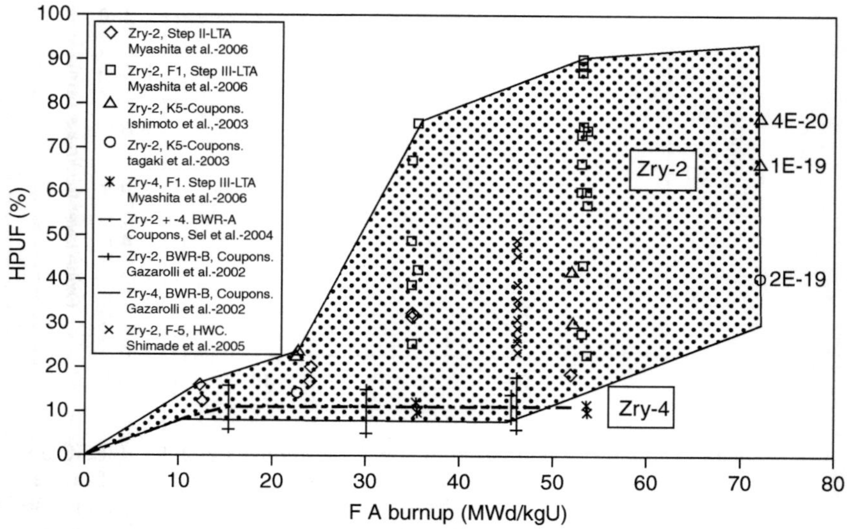

4.50 Hydrogen pickup behaviour of Zircaloy-2 and Zircaloy-4 under isothermal irradiation in different Japanese studies. Particularly important is the data of Miyashita *et al.* (2006); ZIRAT 11 compilation (Adamson, *et al.*, 2006). ([1]Hoffmann & Manzel, 1999; [2]Potts, 2000; [3]Zwicky *et al.*, 2000)

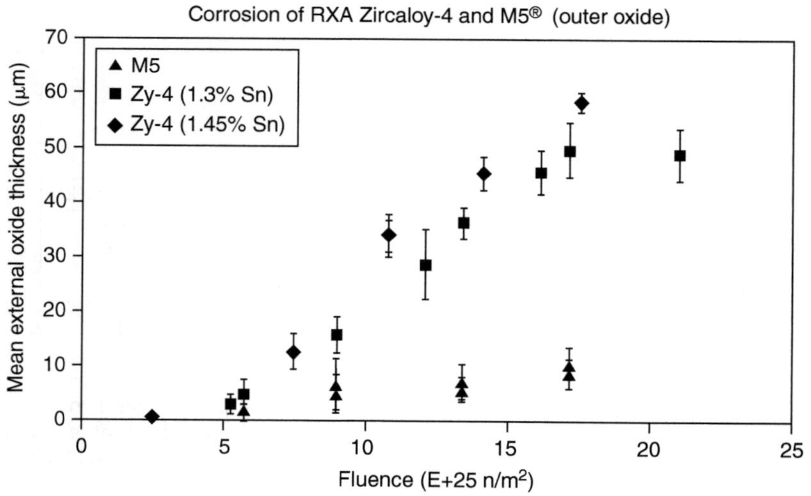

4.51 Corrosion of RXA Zircaloy-4 and M5 at 315°C (588K) in PWR water chemistry. (Source: Reprinted, with permission, from Bossis *et al.* (2007), copyright ASTM International, 100 Barr Harbor Drive, West Conshohocken, PA 19428.)

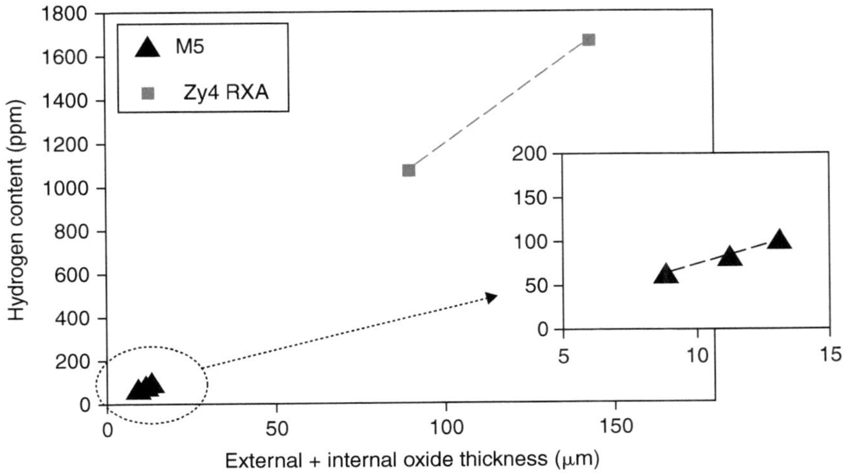

4.52 Hydrogen pickup fraction (HPUF) of RXA Zircaloy-4 and M5 at 315°C (588K) in PWR water chemistry. (Source: Reprinted, with permission, from Bossis *et al.* (2007), copyright ASTM International, 100 Barr Harbor Drive, West Conshohocken, PA 19428.)

be dissolved into the matrix, and SPPs of βNb will exist without appreciable dissolution. Again it appears that it is the existence of the SPPs which is important.

Mechanism implications

Nodular corrosion has been shown to be very sensitive to the concentration of Fe and Ni in the zirconium matrix. For a recent complete review see Franklin (2010). For unirradiated Zircaloy-2 annealing in the high alpha temperature range (<800°C, 1073K) causes a small increase in solute matrix concentration which is correlated to a sharp decrease in nodular corrosion, without a significant change in size of the SPPs (Kruger *et al.*, 1992) and without a significant change in the Fe/Cr ratio in the $Zr(Fe,Cr)_2$ SPP, but with a significant decrease in the Ni/Fe ratio in the $Zr_2(Fe,Ni)$ SPP (Cheng *et al.*, 1987). The mechanistic interpretation of these phenomena, as gleaned from Sections 4.2 and 4.3.1, is either that oxide conductivity is markedly increased by the solutes, or that the Galvanic potential between the SPPs and the matrix is decreased by increased solutes and chemistry changes in the SPPs. The importance of Ni in decreasing nodular corrosion by high alpha annealing is shown by the fact that such treatments for Zircaloy-4, with no alloying Ni, unlike those for Zircaloy-2 do not result in any decrease in nodular corrosion in the 520°C (793K) steam tests used in those studies.

Uniform corrosion appears not to be so sensitive to solute content, as even low fluence irradiation results in large changes in the chemistry of the SPPs and significant increases in solute matrix concentration without a significant change in oxide thickness. The experiments described above clearly indicate, however, that disappearance of the SPPs is correlated to oxide thickness increase.

The HPUF at high fluence appears to be correlated to both SPP disappearance and solute concentrations in the matrix. HPUF increases earlier in fluence compared to oxide thickness, indicating a connection to solutes, but there does not appear to be a clear correlation to Fe concentration in the matrix. When the SPPs do disappear (dissolved by irradiation effects, dispersing Fe and Ni in the matrix and leaving a local concentration of Cr) however, the HPUF sharply increases.

The role of Ni in HPUF issues is emphasized by the observations that HPUF in Zircaloy-4 does not increase at the high fluences reported thus far. A mechanism has been proposed (Garzarolli *et al.*, 2011b) whereby Ni distribution in the growing Zircaloy-2 metal/oxide interface acts as an easy path for hydrogen ingress to the metal. This hypothesis and the data of Ishimoto *et al.* (2006) suggest a high value of the Fe/Ni ratio in Zircaloy-2 is desirable.

4.5.2 Shadow corrosion

Enhanced corrosion of zirconium alloys may occur when the corroding surface is close to, or in contact with, certain other metallic components. The shape of the component is often reproduced in the shape of an area of enhanced corrosion, suggestive of a shadow cast by the component on the zirconium alloy surface. The term 'shadow corrosion' is therefore often used to describe the phenomenon. Observations of shadow corrosion on water reactor components have been noted for many years. In 1974, Johnson *et al.* (1974), reported enhanced corrosion in Zircaloy coupons located near, but not touching, small pieces of platinum in the advanced test reactor (ATR). Also, Trowse *et al.* (1977) reported enhanced fuel rod corrosion beneath steel spacers in steam generating heavy water reactors (SGHWRs).

Most commonly observed are the control blade shadows on BWR channel surfaces adjacent to the control blade handles, such as shown in Fig. 4.53. In this case the stainless steel control blade handle is imaged as a black shadow on a light background, but the reverse is sometimes also observed. It was shown (Chen & Adamson, 1994) that the handle image is faithfully reproduced on the channel surface, but is larger than the actual handle, shown schematically in Fig. 4.54. Hot cell examination of a similar channel shows that oxide thickness within the shadow area can be much higher than outside the shadow (Adamson *et al.*, 2000).

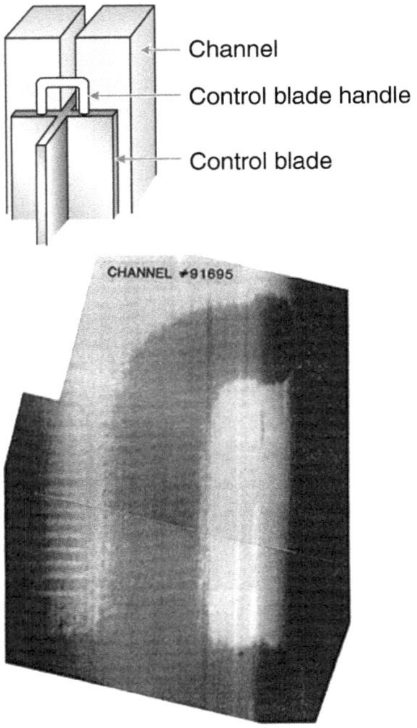

4.53 Control blade shadow on a BWR channel. Upper diagram shows the relative arrangement of blade handle and shadow (Chen & Adamson, 1994). Copyright 1994 by the American Nuclear Society, La Grange Park, Illinois.

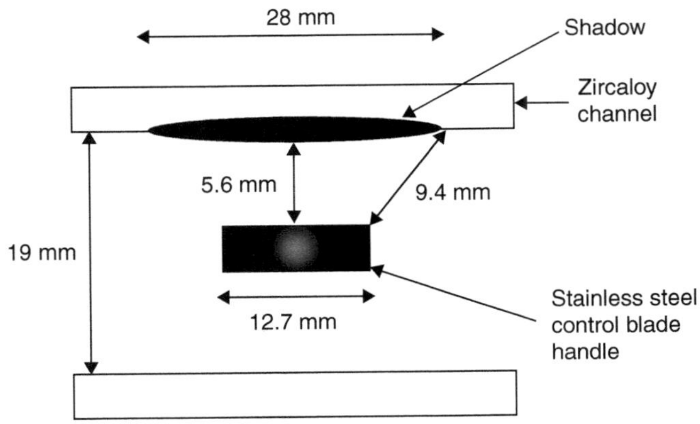

4.54 Geometrical relationships between control blade handle and channel for the shadow shown in Fig. 4.53 (Chen & Adamson, 1994). Copyright 1994 by the American Nuclear Society, La Grange Park, Illinois.

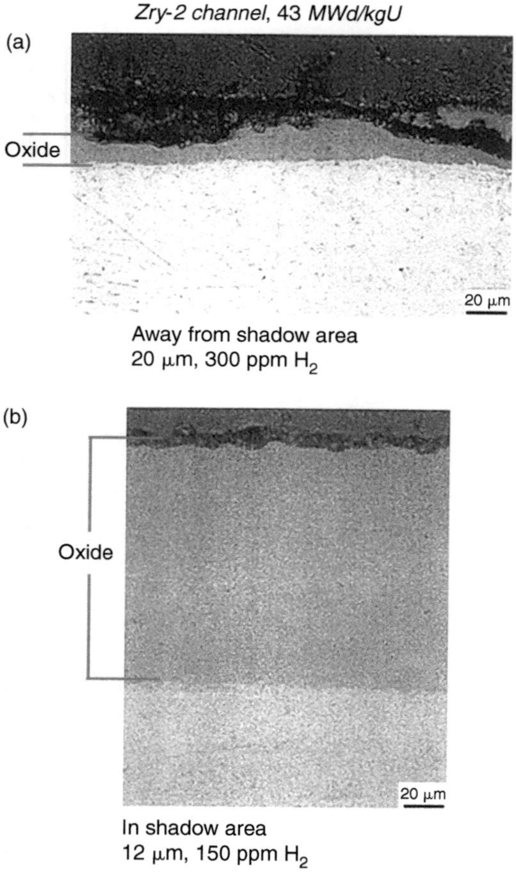

4.55 Oxide thickness (a) away from and (b) within a control blade handle shadow (Adamson *et al.*, 2000).

Figure 4.55 illustrates a 6× difference in oxide thickness within and without of the shadow. It is noteworthy, however, that in this case, hydrogen content in the shadow is 150 ppm, while outside the shadow it is 300 ppm. Since temperature gradients are small in a channel wall, redistribution of hydrogen by thermal-gradient driven diffusion is also small, indicating in this case a much reduced hydrogen pickup rate in the shadow. More recent data indicate variability in pickup fraction (HPUF), as (Mahmood *et al.*, 2010) reported normal HPUF. It is also observed that control blade shadows preferentially form during the first cycle of operation and that the thickness tends to saturate with burnup.

Shadows on fuelled or non-fuelled rods have been observed under Zircaloy spacers with Inconel springs or under all-Inconel spacers (Fig. 4.56). Again,

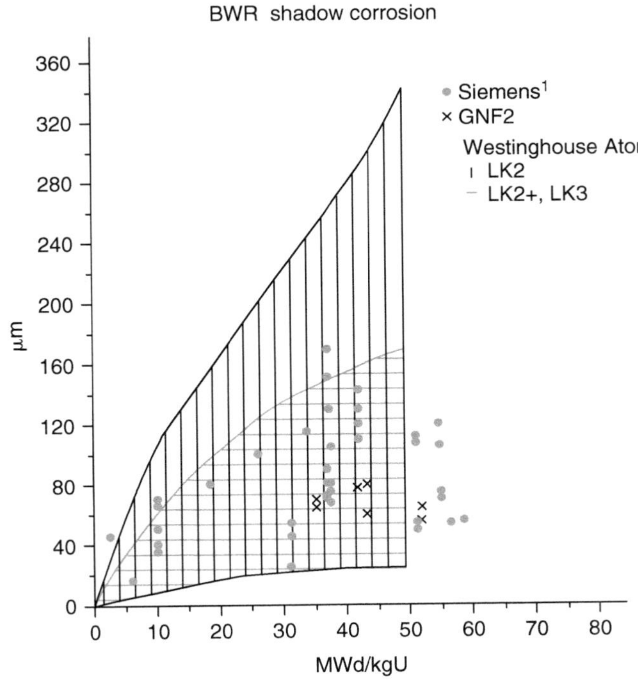

4.56 Shadow corrosion data of various BWR fuel vendors' claddings. (Source: Figure modified according to Hoffmann and Manzel (1999); Potts (2000); Zwicky *et al.* (2000). Copyright 2000 by the American Nuclear Society, La Grange Park, Illinois.)

this illustrates the trend for saturation of the oxide thickness with fluence or burnup. In most cases, shadows have not caused fuel performance problems. The upper curve in Fig. 4.50, is an exception for a specific cladding condition, called Enhanced Spacer Shadow Corrosion (Zwicky *et al.*, 2000).

A number of experiments have been conducted to elucidate the details of the shadow corrosion mechanism. Combined with the commercial reactor observations, these experiments reveal:

1 A variety of metals are observed to cause shadows on Zircaloy. These are:
   (a) Stainless steel (many)
   (b) Pt (Shimada *et al.*, 2002; Johnson *et al.*, 1974)
   (c) Hf (Shimada *et al.*, 2002)
   (d) Inconel X750, X718 (many)
   (e) Inconel 600 (Adamson *et al.*, 2000)
   (f) Welded regions on Zircaloy (Chen & Adamson, 1994; Shimada *et al.*, 2002).

2 Nitronic 32 was observed not to cause shadows (Andersson *et al.*, 2002).
3 Resistance to shadow formation depends upon inherent corrosion resistance (Andersson *et al.*, 2002; Garzarolli *et al.*, 2002; Shimada *et al.*, 2002).
4 The distance between components is critical. Oxide thickness is a function of distance. There is a maximum distance above which there is no effect (a few mm) but there is no minimum distance, including touching (Chen & Adamson, 1994; Lysell *et al.*, 2001; Andersson *et al.*, 2002).
5 In general, the shadow forms at low fluence or exposure and thickness of the shadow saturates with fluence (Fukuya *et al.*, 1994; Hoffmann & Manzel, 1999; Andersson, 2000). In unusual cases, perhaps due to special microstructure and water chemistry, accelerated shadow corrosion begins at high fluence or burnup (Zwicky *et al.*, 2000; Wikmark & Cox, 2001; Andersson *et al.*, 2002).
6 Shadow formation requires a nuclear reactor environment and it has not been possible to reproduce it in laboratory autoclaves (Andersson, 2000; Garzarolli *et al.*, 2001). However, use of ultraviolet light in the laboratory has been shown to increase electro-chemical potentials between common components, believed to be related to shadow formation (Kim *et al.*, 2010).
7 No reports of shadow formation have been made for PWRs or high-hydrogen cases. BWR hydrogen water chemistry conditions, however, do allow shadows (Lefebvre & Lemaignan, 1997; Adamson *et al.*, 2000).
8 Thick oxide shadows do not necessarily result in proportionally high hydrogen pickup and can result in unusually low hydrogen pickup (Adamson *et al.*, 2000) or normal pickup (Mahmood *et al.*, 2010).
9 Pre-oxidation autoclaving of Zircaloy does not prevent shadows, but applying a zirconia layer to Inconel does (Andersson *et al.*, 2002).
10 Shadow corrosion has been observed when the two metals are not in contact physically and are nominally electrically insulated from each other. However, in making such observations it has been assumed that the radiation field has no effect on the conductivity of the insulating medium.
11 Shadow formation has been reported (Châtelain *et al.*, 2000; Andersson *et al.*, 2002) in a reactor position outside and downstream of the MIT test reactor core where the neutron and gamma fluxes are reported to be near zero. On the other hand, no shadows were reported (Lysell *et al.*, 2001) in a reactor position outside and upstream of the R2 test reactor core where the neutron flux and gamma power (flux) were also near zero. However, there does appear to be some uncertainty in the actual gamma intensities in the MIT experiment, so there may need to be a re-evaluation of the out-of-core results.
12 Shadow formation can be prevented if electrical connection between the two components is prevented (Lysell *et al.*, 2005).

The early thoughts on the mechanism of shadow corrosion centred on it being a form of galvanic corrosion. As such, the mechanism required a path for electron transfer between the cathodic shadower (the material that causes the shadow) and the anodic component (the Zircaloy or zirconium alloy component on which the enhanced corrosion occurs) and a conductive path through the water separating the two parts. But since the shadow phenomenon could not be reproduced in the laboratory, it was clear that some sort of radiation effect was also required. Problems with the galvanic hypothesis included lack of evidence, in some cases, of any electrical connection between the shadower and component (although for commercial reactor components an obscure path can always be suspected) and concern that the zirconium oxide, which is always present on component surfaces, was not conductive enough to allow the postulated conductive paths to operate.

Another hypothesis arose when Chen and Adamson (1994) noted that the range in water of beta particles from Mn-56 and Zr-97 (originating in the shadower material) could explain the shape and size of observed shadows if a beta-damage mode could be found. Lemaignan (1992) proposed that extra radiolysis caused by the imposed local beta flux could result in accelerated corrosion. However, additional calculations by Andersson *et al.* (2002) and Shimada *et al.* (2002) indicated that the extra beta flux from the shadower does not make a significant change to the overall beta flux in the reactor, so this hypothesis has been discounted. Also, as noted above, it has been shown that the alloy Nitronic 32 does not cause shadows, even though the flux of beta particles from activated Mn-56 from that alloy is much higher than for the known shadowers Inconel and stainless steel.

The latest view of the mechanism is that it is indeed a form of irradiation-assisted galvanic corrosion. Points which support this hypothesis include:

1 It is known that there is a corrosion potential difference between, for instance, stainless steel or Inconel and Zircaloy in non-hydrogenated water (BWR type) (Table 4.9). Also, this potential difference is enhanced in-reactor (Lysell *et al.*, 2001) and by ultraviolet light outside the reactor (Kim *et al.*, 2010).
2 The observed relationship between component separation distance and shadow oxide thickness is roughly as expected (Adamson *et al.*, 2000; Lysell *et al.*, 2001; Andersson *et al.*, 2002).
3 A true stainless steel/Zircaloy galvanic couple in-reactor produced thick oxide and low hydrogen pickup in Zircaloy, similar to that observed for a control blade handle/Zircaloy shadow (Adamson *et al.*, 2000; Lysell *et al.*, 2005).
4 A radiation enhancement of electrical conductivity of oxides has been reported. Electrical conductivity of oxide films on Zircaloy markedly

Table 4.9 Corrosion potential differences (mV) between Zircaloy-4 and Inconel in BWR and PWR environments

Environment	Material coupling	
	Inconel-Zry-4 ground	Inconel-Zry-4 pickled
PWR type	−60	−75
BWR type	−420	−640

Source: A.N.T. International (2011) and Garzarolli *et al.* (2001b).

increased during electron irradiation (Howla *et al.*, 1999), gamma irradiation (Kang *et al.*, 1994) and in-reactor (Shannon, 1962). Also, the conductivities of various ceramics, including Al_2O_3 were reported to increase dramatically under proton or x-ray irradiation (Hobbs *et al.*, 1994). So perhaps a way to allow closing of the galvanic circuit is indicated. Note that Al_2O_3 was used as insulation in the MIT experiment (Andersson *et al.*, 2002) and ZrO_2 as insulation in others (Chen & Adamson, 1994; Shimada *et al.*, 2002). Some distinction must be made between surface conduction and bulk conduction in thick ceramics, but little is known.
5 Shadows were not formed when the components were confirmed to be electrically insulated (Lysell *et al.*, 2005).

Points which bring doubt to the irradiation-assisted galvanic mechanism hypothesis include:

6 It is not certain that an electrically-conducting path truly exists between the shadower and component.
7 Conventional galvanic reactions are inhibited when the cathode is small and the anode is large, as is often the case with observed shadows.
8 The MIT experiments (Andersson *et al.*, 2002) appear to produce shadows in a region of very low radiation.

It is also quite possible that several different mechanisms contribute to shadow corrosion depending on specific conditions.

It is obvious that the mechanism of shadow corrosion is not precisely defined, although the number of observations which help to provide a working hypothesis are many. It is clear that enhanced corrosion can and will occur in cases where Zircaloy and a variety of other metals or alloys are in close contact. Potential problems include loss of strength and integrity due to wall thinning and oxide spallation. In most cases, the enhanced corrosion does not cause serious operational problems. The oxide thicknesses are moderate, hydrogen absorption is not increased by straightforward shadows and spalling usually does not occur. The severe problem of enhanced spacer shadow corrosion (ESSC)) can apparently be mollified by control of both Zircaloy microstructure and by reactor water chemistry. No harmful effects have been reported or been related to the use of Pt in noble water chemistry treatments in BWRs, but further studies are underway. BWR channel bowing has been attributed to manifestations of shadow corrosion, and has produced major in-reactor problems to be discussed later in this chapter, but further study is required to understand the details (Mahmood *et al.*, 2007; Blavius *et al.*, 2008; Münch *et al.*, 2008). There does not appear to be an immediate remedy for shadow corrosion – for instance an oxide prefilm on Zircaloy does not prevent shadows. Coatings on the shadower (such as Inconel springs) would appear to be promising, but may not be practicable.

4.5.3 Classical crud-induced localized corrosion (CILC)

The major sources of crud formation on fuel rod surfaces in BWRs (as in PWRs) are the metallic impurities in the coolant that result from the corrosion of reactor coolant system materials. Other sources are leakage of solids, liquids and gases into the system and impurities from the surface of components placed in the core (fuel assemblies, etc.).

Crud types identified on BWR fuel surfaces have generally been:

- tightly-adherent, dense Fe_2O_3 – generally a thin layer
- loosely-adherent Fe_2O_3 – various levels of 'looseness'
- a combination of the two types – tight and loose layers
- tightly-adherent, dense Fe_2O_3 + Cu or CuO
- tightly-adherent, dense Fe_2O_3 + ZnO ($ZnFe_2O_4$).

The effect of the thin, tightly adherent Fe_2O_3 layers on clad surface temperatures is minimal. The loosely adherent layers are generally very porous, filled with water and, as a result, have high thermal conductivity and only a small effect on clad surface temperatures.

Zn, injected to reduce activity transport, tends to form the spinel with Fe, as listed above, but is not known to have been the cause of fuel failures. A particularly detrimental crud type is Fe_2O_3 infiltrated with Cu. This has been associated with CILC cladding failures. The topic has been thoroughly reviewed in a ZIRAT report (Wikmark & Cox, 2001) and by Marlowe *et al.* (1985). An overview of CILC mechanisms is presented here.

In 1979 and the early 1980s, localized fuel cladding corrosion failures occurred in some plants that had copper alloy (Admiralty brass) condenser tubes and filter-demineralizer condensate clean-up systems. Such plants have higher levels of soluble copper in the water than those with stainless steel or titanium condenser tubes or with deep-bed resin clean-up systems. A few rods per bundle failed in susceptible bundles at burnups >15 GWd/MT. Over

Table 4.10 Typical elemental analysis of crud composition in a CILC-susceptible plant

	Standard crud	CILC crud
Major phase	Fe_2O_3	CuO
Iron	87%	21.1%
Copper	2.0	52.8
Zinc	4.4	11.1
Nickel	3.3	2.5
Manganese	2.2	3.3
Chromium	1.1	2.5
Cobalt	0.3	0.6

Source: A.N.T. International (2011) and Baily *et al.* (1985).

90% of the failed rods contained $(U,Gd)O_2$ fuel (i.e. they were 'gadolinia rods'). However, most fuel reloads and fuel bundles were not affected, even in susceptible plants. Poolside and hot cell examinations revealed unusual crud scale deposits, with high copper concentrations, rather than the typical fluffy, Fe_2O_3 crud. Table 4.10 gives a breakdown of crud elemental composition in a CILC-susceptible plant (Baily *et al.*, 1985).

It was reported (Marlowe *et al.*, 1985) that three factors interacted to cause CILC fuel failures: reactor water chemistry, fuel duty and Zircaloy resistance to nodular corrosion. Marlowe *et al.* (1985) and Wikmark and Cox (2001) provide details and analysis. An interpretive summary is given here as an illustration of a crud-induced failure process. Failure in the gadolinia rods proceeds by the following steps:

1 Incubation phase (low to moderate power)
   (a) extensive nodular corrosion occurs early in first cycle
   (b) oxide nodules grow on some fuel rods to produce 90–100% coverage
   (c) copper, in reactor water, deposits between oxide nodules
   (d) copper deposition continues, crud grows within oxide nodules to form a thick sandwich structure (ZrO_2/*crud*/ZrO_2)
2 Failure phase (moderate to high power)
   (a) cracks form within sandwich structure producing local, steam-insulated regions
   (b) insulating effect accelerates cladding corrosion and hydriding
   (c) cladding penetrations occur locally by formation of auto-catalytic corrosion pits or by cracking of hydrided Zircaloy in spalled regions.

Gadolinia rods are at low power during the first cycle and never become the highest power rod in a bundle (see Fig. 4.57). For reasons that are still not fully

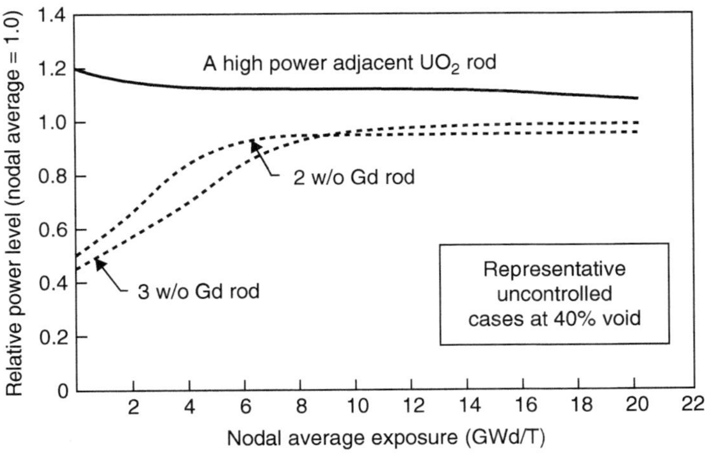

4.57 Relative power history of $(U,Gd)O_2$ and nearby high power UO_2 rods (Marlowe *et al.*, 1985).

understood, but perhaps related to the specifics of the boiling phenomena in a BWR core, low heat flux rods, or regions of rods, have been shown to be particularly susceptible to nodular corrosion (Marlowe *et al.*, 1985). Therefore, in gadolinia rods, Step 1b (in the incubation phase) is reached quickly. Figure 4.58 shows that when the nodular coverage reaches about 90%, copper deposits copiously on the crudded rod surface. The most likely deposition location is between nodules, as it has been shown that microscopic heat flux increases between nodules and, therefore, good conditions for wick boiling are established (Wikmark & Cox, 2001). At this point, copper also begins to deposit in lateral cracks in the thick nodules. Figure 4.59 is an elemental X-ray map of a typical nodule, clearly showing copper (and zinc) in cracks in the nodule. At this point the ZrO_2/CuO layer can be more than 100 μm thick and heat transfer through the nodule is inhibited. However, the conductivity of ZrO_2/CuO is still quite high (Wikmark & Cox, 2001, Table 12) and gross overheating should not occur. The failure phase accelerates dramatically at step 2a, when new cracks in the oxide 'sandwich' form and become steam-filled. This can be facilitated by the deposited copper blocking normal ingress of coolant and egress of steam from the cracks, and by expansion of the CuO_2 as the temperature increases. Once steam-blanketed regions form, clad temperatures can become very high, as steam has a very low conductivity compared to the crud or oxides present. Step 2c follows, as the steam blanketing and the resulting high temperatures dooms the cladding.

Two-step steam testing (Cheng *et al.*, 1987) showed that only a small percentage of material in failed fuel bundles is susceptible to nodular corrosion, thus explaining why most rods do not fail.

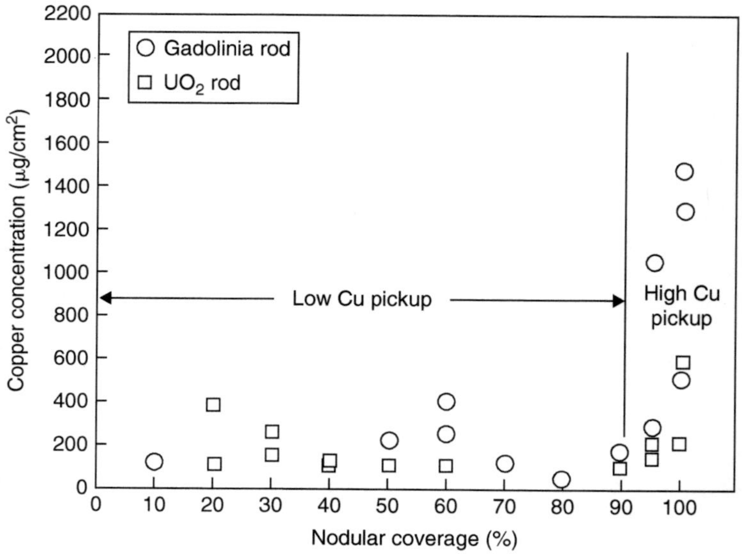

4.58 Effect of extent of nodular coverage on copper bearing and deposition (Marlowe *et al.*, 1985). Copyright 1985 by the American Nuclear Society, La Grange Park, Illinois.

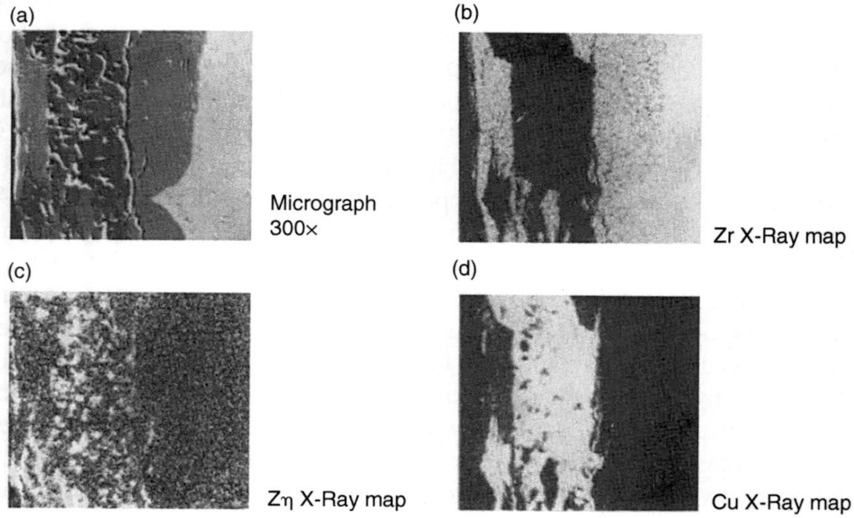

4.59 Copper-rich crud deposited in laminations (cracks) of zirconium oxide. In the X-ray map, a light region indicates the presence of the element in question (Marlowe *et al.*, 1985). Copyright 1985 by the American Nuclear Society, La Grange Park, Illinois.

Experience of CILC failures drove the BWR industry to develop nodule-resistant Zircaloy microstructures, to refine the 500°C-type steam test as a tool to predict in-reactor corrosion resistance and to tighten-up water chemistry specifications, particularly for copper.

It should be noted that there is no evidence that the presence of copper in the water enhances nodule nucleation. In fact at least one careful study indicates that copper either has no effect or improves nodular corrosion resistance (Ito *et al.*, 1994; Shimada *et al.*, 1997). Since nodular corrosion by itself has never been shown to affect fuel performance (assuming oxide spallation does not occur), elimination of either copper in the water or of nodular corrosion is claimed, with considerable justification, to eliminate CILC-type failures. In fact, the industry has been free of classical CILC failures for the past 15 years. Recent fuel failures in the United States may be caused by a crud-induced process, but the characteristics are different from CILC. Both, however, induce fuel rod failures by temperature-driven corrosion processes.

4.6 Dimensional stability of zirconium alloys

One of the most unique aspects of material behaviour in a nuclear power plant is the effect of radiation (mainly neutrons) on the dimensional stability of the reactor components. In fast breeder reactors the Fe and Ni-based alloys creep and swell, that is, they change dimensions in response to a stress and change their volume in response to radiation damage. In LWRs, zirconium alloy structural components creep, do not swell, but do change their dimensions through the approximately constant volume process called irradiation growth. Radiation effects are not unexpected since during the lifetime of a typical component every atom is displaced from its normal lattice position at least 20 times (20 dpa). With the possible exception of elastic properties like Young's Modulus, the properties needed for reliable fuel assembly performance are affected by irradiation. A summary of such effects is given by Adamson (2000).

Practical effects of dimensional instabilities are well known and it is rare that a technical conference in the reactor performance field does not include discussions on the topic. Because of the difference in pressure inside and outside the fuel rod, cladding creeps down on the fuel early in life, and then creeps out again later in life as the fuel begins to swell. A major issue is to have creep strength sufficient to resist outward movement of the cladding if fission gas pressure becomes high at high burnups. PWR guide tubes can creep downward or laterally due to forces imposed by fuel assembly hold down forces or cross flow hydraulic forces – both leading to assembly bow which can interfere with smooth control rod motion. BWR channels can creep out or budge in response to differential water pressures across the channel wall, again leading toward control blade interference. Fuel rods, water rods or boxes, guide tubes and tie rods can lengthen, possibly leading to bowing problems. (For reference, a recrystallized (*RX* or *RXA*) Zircaloy water rod or guide tube could lengthen due to irradiation growth more than 2 cm during service; a CWSR component could lengthen more than 6 cm.) Even RX spacer/grids could widen enough due to irradiation growth (if texture or heat treatment was not optimized) to cause uncomfortable interference with the channel. In addition, corrosion leading to hydrogen absorption in Zircaloy can contribute to component dimensional instability due, at least in part, to the fact that the volume of zirconium hydride is about 16% larger than zirconium.

The above discussion leads to the concept that understanding the mechanisms of dimensional instability in the aggressive environment of the nuclear core is important for more than just academic reasons. Reliability of materials and structure performance can depend on such understanding.

Comprehensive reviews of dimensional stability have been given in the *ZIRAT* Special Topical Reports (Adamson & Rudling, 2002; Adamson *et al.*, 2009; Cox *et al.*, 2005). The sources of dimensional changes of reactor components (in addition to changes caused by conventional thermal expansion and contraction) are: irradiation growth, irradiation creep, thermal creep, stress relaxation (which is a combination of thermal and irradiation creep), and hydrogen and hydride formation.

Irradiation effects are primarily related to the flow of irradiation-produced point defects to sinks such as grain boundaries, deformation-produced dislocations, irradiation-produced dislocation loops, and alloying and impurity element complexes. In zirconium alloys, crystallographic and diffusional anisotropy are key elements in producing dimensional changes.

In the past, hydrogen effects have been considered to be additive to and independent of irradiation. Although this independency has yet to be definitively proven, it is certain that corrosion-produced hydrogen does cause significant dimensional changes simply due to the 16–17% difference in density between zirconium hydride and zirconium. A length change in the order of 0.20% can be induced by 1000 ppm hydrogen in an unirradiated material (Fig. 4.60) (King *et al.*, 2002; Seibold *et al.*, 2000). That the presence of hydrides contributes to the mechanisms of irradiation creep and growth is highly suspected but yet to be determined in detail.

Fuel rod diametral changes are caused by stress dependent creep processes. Fuel rod length changes are caused by several phenomena:

- Stress free axial elongation due to irradiation growth.
- Anisotropic creep (before pellet/cladding contact) due to external reactor system pressure. Because of the tubing texture, axial elongation generally results from creep down of the cladding diameter; however for

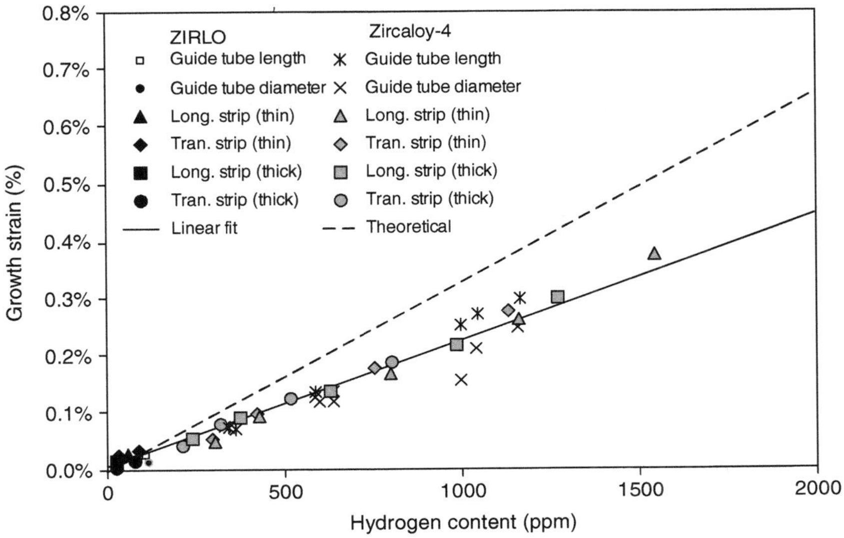

4.60 Dimensional changes in unirradiated ZIRLO and Zircaloy-4 tubing and strip for different sample orientations as a function of hydrogen content. (Source: Reprinted, with permission, from King *et al.* (2002), copyright ASTM International, 100 Barr Harbor Drive, West Conshohocken, PA 19428.)

heavily cold worked material, it has been reported that some shrinkage may occur. In a non-textured material such as SS, creep down of the cladding would only result in an increase in cladding thickness, with no change in length.

- Creep due to PCMI after hard contact between the cladding and fuel. This occurs in mid-life, depending on the cladding creep properties and the stability of the fuel.
- Hydriding of the cladding due to corrosion.

Bow of a component such as a BWR channel or PWR control rod assembly can occur if one side of the component changes length more than the other side. Such differential length changes occur due to differential stress and creep, to relaxation of differential residual stresses or to differential growth due to differences in flux-induced fluence, texture, material cold work and hydrogen content (and, although not usually present, differences in temperature or alloying content). This is described more in the *ZIRAT*10 Special Topics Report on Structural Behaviour of Fuel and Fuel Components (Cox *et al.*, 2005).

The next section discusses the effect of irradiation on dimensional stability.

4.6.1 Irradiation growth

Irradiation growth occurs simultaneously with irradiation creep if there is an applied stress. The two processes are considered to be independent and additive, even though they compete for the same irradiation-produced defects mechanistically. Earlier ZIRAT reviews providing more detail can be found in the ZIRAT7 STR (Adamson & Rudling, 2002) and the Fuel Material Technology Report, Vol. 2 (Rudling *et al.*, 2007).

Irradiation growth is a change in the dimensions of a zirconium alloy reactor component even though the applied stress is nominally zero. It is an approximately constant volume process, so if there is, for example, an increase in the length of a component, the width and/or thickness must decrease to maintain constant volume. Understanding of the detailed mechanism is still evolving; however a clear correlation of growth to microstructure evolution exists, and many empirical observations have revealed key mechanistic aspects. The inherent anisotropy of the Zr crystallographic structure plays a strong role in the mechanism, as materials with isotropic crystallographic structure (like stainless steel, copper, Inconel, etc.) do not undergo irradiation growth. It should not be confused with irradiation swelling, which does not conserve volume and does not occur in zirconium alloys under normal reactor operating conditions.

Irradiation growth is strongly affected by fluence, CW, texture, irradiation temperature and material chemistry (alloying and impurity elements).

Growth characteristics and rate

Figure 4.61 gives schematic growth curves for Zircaloy illustrating several points. Note that L-textured (longitudinal, or in the original rolling direction) material grows, while T-textured (transverse to the rolling direction) material shrinks; when taking into account shrinkage of a component in the third direction (N, normal to the rolling direction), this behaviour results in approximately constant volume. The long direction (L) of a component is the most important: for instance the length of a fuel rod, channel box or GT. Note that cold worked (CW or SRA) material grows at a high and almost linear rate, while recrystallized (RXA) material grows in a 3-stage process, with the final high rate being called 'breakaway' growth. The various stages can be directly correlated to the irradiation-produced microstructure described earlier. For RXA Zircaloy, at low fluences where only <a> component loops exist, growth is small (~0.1%) and saturates. When <c> component loops begin to appear the growth rate increases and becomes nearly linear with fluence in the range 6–10 × 10^{25} n/m^2, E >1 MeV. For L-texture material growth can reach 1% at 20 × 10^{25} n/m^2. In initially cold worked (CW) or stress relieved material (SRA), <c> component dislocations occur as part of the deformation-induced structure and more are formed during irradiation (Holt *et al.*, 1996). The growth

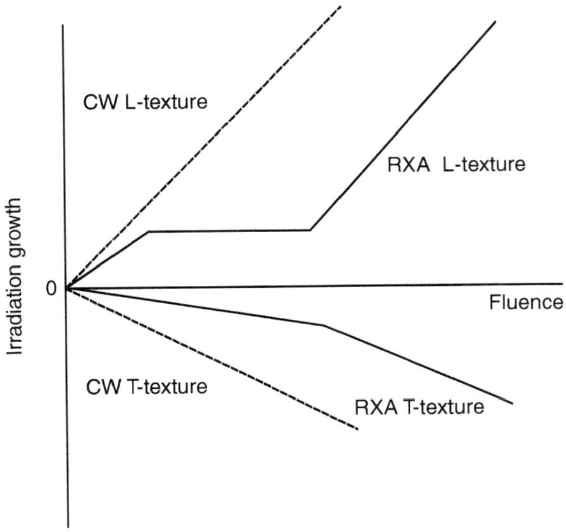

4.61 Schematic curves for irradiation growth as a function of fluence for recrystallized (RXA) and CW Zircaloy having textures characterized as L (f ≈ 0.1) and T (f ≈ 0.4) and an irradiation temperature near 300°C (573K).

rate is nearly linear with fluence and the magnitude is almost linear with the amount of initial CW. In heavily-worked material (typically 70–80% in a fuel rod) a growth of 2% can be reached by 20 × 10^{25} n/m^2 (corresponding to a burnup of about 100 MWd/kgU). Figure 4.62 gives some values of irradiation growth for Zircaloy materials of different heat treatments, reflecting the amount of residual CW and dislocation density. An overview of factors affecting growth is given by Fidleris *et al.* (1987).

Texture

It can be argued (Hesketh *et al.*, 1969; Alexander *et al.*, 1977) that the magnitude of growth strain in any given direction of a polycrystalline material can be related to the crystallographic texture and is proportional to a growth anisotropy factor G_d, given by

$$G_d = 1 - 3f_d^c, \qquad [4.1]$$

where f_d^c is the resolved fraction of basal poles, f^c, in the d-direction. The anisotropy factor depends on the assumptions that each grain behaves as an independent single crystal and that the volume change due to irradiation growth is zero.

At high burnup and high temperature (greater than about 360°C, 633K) and perhaps also in a heavily cold worked material, the familiar (1–3f) and

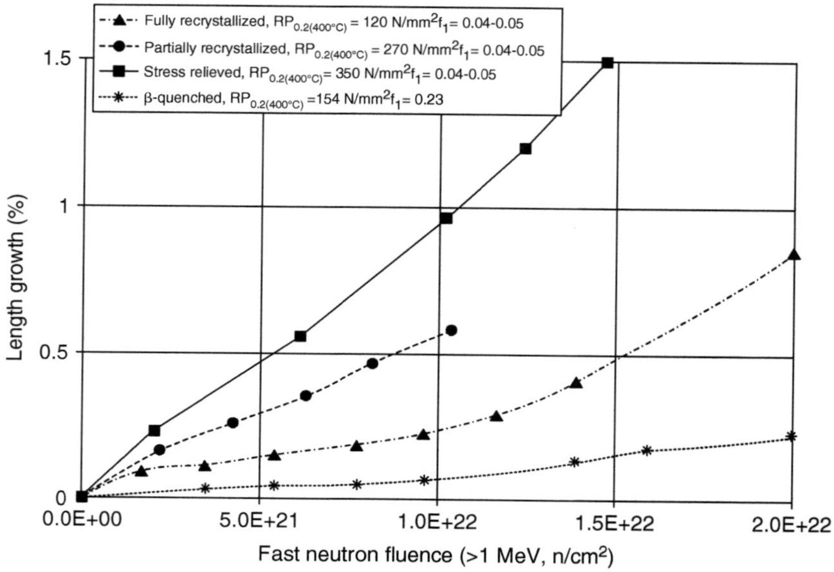

4.62 Irradiation growth of Zircaloy at 300°C measured on samples with different yield strength (CW, recrystallization) and different textures (f1). (Source: Reprinted, with permission, from Garzarolli *et al.* (1989, 1996), copyright ASTM International, 100 Barr Harbor Drive, West Conshohocken, PA 19428.)

constant volume assumptions may not be valid. At low fluence the two assumptions are reasonable, but at high fluence the transverse strain is not zero (as would be predicted by the f_x value) and the sum of the strains is strongly positive. It is also noted that cold worked material and recrystallized material have similar growth behaviour at high temperatures. It is further noted that the temperature of this irradiation is at the upper range (>378°C) expected for even a hot PWR.

For the high temperature data presented in Fig. 4.64, STEM studies revealed grain boundary cavities and occasional IG voids (Tucker *et al.*, 1984) which may explain the observed change of volume. Other studies have not reported cavities or voids at very high fluence at 290°C (363K) (Mahmood *et al.*, 2000) or high fluence at 350°C (623K). Holt & Causey (2004) reported that for Zr-2.5Nb there is a small volume increase (0.05–0.1%) at low fluence, but at high fluence the volume change was close to zero.

Materials chemistry of the alloy

Irradiation growth of RXA Zr-Nb alloys (E110, E635, NSF, M5, ZIRconium Low Oxidation (ZIRLO)) all exhibit a resistance to formation of <c> loops at low or intermediate fluence and as a result have lower growth than

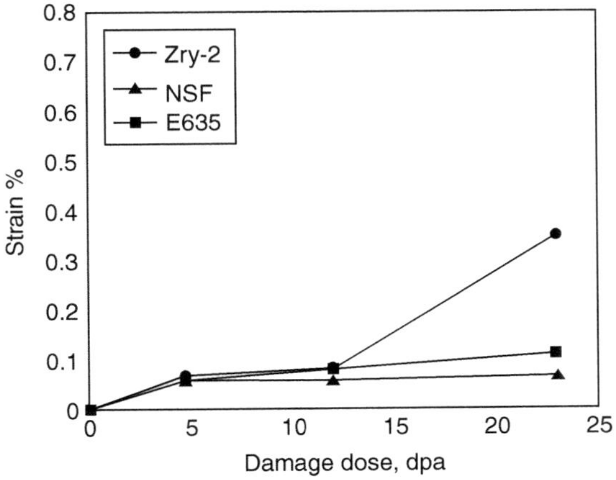

4.63 Irradiation growth of specimens at 320°C (593K) in the BOR 60 reactor. 20 dpa is equivalent to about 13 × 10^{21} n/cm^2 in a BWR/PWR. NSF and E635 alloys are nominally Zr1Nb1Sn0.35Fe alloys (Kobylyansky *et al.*, 2007).

Zircaloys. The fluence to breakaway is not yet well defined but is probably at least 1.5 × 10^{22} n/cm^2, E>1 MeV (see Fig. 4.63) corresponding to a burnup of about 75 MWd/kgU.

Temperature during irradiation

The temperature of a component during irradiation is an important variable. In a BWR, temperature variation along the length of the core is relatively small: water rods and spacers are near the temperature of the boiling water (288°C, 561K); fuel cladding material operates at slightly higher temperatures due to heat generation from the fuel and buildup of oxide and crud; but the range is between 288°C (561K) and 320°C (593K). However, in a PWR the components all operate with a substantial axial temperature gradient due to increase in the water temperature as it rises through the core. Depending on core design and duty, material temperatures could be as low as 280°C (553K) at the bottom and nearly 400°C (673K) at the top of the core. Therefore, the temperature dependence of irradiation growth and creep must be accounted for.

Growth as a function of temperature is not straightforward, as shown schematically in Fig. 4.65 (Holt, 1988) for RXA Zircaloy. In general for Zircaloy at low fluence, growth decreases with increasing temperature and at high fluences growth increases with temperature, with the critical temperature (T_1 in Fig. 4.65) being near 360°C (633K). It is seen that at less than about 2 × 10^{25} n/m^2 (E>1MeV) (which is before the region of breakaway growth and before c-component dislocations form) growth peaks at about 300°C (573K) and then steadily decreases at higher temperatures. At post-breakaway fluences, growth rate (the slope of the growth vs fluence

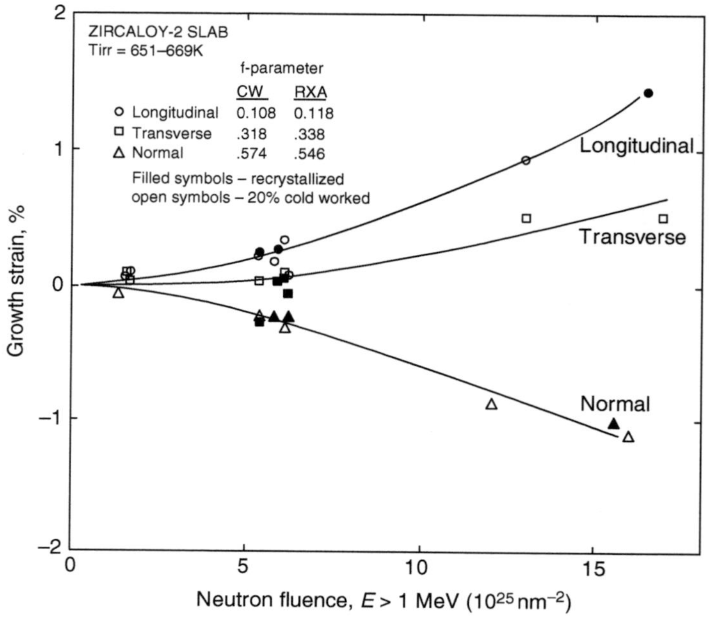

4.64 Dependence of growth on neutron fluence in three orthogonal directions at irradiation temperatures in the range 651–669K (Tucker *et al.*, 1984).

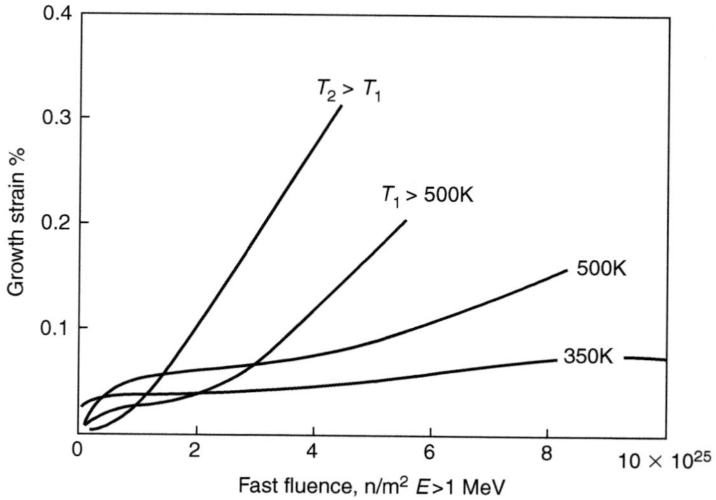

4.65 Schematic diagram showing the growth of annealed Zircaloy in the longitudinal direction (FL <0.1) as a function of temperature (Holt, 1988).

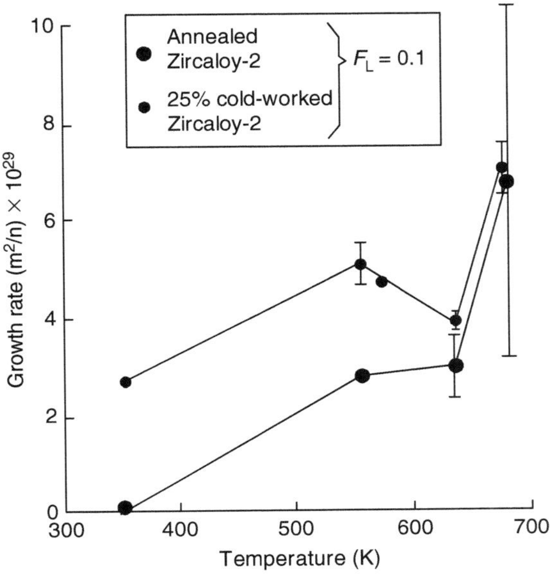

4.66 Temperature dependence of growth in Zircaloy-2 at high fluence (Rogerson, 1988).

curve) steadily increases with increasing temperature. Also, as indicated in Fig. 4.65, the fluence at which growth breakaway occurs gets smaller as temperature increases. Also note in Fig. 4.66 (Rogerson, 1988) that at temperatures above about 360°C (633K) the growth rate is the same for recrystallized (RXA) and cold worked (SRA or CWSRA) materials.

Temperature effect information is given by Gilbon & Simonot (1994) and Gilbon *et al.* (2000) and confirms that growth dramatically increases between the temperatures 350°C and 400°C (623–673K). Gilbon *et al.* (2000) also give data showing that growth of the M5 alloy at 5×10^{25} n/m^2 decreases from 0.08% at 280°C (553K) to 0.04% at 350°C (623K).

The effect of temperature on growth rates of cold-worked Zr-2.5Nb is given by Holt *et al.* (2000). In the range of interest for CANDU reactors, that is 290–317°C (563–590K), the growth rate decreases with increasing temperature at high or low fluences. This is slightly different and opposite to that observed for Zircaloy, but the range of temperature data available is smaller than for Zircaloy.

The practical implications of the discussed temperature dependence should be clear. For BWRs and CANDUs variations in growth should be small and predictable, at least for the alloys currently used, and for PWRs variation may be large and, particularly above about 350°C (623K), may be unpredictable with our current state of knowledge.

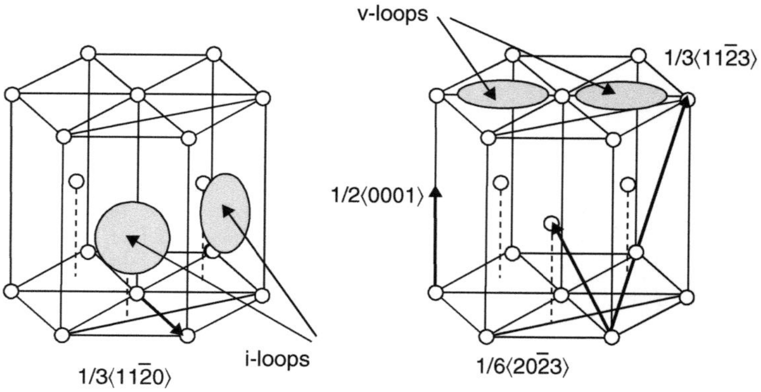

4.67 Schematics showing the irradiation growth process in a simplified manner (Holt, private communication).

Basic mechanism

The most simple view of the growth mechanism is that it is due to interstitial <a> loops (with displacement vectors or Burgers vectors pointing in the a-direction) lying on prism planes, and vacancy <c> loops (with displacement vectors or Burgers vectors pointing in the c-direction) lying on basal planes (Buckley 1961). Figure 4.67 gives a schematic illustration.

The simple vision is that planar arrays of vacancies cause shrinkage in the direction normal to the plane, and planar arrays of interstitials cause expansion. However, all vacancies and interstitials do not end up as loops. As discussed in earlier sections on irradiation creep, they can also be deposited at grain boundaries, solute atoms and network dislocations (that is, dislocations introduced by deformation rather than irradiation) having <a> or <c+a> Burgers vectors (i.e., line dislocations having a net strain in the <a> or <c+a> lattice directions). In the final analysis, irradiation growth is due to anisotropic deposition of vacancies and interstitials at these sinks and is strongly influenced by the anisotropic diffusion of self interstitial atoms (SIAs) in the basal plane, or more exactly, in the directions normal to the c-axis.

The mechanisms for irradiation growth parallel those for irradiation creep, with the notable exception of any applied stress.

> In general, growth is now believed to occur when there is an anisotropic distribution of sinks receiving a net flux of vacancies and this anisotropy is different from that of the distribution of sinks receiving a net flux of the self interstitial atoms (SIAs). The dilations associated with the addition of lattice planes accompanying the precipitation of SIAs then do not cancel the contractions associated with the removal of lattice planes associated with the condensation of vacancies. For a complete understanding of irradiation growth one must identify the possible sinks and explain their evolution, identify the source

> of the anisotropy and explain the partitioning of the PDs amongst the sinks. (Holt, 1988)

The key unique mechanistic feature of irradiation growth is the initiation and growth of <c> component dislocations (either as irradiation-produced loops or deformation-induced networks). Without them breakaway growth does not occur. It also appears to be essential that diffusion of SIAs be anisotropic and favour the directions in the basal plane. More details are given in Holt *et al.* (1996), Woo (1988), Holt *et al.* (2000), Holt (2008) and Christien and Barbu (2009).

Irradiation growth will occur to some extent in all zirconium alloys. To minimize the effects of growth several actions are important:

- minimize residual stresses,
- plan for the effects of cold work,
- minimize hydrogen absorption,
- plan for the effects of temperature,
- understand and control alloy chemistry,
- control texture.

4.6.2 Irradiation creep

Creep is plastic deformation occurring as a constant volume process, normally at low stresses below the yield stress. For materials in the irradiation field of a nuclear reactor, the deformation occurs by the motion of dislocations and irradiation-produced defects under the influence of stress. Neutron irradiation produces large quantities of point defects – vacancies and SIAs – which migrate to and collect at various sinks. Due to the anisotropy of the zirconium crystal lattice, motion of both dislocations and SIAs is anisotropic, preferring to occur parallel to the basal plane in the <a> direction of the lattice. Dislocations are sinks for both vacancies and SIAs, but it is normally considered that an edge dislocation attracts SIAs more than vacancies. Dislocations produced by deformation and by irradiation lie on both basal and prism planes. Because of the diffusional anisotropy of SIAs, they tend to be absorbed by dislocations lying on prism planes. The diffusion of vacancies is isotropic, and they tend to be absorbed preferentially by dislocations lying on basal planes. Similarly, SIAs tend to be absorbed at grain boundaries oriented parallel to prism planes and vacancies on boundaries parallel to basal planes. Absorption of either vacancies or SIAs at dislocations of grain boundaries causes plastic strain: positive for SIAs and negative for vacancies. If the absorptions occurred randomly and in a non-biased fashion, the net strain would be zero; however in zirconium alloys the built-in anisotropy results in separate positive and negative strains, and in constant volume deformation. Also, in addition to the natural anisotropy of the zirconium crystal lattice, another factor is the concept that anisotropic diffusion is enhanced by stress.

Basic mechanism

Many mechanisms of irradiation creep have been proposed. No single mechanism has been accepted as *the* dominating mechanism, and it seems very likely that several processes contribute simultaneously. The most comprehensive review of creep mechanisms is quite old (Matthews & Finnis, 1988) with newer reviews given by Holt (2008) and Adamson *et al.* (2009). The two most prominent mechanisms are stress induced preferential absorption (SIPA) and climb and glide. SIPA assumes a bias of the motion of vacancies and SIAs to dislocations depending on the orientation of the Burgers vectors with respect to the applied shear stress. There are several variations of SIPA, including the elasto-diffusion modification which invokes the effect of stress on the diffusion anisotropy itself. Elasto-diffusion appears to have the strongest effect on creep within the SIPA 'family'.

The most straightforward irradiation creep mechanism is the climb and glide mechanism, by which deformation-producing dislocations are aided in bypassing obstacles to their motion by irradiation-produced point defects. As long as an individual dislocation attracts a net flux of either vacancies or SIAs, it can 'climb around' a barrier and under the influence of an applied stress, glide to the next barrier, thereby producing strain and eventually causing a slip step at the material surface. The weak dependence of creep rate on dislocation density suggests that glide may not be the main strain-generating process.

A further contributor to the strain measured in a creep experiment is irradiation growth. Although not strictly in the 'creep' category because it occurs in the absence of an applied stress, it is inevitably measured as part of the overall strain in all in-reactor experiments, except for most bent beam stress relaxation tests. Irradiation growth results from mechanisms similar to irradiation creep in that it is dependent on the anisotropic properties of the zirconium crystal lattice. It was discussed in the previous section.

Parameters

Creep experiments in a neutron irradiation environment have been conducted since the early 1960s and continue today. Since creep without irradiation tends to have a relatively high initial rate, and since irradiation damage builds up with time, the creep curve (strain vs time or fluence) is usually divided into primary and secondary stages, as shown in Fig. 4.68. Whether or not a true 'steady state' creep rate is obtained in-reactor is problematic to prove, but in any case it is assumed for the analysis of the

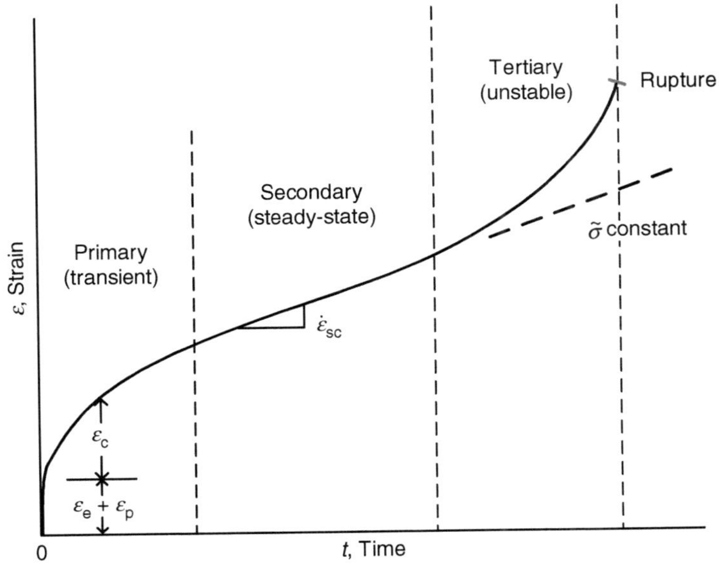

4.68 Strain vs time behaviour during creep under constant load, and the three stages of creep.

important parameters influencing creep strains and rates. A tertiary stage is not reached except in very rare, localized high stress cases.

In-reactor creep of any component or specimen consists of three parts: (1) thermal creep (which in most practical cases is small, and in all cases is different from thermal creep of unirradiated material), (2) true irradiation creep and (3) irradiation growth. The most common practice is to assume that the three parts are independent and additive, although from a mechanistic view this is questionable. In data analyses, only in-reactor creep has been analysed typically, without any separation of thermal and irradiation creep components; if possible, any irradiation growth should be subtracted from the experimental creep measurements.

The steady state creep rate, $\dot{\varepsilon}$, is most often expressed as a function of variables:

$$\dot{\varepsilon} = \frac{d\varepsilon}{dt} = f(\phi^p, \sigma^n, e^{-\frac{Q}{RT}}, f, G, \rho, A) \qquad [4.2]$$

and in-reactor strain, ε, is often expressed for a specific alloy and alloy condition as:

$$\varepsilon = A(\phi t)^m \sigma^n e^{-\frac{Q}{RT}} \qquad [4.3]$$

where p, n and m are constants and

ϕ is the neutron flux, n/m²/s or dpa/s
σ is the applied stress, MPa
t is the time, s
Q is the activation energy, cal/mole
Q/R is the activation temperature, K
R is the universal gas constant, 1.98 cal/K/mole
T is the temperature, K
f is a texture parameter
G is the grain size
ρ is a parameter related to the dislocation density and degree of CW
A is a metallurgical factor for a specific alloy.

Recent extensive reviews of creep data have been given by Holt (2008) and Adamson *et al.* (2009, noted below as AGP, 2009).

Using large amounts of data, it was determined (AGP, 2009) that in the temperature range 275–390°C (548–663K), with flux in the range $0.50–50 \times 10^{13}$ n/cm²/s, E > 1 MeV and independent of the direction of creep, material composition and material condition, the flux dependency, p, is 0.85. Purely irradiation creep mechanisms would predict $p = 1$, but the inevitable contributions of thermal creep tend to reduce the value below unity. For the very low neutron fluxes that occur at fuel bundle extremities, lower values of p would be expected. For the value of m, giving the fluence dependency, Soniak *et al.* (2002) obtained 'm' near 0.5.

Continued analyses conclude that the stress dependency of in-reactor creep is linear, $n = 1$, in the most relevant stress range of <10–200 MPa. This applies for temperatures in the range of 275–390°C (548–663K), independent of the direction of creep, flux, material composition and material condition. Mechanistic analyses, described briefly above, predict a range of n-values from 1 to higher than 2. SIPA and elastodiffusion predict a linear stress dependency, but the favoured climb and glide mechanisms predict higher values of n.

The temperature dependence of in-reactor creep also has components based on thermal and irradiation processes. It is generally assumed that pure irradiation creep has only a small temperature dependence (often the process is called 'athermal'), but thermal processes play an increasingly large role at temperatures above about 300°C (573K). At temperatures above about 350°C (623K) thermally produced vacancies and annealing of small vacancy clusters compete with irradiation-produced PDs, and thermal processes dominate in-reactor creep by 400°C (673K). In all cases increasing the temperature increases the creep rate. Analysis of a large amount of normalized data by Garzarolli (in AGP, 2009) indicates that temperature dependency in the most interesting range of 270–400°C varies and depends on several important variables include cold work, Sn content, alloy content and fabrication history. The lowest temperature dependency is ascribed to heavily cold-worked Zircaloy-2 and -4, while the highest is for Zr-2.5Nb CANDU pressure tubes. (It should be noted that the operating temperature of CANDU pressure tubes is generally below 310°C, 583K.)

Several metallurgical factors influence creep. Disturbances in the regularity of the lattice such as those caused by under- or over-sized atoms interfere with dislocation motion and act as sinks for irradiation-produced vacancies and SIAs. Of particular importance are the elements disturbing the regularity of the lattice and having appreciable solubility in Zr: Sn, Nb and O, all of which increase the creep strength (Seibold & Garzarolli, 2002). On the other hand, an element having low solubility, *S*, has been proposed to increase the creep strength of Zr-1Nb alloys (Soniak *et al.*, 2002; Mardon and Bordy, 2004; Rebeyrolle *et al.*, 2004). The mechanism is not yet well defined, but is proposed to be a dislocation interaction effect.

Cold working affects three metallurgical parameters: dislocation density, grain shape and texture (anisotropy), all of which can affect creep rate. The general observation is that cold worked or cold worked/stress relieved (CWSRA) zirconium alloys have higher creep rates than RXA alloys. An example is shown in Fig. 4.69 (Soniak, *et al.*, 2002). Mechanistic analyses are, once again, not firm on a quantitative effect of dislocation density, but most, including varieties of the SIPA mechanism, predict higher in-reactor creep rates with higher dislocation density, ρ. Stress relaxation experiments

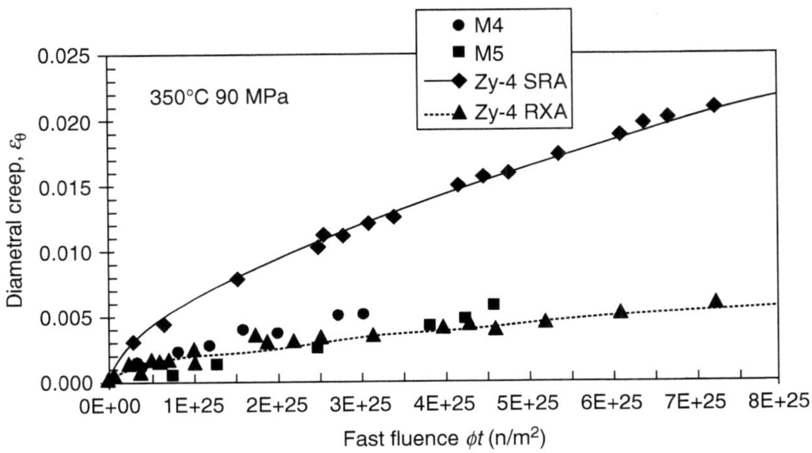

4.69 Effect of metallurgical condition and alloy composition on hoop creep strain vs fluence for SRA and RXA Zircaloy-4 and RXA M4 and M5. (Source: Reprinted, with permission, from Soniak *et al.* (2002), copyright ASTM International, 100 Barr Harbor Drive, West Conshohocken, PA 19428.)

suggest a moderate dependency ($\rho^{0.2}$), whereas for irradiation growth it is about $\rho^{0.8}$, summarized by Holt (2008).

Since the HCP crystal structure is anisotropic, many properties are expected to be anisotropic, and so it proves to be. The often-preferred slip (glide) direction is <a> (prism slip), as is the preferred diffusion direction of SIAs, both occurring parallel to the basal plane. The prevalence of basal slip increases in irradiated material, but even so prism slip predominates at low stress, so the climb and glide mechanism predicts higher creep strain in the <a> direction (in Zircaloy components this is called the longitudinal direction), as do conventional and modified SIPA mechanisms. Although texture (as defined by basal pole figures) is nearly the same for RXA and SRA materials, the anisotropy as expressed by anisotropy coefficients are different for the two. Forfuel rods, for example, creep in the rod results in a rod length decrease for SRA material and an increase for RXA material (AGP, 2009). Analyses confirm that for uniaxial stress conditions, creep of Zircaloy components in the longitudinal direction is greater than in the transverse direction (Fidleris, 1988). Zr-2.5Nb pressure tubes have a different texture to most Zircaloy components, so the anisotropy is different to that for typical Zircaloy components, but the basic principles are the same (Holt, 2008).

4.7 Future trends and research needs

- In PWRs it is found the Zircaloy-4 no longer meets corrosion and hydriding needs; therefore virtually all current PWR cladding use a zirconium alloy containing niobium. Table 4.11 lists materials currently being explored for use as cladding and structural materials, with the most widely used to date in the West, in addition to Zircaloy-4, being M5 and ZIRLO.
- For BWRs, Zircaloy-2 cladding, with various heat treatments to optimize the second phase precipitate size and distribution, remains the standard for BWR components. However, new channel materials are currently being explored to meet the challenges of channel bowing (described in the next chapter and in Garzarolli *et al.*, 2011a). Table 4.12 lists some of those materials and heat treatments currently under development and early usage.
- Other trends and needs are given in the appropriate section of the next chapter.

4.8 Sources of further information

Major sources are many, including:

- ZIRAT Annual Reports and ZIRAT Special Topic Reports, A.N.T. International, Mölnlycke, Sweden (www.antinternational.com).

Table 4.11 Material variations being used in or considered for PWR fuel bundle components

Alloy	Sn (%)	Nb (%)	Fe (%)	Cr (%)	Others	Cond.*	FF**	HPUF (%)***
Modif. Zry-4	1.3		0.3	0.2		PR/RX	~1.8/2.3	–/12
NDA	1	0.1	0.3	0.2		SR	~2.4	15/–
S2	0.8	0.1	0.3	0.1		RX	~2.2	
DX ELS 0.8B	0.8		0.3	0.2		SR	~1.1	–/24
DX D4	0.5		0.5	0.2		SR	~1	–/20
HPA-4	0.5		0.5		0.3V	SR/RX	~1/1.3	–/10
E635	1.3	1	0.4			RX		
ZIRLO	1	1	0.1			SR/RX	~2.4	15/–
MDA	0.8	0.5	0.2	0.1		SR	~2.4	15/–
Low-Sn-ZIRLO	0.7	1	0.1			SR	~1.4	15/–
Optim. ZIRLO	0.67	1	0.1			PR	~1.9	15/–
M-MDA	0.5	0.5	0.4	0.3		SR/RX	~1.4/2.5	
Quart-NbSnFe	0.5	1	0.1			RX	~2	–/>30
AXIOM X5A	0.45	0.3	0.35	0.25		50%PR	~1.6	
HANA-4	0.4	1.5	0.2	0.1		PR	~1.6	
AXIOM-X1	0.3	0.7–1	0.05		0.12Cu, 0.2V	80%PR	~1.3	
AXIOM-X5	0.3	0.7	0.35	0.25		50%PR	~0.9	
Quart-NbFeSn	0.3	1	0.1			RX	~1.4	10
Quart-NbFeSn	0.3	1	0.2			RX	~1.2	10
E110		1	0.01			RX	~0.8	≤10
M5		1	0.04		20ppm S	RX	~0.8	≤10
Quart-NbFe		1	01			RX	~0.6	
AXIOM-X2		1	0.06			RX	~1	
AXIOM-X4		1	0.06	0.25	0.05Cu	80%PR	~0.8	
HANA-6		1.1			0.05Cu	PR	~1.3	
J-Alloys		1.6–2.5		0–0.01	βNb	RX	~0.9	

* PR, partially recrystallized; RX, recrystallized; SR, stress relieved.
** Fitting factor (calculated by Garzarolli *et al.*, 2011a).
*** Hydrogen pickup fraction (calculated by Garzarolli *et al.*, 2011a).
Source: Garzarolli *et al.* (2011a).

Table 4.12 Material variations being used or considered for BWR channels

Material	Composition	Proposed advantage					
		C	SC	HPUF	HPUF HB	G	Reference
Zry-4	Zr-1.3Sn-0.2Fe-0.1Cr		x	x	x		Cantonwine *et al.*, 2008; AREVA
NSF	Zr-1Nb-1Sn-0.35Fe		x	x	x	x	Ledford *et al.*, 2010; Kobylyansky *et al.*, 2010
VB	Zr-0.5Sn-1Cr-0.5Fe			x	x	x	Vaidyanathan *et al.*, 2000
ZIRLO	Zr-1Nb-1Sn-0.1Fe		x	x	x	x	Helmersson & Dag, 2008
βQ Zry-4	Zr-1.3Sn-0.2Fe-0.1Cr		x			x	Sedano *et al.*, 2010
βQ Zry-2	Zr-1.3Sn-0.17Fe-0.1Cr					x	Dahlbäck *et al.*, 2005; Möckel *et al.*, 2008

Notes: Proposed advantage relative to current Zry-2; C – corrosion; SC – shadow corrosion; HPUF – hydrogen pickup fraction; HPUF HB – at high burnup; G – irradiation growth; βQ – beta quenched.
Source: A.N.T. International (2011) and Garzarolli *et al.* (2011a).

- The series of *Zirconium in the Nuclear Industry, International Symposiums*, ASTM International, West Conshohocken, PA, USA, held every 2–3 years.
- *Zirconium Production and Technology: The Kroll Medal Papers 1985–2010*, editor, R. B. Adamson, ASTM International RPS2, ASTM I, West Conshohocken, PA, USA, 2010.
- Proceedings of the LWR Fuel Performance Meeting/Top Fuel/WRFPM, held annually in the United States, Europe or Asia.
- References given in Section 4.10
- The next chapter of this book – 'Performance and Inspection of Zirconium Alloy Components in Nuclear Power Light Water Reactors' P. Rudling, ANT International, Mölnlycke, Sweden and R. B. Adamson, Zircology Plus, Fremont, CA, USA.

4.9 Acknowledgements

The authors sincerely thank our colleagues in the expert network staff of ANT International: Brian Cox, Friedrich Garzarolli, Charles Patterson and Alfred Strasser. Their discussions, their expertise, their comments and their contributions to the ZIRAT programme reports have greatly contributed to this chapter.

4.10 References

Adamson R., Cox B., Davies J., Garzarolli F., Rudling P. and Vaidyanathan S., '*Pellet-Cladding Interaction (PCI and PCMI)*', ZIRAT11, Special Topics Report, ANT International, Mölnlycke, Sweden, 2006.

Adamson R., Garzarolli F., Cox B., Strasser A. and Rudling P., '*Corrosion Mechanisms in Zirconium Alloys*', ZIRAT12 Special Topics Report, ANT International, Mölnlycke, Sweden, 2007.

Adamson R. B., 'Cyclic Deformation of Neutron Irradiated Copper', *Phil. Mag.* **17**, p. 681, 1968.

Adamson R. B., '*Effects of Neutron Irradiation on Microstructure and Properties of Zircaloy*', Zirconium in the Nuclear Industry; Twelfth International Symposium, ASTM STP 1354, pp. 15–31, West Conshohocken, PA, 2000.

Adamson R. B., '*Recovery of Irradiation Damage by Posts-Irradiation Thermal Annealing-Relevance to Hydrogen Solubility and Dry Storage Issues*', EPRI Technical Report 1013446, June 2006.

Adamson, R. B. editor, *Zirconium Production and Technology*, The Kroll Medal Papers 1975–2010, RPS2, ASTM International, 2010.

Adamson R. B. and Bell W. L., '*Effects of Neutron Irradiation and Oxygen Content on the Microstructure and Mechanical Properties of Zircaloy*', Microstructure and Mechanical Behaviour of Materials, Proceedings: International Symposiums, Xian, China, October, 1985, EMAS, pp. 237–246, Warley, UK, 1986.

Adamson R. B., Cox B., Garzarolli F., Strasser A., Rudling P. and Wikmark G., '*Corrosion of Zirconium Alloys'*, ZIRAT7 Special Topics Report, ANT International, Mölnlycke, Sweden, 2002.

Adamson R. B, Garzarolli F. and Patterson C., '*In-Reactor Creep of Zirconium Alloys*', ZIRAT14/IZNA9 Special Topical Report, ANT International, Mölnlycke, Sweden, 2009.

Adamson R. B, Garzarolli F., Patterson C., Rudling P. Strasser A. and Coleman K., '*ZIRAT15 Annual Report'*, ANT International, Mölnlycke, Sweden, 2010.

Adamson R. B., Lutz D. R. and Davies J. H., '*Hot Cell Observations of Shadow Corrosion Phenomena'*, Proceedings Fachtagung der KTG-Fachgruppe, Brennelemente und Kernbautelle, 29 February–1 March 2000, Forschungszentrum Karlsruhe, 2000.

Adamson R. B. and Rudling P., '*Dimensional Stability of Zirconium Alloys'*, ZIRAT7 Special Topics Report, ANT International, Mölnlycke, Sweden, 2002.

Adamson R. B., Wisner, S. B., Tucker, R. P. and Rand R. A., '*Failure Strain for Irradiated Zircaloy Based on Subsided Specimen Testing and Analysis'*, The Use of Small-Scale Specimens for Testing Irradiated Material, ASTM STP 888, W. R. Corwin and G. E. Lucas, Eds, American Society for Testing and Materials, Philadelphia, pp. 171–185, 1986.

Alexander W. K., Fidleris V. and Holt R. A., '*Zry-2 Pressure Tube Elongation at the Hanford N Reactor'*, Zirconiumin the Nuclear Industry, ASTM Spec. Techn. Publ. 633, A.L. Lowe and G.W. Parry, Eds, American Society for Testing and Materials, pp. 344–364, 1977.

Andersson B., '*The Enhanced Spacer Shadow Corrosion Phenomenon'*, Fachtagung der KTG-Fachgruppe 'Brennelemente und Kernbauteile', 29 February/1 March 2000, Forschungszentrum Karlsruhe, 2000.

Andersson B., Limbäck M., Wikmark G., Hauso E., Johnsen T., Ballinger R. G. and Nystrand A-C., '*Test Reactor Studies of the Shadow Corrosion Phenomenon'*, Zirconium in the Nuclear Industry: Thirteenth International Symposium, ASTM STP 1423, Moan G. D. and Rudling P., Eds, ASTM, 2002.

Averin S. A., Panchenko V. L., Kozlov A. V., Sinelnikov L. P. Shishov V. N. and Nikulina A. V., '*Evolution of Dislocation and Precipitate Structure in Zr Alloys under Long Term Irradiation'*, Zirconium in the Nuclear Industry: Twelfth International Symposium, ASTM STP 1354, G. P. Sabol and G. D. Moan, Eds, American Society for Testing and Material, West Conshohocken, PA, pp. 105–121, 2000.

Azzarto F. J., Baldwin E. E., Wiesinger F. W. and Lewis D. M., 'Unirradiated, In-Pile and Post-Irradiation Low Strain Rate Tensile Properties of Zircaloy-4', *J. Nucl. Mater.*, 30, pp. 208–218, 1969.

Bai J., 'Effect of Hydriding Temperature and Strain Rate on the Ductile-Brittle Transition in β Treated Zircaloy-4', *J. Nucl. Sci. Technol.*, **33**(2), pp. 141–146, February, 1996, Taylor and Francis, www.tandfonline.com.

Baily W. E., Marlowe M. O. and Probstle, R. A., '*Trends in Nuclear Fuel Performance'*, Proceedings of ANS International Topical Meeting LWR Fuel Perform., Orlando, FL, USA, April 21–24, 1985, 1-13-1-15, 1985.

Bajaj R., Kammenzind B. F. and Farkas D. M., '*Effects of Neutron Irradiation on the Microstructure of Alpha-Annealed Zircaloy-4'*, Zirconium in the Nuclear Industry: 13th International Symposium, ASTM STP 1423, G. D. Moan and P. Rudling, Eds, ASTM International, pp. 400–426, West Conshohocken, PA, 2002.

Bell W. L., *Zirconium in Nuclear Applications*, ASTM STP 551, ASTM, 199–200, 1974.

Bement A. L., Tobin J. C. and Hoagland R. G., '*Effects of Neutron Irradiation on the Flow and Fracture Behavior of Zircaloy-2*', Flow and Fracture, ASTM-STP 380, ASTM, 364–383, 1965.

Bertsch J., Valance S. and Zubler R., '*Crack Resistance Determination of Irradiated Fuel Cladding using the Cladding Tensile Fracture Test (CTFT)*', 2010 LWR Fuel Performance Meeting/Top Fuel/WRFPM, Orlando, Florida, paper 047, Sept. 2010.

Blavius D., Münch C-J. and Garner N., '*Dimensional Behavior of Fuel Channels – Update on the Operational Experiences and Evaluation Results*', KTG Jahrestagung Kerntechnik, Hamburg, 2008.

Bossis P., Thomazet J. and Lefebvre F., '*Study of the Mechanisms Controlling the Oxide Growth Under Irradiation: Characterization of Irradiated Zircaloy-4 and Zr-1Nb-O Oxide Scales*', Zirconium in the Nuclear Industry: Thirteenth International Symposium: ASTM STP 1423, G. D. Moan and P. Rudling, Eds, ASTM International, pp. 190–221, West Conshohocken, PA, 2002.

Bossis P., Verhaeghe B., Doriot S., Gilbon D., Chabretou V., Dalmais A., Mardon J. P., Blat M. and Miquet A., '*In PWR Comprehensive Study of High Burn-up Corrosion and Growth Behavior of M5 and Recrystallized Low-Tin Zircaloy-4*', 15th ASTM International Symposium: Zirconium in the Nuclear Industry – Sunriver, OR, June 25–27, 2007.

Bourdiliau B., *et al.*, '*Impact of Irradiation Damage Recovery during Transportation on the Subsequent Room Temperature Tensile Behavior of Irradiated Zirconium Alloys*', 16th International Symposium on Zirconium in the Nuclear Industry, ASTM, Chengdu, China, May, 2010.

Buckley S. N., '*Properties of Reactor Materials and the Effects of Irradiation Damage*', Butterworths, London, p. 443, 1961.

Cann C. D., Puls M. P., Sexton E. E. and Hutchings W. G., 'The Effect of Metallurgical Factors on Hydride Phases in Zirconium', *J. Nucl. Mater.* **126**, pp. 197–205, 1984.

Cantonwine P. E., Karve A. A., Lin Y. P., Mahmood S. T., White D. W. and Crawford D. C., '*GNF channel performance and success in mitigating channel distortion and cell friction*', 2008 Water Reactor Fuel Performance, Seoul, Korea, paper 8078, October, 2008.

Châtelain A., Anderson B., Ballinger R. G., and Wikmark G., '*Enhanced Corrosion of Zirconium Base Alloys in Proximity to Other Metals: the Shadow Effect*', Int. Topical Meeting on *LWR* Fuel Performance, Park City, UT, pp. 485–498, April, 2000.

Cheadle B. A., Ells C. E. and van der Kuur J., '*Plastic Instability in Irradiated Zr-Sn and Zr-Nb Alloys*', Zirconium in Nuclear Applications, ASTM STP 551, 370–384, 1974.

Chen J. S. F. and Adamson R. B., '*Observations of Shadow Phenomena on Zirconium Alloys*', Proceedings of the International Topical Meeting on LWR Fuel Performance, ANS, West Palm Beach, Florida, pp. 309–317, 1994.

Cheng B. and Adamson R. B., '*Mechanistic Studies of Zircaloy Nodular Corrosion*', Zirconium in the Nuclear Industry: Seventh International Symposium, ASTM STP 393, Adamson R. B. and Van Swam L. F. P., Eds, ASTM, Philadelphia, pp. 387–416, 1987.

Cheng B., Levin H. A., Adamson R. B., Marlowe M. O. and Monroe, V. L., '*Development of a Sensitive and Reproducible Steam Test for Zircaloy Nodular Corrosion*', Zirconium in the Nuclear Industry, Seventh International Symposium, ASTM

STP 939, Adamson R. B. and Van Swam L. F. P., Eds, ASTM, Philadelphia, pp. 257–283, 1987.

Cheng B-C., Krüger R. M. and Adamson R. B., '*Corrosion Behaviour of Irradiated Zircaloy'*, Proc. 10th Int. Symp. in the Nucl. Ind., ASTM STP 1245, pp. 400–418, 1994.

Christien F. and Barbu A., 'Cluster Dynamics Modeling of Irradiation Growth of Zirconium Single Crystals', *J. Nucl. Mater.*, **393**, pp. 153–161, 2009.

Chu H., Wu S. and Kuo R., 'Hydride Reorientation in Zircaloy-4 Cladding', *J. Nucl. Mater.*, **373**, pp. 319–327, 2008.

Chu H. C., Wu S. K., Kuo R. C. and Cheng S. C., '*Effect of Radial Hydrides on Mechanical Properties of Zircaloy-4 Cladding'*, Proc. Water Reactor Fuel Performance Meeting, Kyoto, Japan, October 2–6, 2005.

Coleman C. E., 'Cracking of Hydride-Forming Metals and Alloys', *Comprehensive Structural Integrity*, Vol. **6**, Environmentally Assisted Failure, F. Petit and P. Scott, Eds, chapter 6–03, pp. 103–161, Oxford, UK, 2003.

Coleman C. E. and Theaker J. R., '*Effect of Ingot Preparation Method on Irradiation Response of Zr-2.5Nb'*, Proc. 14th ASTM Symposium on Zirconium in the Nuclear Industry, Stockholm, 2004.

Comstock R. J., Schoenberger G. and Sabol G. P., '*Influence of Processing Variables and Alloy Chemistry on the Corrosion Behavior of ZIRLO Nuclear Fuel Cladding'*, Zirconium in the Nuclear Industry: 11th Int'l Symposium, ASTM STP 1295, E. R. Bradley and G. P. Sabol, Eds, American Society for Testing and Materials, pp. 710–725, 1996.

Cox B., 'Some Thoughts on the Mechanisms of In-Reactor Corrosion of Zirconium Alloys', *J. Nucl. Mater.*, **336**, pp. 331–368, 2005.

Cox B., Garzarolli F., Adamson R., Rudling P. and Strasser A., '*Fuel Material Technology Report Vol. I'*, ANT International, Mölnlycke, Sweden, 2006.

Cox B., Garzarolli F. and Rudling R., '*Corrosion of Zr-Nb Alloys'*, ZIRAT9 Special Topics Report, ANT International, Mölnlycke, Sweden, 2004.

Cox B., Garzarolli F., Strasser A. and Rudling P., '*Structural Behavior of Fuel and Fuel Channel Components'*, ZIRAT10 Special Topics Report, ANT International, Mölnlycke, Sweden, 2005.

Cox B. and Rudling P., '*Hydriding Mechanisms and Impact on Fuel Performance'*, ZIRAT5, Special Topics Report, ANT International, Mölnlycke, Sweden, 2000.

Dahlbäck M, Limbäck M., Hallstadius L., Barberis P., Bunel G., Simonot C., Andersson T., Askeljung P., Flygar J., Lehtinen B. and Massih A. R., 'The Effect of Beta Quenching in Final Dimension on the Irradiation Growth of Tubes and Channels', *J. ASTM Int.*, **2**, June 2005, paper JAI12337, 2005.

Daum RS., Majumdar S., Liu Y. and Billone MC., '*Mechanical Testing of High-Burnup Zircaloy-4 Fuel Cladding under Conditions Relevant to Drying Operations and Dry-Cask Storage*', Proc. Water Reactor Fuel Performance Meeting, Kyoto, Japan, pp. 498–531, October 2–6, 2005.

Davies P. H., Hosbons R. R., Griffiths M. and Chow C. K., '*Correlation Between Irradiated and Unirradiated Fracture Toughness of Zr-2.5Nb Pressure Tubes*', Zirconium in the Nuclear Industry: Tenth International Symposium, ASTM STP 1245, A. M. Garde and E. R. Bradley, Eds, American Society for Testing and Materials, Philadelphia, 135–167, 1994.

Doriot Sylvie, Gilbon Didier, Béchade Jean-Luc, Mathon Marie-Hélène, Legras Laurent, and Mardon Jean-Paul, 'Microstructural Stability of M5TM Alloy

Irradiated up to High Neutron Fluences', *J. ASTM Int.*, **2**(7), paper JAI12332, July/August, 2004.

Ells C. E., 'Hydride Precipitates in Zirconium Alloys', *J. Nucl. Mater.*, **28**, pp. 129–151, 1968.

Etoh Y. and Shimada S., 'Neutron Irradiation Effects on Intermetallic Precipitates in Zircaloy as a function of Fluence', *J. Nucl. Mater.*, **200**, pp. 59–69, 1993.

Fidleris V., 'The Irradiation Creep and Growth Phenomena', *J. Nucl. Mater.*, **159**, pp. 22–42, 1988.

Fidleris V., Tucker R. P. and Adamson R. B., '*An Overview of Microstructural and Experimental Factors That Affect the Irradiation Growth Behaviour of Zirconium Alloys*', Zirconium in the Nuclear Industry: 7th International Symposium, Adamson R. B. and Von Swam, Eds, ASTM STP 939, 49–85, 1987.

Franklin D. G., 'Performance of Zirconium Alloys in Light Water Reactors with a Review of Nodular Corrosion', *J. ASTM Intl.*, **7**(6). Doi:10.1520/JAI103032, 2010.

Fregonese M., Régnard C., Rouillon L., Magnin T., Lefebvre F. and Lemaignan C., '*Failure Mechanisms of Irradiated Zr Alloys Related to PCI: Activated Slip Systems, Localized Strains, and Iodine-Induced Stress Corrosion Cracking*', Zirconium in the Nuclear Industry: Twelfth Int'l Symposium, ASTM STP 1354, G. P. Sabol and G. D. Moan, Eds, American Society for Testing and Materials, West Conshohocken, PA, 377–398, 2000.

Fuketa T. *et al*, '*Effects of Pellet Expansion and Cladding Hydrides on PCMI Failure of High Burnup LWR Fuel During Reactivity Transients*', Nuclear Safety Research Conference, Washington, DC, October, 2003.

Fukuya *et al.*, '*BWR Fuel Channel Performance and Localized Corrosion at High Burnups*', Proc. ANS LWR Fuel Performance Conference, pp. 580–586, Palm Beach, 1994.

Garde A. M., Patti S. R., Krammen M. A., Smith G. P. and Endter R. K., '*Corrosion Behavior of Zircaloy-4 Cladding with Varying Tin Content in High-Temperature Pressurized Water Reactors*', Zirconium in the Nuclear Industry: 10th Int'l Symposium, ASTM STP 1245, pp. 760–778, A. M. Garde and E. R. Bradley, Eds, ASTM, Philadelphia, 1994.

Garzarolli F, Adamson R., and Rudling P., '*Optimization of BWR Fuel Rod Cladding Condition for High Burnups*', 2010 ANS LWR Fuel Performance Meeting, Paper 069, Orlando, Florida, 2010.

Garzarolli F., Adamson R. B., Rudling P. and Strasser A., '*BWR Fuel Channel Distortion*', ZIRAT16 Special Topical Report, ANT International, Mölnlycke, Sweden, 2011a.

Garzarolli F, Cox B., and Rudling P., '*Optimization of Zry-2 for High Burnups*', Zirconium in the Nuclear Industry: 16th International Symposium, ASTM STP 1529, M. Limback and P. Barberis, Eds, American Society for Testing and Material, West Conshohocken, PA, pp. 711–727, 2011b.

Garzarolli F., Dewes P., Maussner G. and Basso H-H., '*Effects of High Neutron Fluences on Microstructure and Growth of Zircaloy-4*', Zirconium in the Nuclear Industry: Eighth International Symposium, ASTM STP 1023, 641–657, 1989.

Garzarolli F., Hoffmann P. B. and Seibold A., 'Shadow Corrosion or Crevice Corrosion', *J. Nucl. Mater.*, **289**, pp. 338–341, 2001.

Garzarolli F., Ruhmann H. and Van Swam L., '*Alternative Zr Alloys with Irradiation Resistant Precipitates for High Burnup BWR Application*', Zirconium in the Nuclear Industry: Thirteenth International Symposium, ASTM STP 1423, G. Moan and P. Rudling, Eds, ASTM International, pp. 119–132, West Conshohocken, PA, 2002.

Garzarolli F., Schumann R. and Steinberg E., '*Corrosion Optimized Zircaloy for Boiling Water Reactor (BWR) Fuel Elements*', Zirconium in the Nuclear Industry, Tenth Int'l Symposium, ASTM STP 1245, A. M. Garde and E. R. Bradley, Eds, American Society for Testing and Materials, Philadelphia, pp. 709–723, 1994.

Garzarolli F., Stehle H. and Steinberg, E., '*Behavior and Properties of Zircaloys in Power Reactors: A Short Review of Pertinent Aspects in LWR Fuel*', Zirconium in the Nuclear Industry: 11th Int'l Symposium, ASTM STP 1295, Bradley E. R. and Sabol G. P., Eds, American Society for Testing and Materials, pp. 12–32, 1996. (Also, Adamson *et al.* 2006).

Gilbon D. and Simonot C., '*Effect of Irradiation on the Microstructure of Zircaloy-4*', Zirconium in the Nuclear Industry: 10th Int'l Symposium, ASTM STP 1245, A. M. Garde and E. R. Bradley, Eds, ASTM, Philadelphia, pp. 521–548, 1994.

Gilbon D., Soniak A. Doriot S., and Mardon J. P., '*Irradiation Creep and Growth Behavior, and Microstructural Evolution of Advanced Zr-base Alloys*', Zirconium in the Nuclear Industry, Twelfth International Symposium, ASTM STP 1354, ASTM, West Conshohocken, PA, pp. 51–73, 2000.

Griffiths M., Gilbert R. W. and Carpenter C. J. C., 'Phase Instability, Decomposition and Redistribution of Intermetallic Precipitates in Zircaloy-2 and -4 During Neutron Irradiation', *J. Nucl. Mater.*, **150**, pp. 53–66, North-Holland, Amsterdam, 1987.

Griffiths M., Mecke J. F. and Winegar J. E., '*Evolution of Microstructure in Zirconium Alloys During Irradiation*', Zirconium in the Nuclear Industry: 11th Int'l Symposium, ASTM STP 1295, E. R. Bradley and G. P. Sabol, Eds, Am. Soc. Test. Mater., pp. 580–602, 1996.

Helmersson S. and Dag A., '*BWR Fuel Design Modifications to Enhance Reliability*', Water Reactor Fuel Performance Meeting Paper No. 8075, Seoul, Korea, 19–23 October 2008.

Hesketh R. V., Harbottle J. E., Waterman N. A. and Lobb R. C., in Radiation Damage in Reactor Materials, Vol. 1, pp. 365-, IAEA, Vienna, 1969.

Hobbs L. W., Clinard F. W., Jr., Zinkle S. J. and Ewing R. C., 'Radiation Effects in Ceramics', *J. Nucl. Mater.*, **216**, pp. 291–321, 1994.

Hoffmann P. B. and Manzel R., '*Shadow Corrosion. Schönheitsfehler oder Gefahr*', Proc. Jahrestagung Kerntechnik 99, pp. 395–398, Karlsruhe, Germany, 1999.

Holt R. A., 'Mechanisms of Irradiation Growth of Alpha-Zirconium Alloys', *J. Nucl. Mater.*, **159**, pp. 310–338, 1988.

Holt R. A., 'In-reactor Deformation of Cold-worked Zr-2.5Nb Pressure Tubes', *J. Nucl. Mater.*, **372**, pp. 182–214, 2008.

Holt R. A. and Causey A. R., 'Volume Conservation during Irradiation Growth of Zr-2.5Nb', *J. Nucl. Mater.*, **335**, pp. 529–533, 2004.

Holt R. A., Causey A. R., Christodoulou N., Griffiths M., Ho E. T. C. and Woo C. H., '*Non-Linear Irradiation Growth of Cold-Worked Zircaloy-2*', Zirconium in the Nuclear Industry: 11th Int'l Symposium, ASTM STP 1295, E. R. Bradley and G. P. Sabol, Eds, ASTM, pp. 623–637, 1996.

Holt R. A., Causey A. R., Griffiths M. and Ho E. T. C., '*High Fluence Irradiation Growth of Cold-Worked Zr-2.5Nb*', Zirconium in the Nuclear Industry; Twelfth International Symposium, ASTM STP 1354, ASTM, West Conshohocken, PA, pp. 86–104, 2000.

Hong S. I. and Lee K. W., 'Stress-Induced Reorientation of Hydrides and Mechanical Properties of Zircaloy-4 Cladding Tubes', *J. Nucl. Mater.*, **340**, pp. 203–208, 2005.

Howlader M. M. R., Kinoshita C., Shiiyama K., Kutsuwada M. and Inagaki M., 'In situ Measurement of Electrical Conductivity of Zircaloy Oxides and Their Formation Mechanism under Electron Irradiation', *J. Nucl. Mater.*, **265**, pp. 100–107, 1999.

Huang P. Y., Mahmood S. T. and Adamson R. B., '*Effects of Thermomechanical Processing on In-Reactor Corrosion and Post-Irradiation Mechanical Properties of Zircaloy-2I'*, Zirconium in the Nuclear Industry: Eleventh International Symposium, ASTM STP 1295, E. R. Bradley and G. P. Sabol, Eds, ASTM, pp. 726–757, 1996.

Ishimoto S., Etoh Y.i, Matsumoto T., Lutz D. and Takagi A., '*Improved Zr Alloys for High Burnup BWR Fuel',* Top Fuel 2006, IM *LWR* FP, Salamanca, Spain, pp. 318–329, October 2006.

Ito K., Kamimura K. and Tsukuda Y., '*Evaluation of Irradiation Effect on Spent Fuel Cladding Creep Properties*', Proceedings of the 2004 International Meeting on LWR Fuel Performance, pp. 1117, Orlando, FL, 19–22 September, 2004.

Ito K., Shimada S., Levin H., Adamson R. B., Chen J. S. F., Oguma M., Cheng B., Ikeda T., Takei K. and Ishii Y., '*Effects of Water Chemistry Impurities on Corrosion of Zr-Alloys under BWR Condition*', Proc. ANS Int'l Topical Meeting LWR Fuel Perform.; West Palm Beach, FL, USA, 17–21 April, 1994, 273–285, 1994.

Johnson A. B., Lesurf J. E. and Proebstle R. A., '*Study of Zr Alloy Corrosion Parameters in Advanced Test Reactor*', Zirconium in the Nuclear Industry, ASTM STP 551, pp. 495–512, 1974.

Kang T. K., Kuk I–H., Katano Y., Igawa N., Ohno H., *J. Nucl. Mater.*, **209**, p. 321, 1994.

Kim Y-J. *et al.*, '*Photoelectrochemical Investigation of Radiation-Enhanced Shadow Corrosion Phenomenon'*, 16th International Symposium on Zirconium in the Nuclear Industry, Chengdu, China, 2010.

King S. J., Kesterson R. L., Yueh K. H., Comstock R. J., Herwig W. M. and Ferguson S. D., '*Impact of Hydrogen on Dimensional Stability of ZIRLO™ Fuel Assemblies*', ASTM 13th Int'l Conference on Zirconium in the Nuclear Industry, ASTM STP 1423, 471–489, Annecy, France, 10–14 June 2002.

Kobylyansky G. P., Novoselov A. E., Ostrovsky Z. E., Obukhov A. V., Shishin V., Shishov V. N., Nikulina A. V., Peregud M. M., Mahmood S.T., White D., Lin Y-P. and Dubecky M., '*Irradiation-Induced Growth Strain and Microstructure of Recrystallized, Cold-Worked and Quenched Alloys Zry-2, NFS and E635*', 15th Symposium 'Zirconium in Nuclear Industry', Sunriver, OR, USA, 24–26 June 2007.

Kruger R. M., '*Precipitate Stability in Zircaloy-2*', EPRI NP 6845-D, 1990.

Kruger R. M., Adamson R. B. and Brenner S. S., 'Effects of Microchemistry and Precipitate Size on Nodular Corrosion Resistance of Zircaloy-2', *J. Nucl. Mater.*, **189**, pp. 193–200, 1992.

Kupalov-Yaropolk A. I., Nikolaev V. A., Cherkashov Yu. M., Pahyushkin A. K. and Fedosov A. M., '*RMBK Fuel Assemblies: Current Status and Perspectives*', Proc. IAEA Tech. Comm. Meeting Vienna, Dec. 1996, IAEA-TECDOC-997, pp. 143–149, 1998.

Ledford K. L. *et al.*, 'GNF Defense in Depth 2010 Update', Proceedings of 2010 LWR Fuel Performance conference, Paper 106, Orlando, Florida, USA, 2010.

Lefebvre F. and Lemaignan C., 'Irradiation Effects on Corrosion of Zirconium Alloy Cladding', *J. Nucl. Mater.*, **248**, pp. 268–274, 1997.

Lemaignan C., 'Impact of Beta Radiolysis and Transient Products on Irradiation-Enhanced Corrosion of Zr Alloys', *J. Nucl. Mater.*, **187**, pp. 122–130, 1992.

Lysell G., Nystrand A–C. and Ullberg M., '*On the Shadow Corrosion Mechanism for Zirconium Alloys*', TopFuel 2001, P2–13, Stockholm, Sweden, 2001.

Lysell G., Nystrand A-C., and Ullberg M., 'Shadow Corrosion Mechanism of Zircaloy', *J. ASTM Int.*, **2**(6), Paper JAI12374, 2005.

Mahmood S. T., Cantonwine P. E., Lin Y-P. and Crawford D. C., '*Shadow Corrosion-Induced Bow of Zircaloy-2 Channels*', 16th International Symposium on Zirconium in the Nuclear Industry, Chengdu, China, 2010.

Mahmood S. T., Edsinger K. W., Farkas D. M. and Adamson R. B., '*Effects of SPP Dissolution on Mechanical Properties of Zircaloy-2*', Proc., Int. Topical Meeting on LWR Fuel Performance, ANS, Portland, pp. 440–451, 2–6 March, 1997.

Mahmood S. T., Farkas D. M., Adamson R. B. and Etoh Y., *Post Irradiation Characterization of Ultra-High-Fluence Zircaloy-2 Plate*, Zirconium in the Nuclear Industry: Twelfth International Symposium, ASTM STP 1354, Sabol G. P. and Moan G. D., Eds, American Society for Testing and Materials, West Conshohocken, PA, pp. 139–169, 2000.

Mahmood S.T., Lin Y-P, Dubecky M. A., Edsinger K. and Mader E. V., '*Channel Bow in Boiling Water Reactors – Hot Cell Examination Results and Correlation to Measured Bow*', Proc. of the 2007 International LWR Fuel Performance Meeting, Paper 1061, American Nuclear Society, San Francisco, CA, pp. 124–133, 2007.

Mardon Jean Paul and Bordy Michel, '*Performance of M5™ Alloy as Cladding Tube and Structural Component Material*', Proc: 14th Pacific Basin Nuclear Conference, Honolulu, Hawaii, 21–25 March 2004.

Marlowe M. O., Armijo J. S., Cheng B. and Adamson R. B., '*Nuclear Fuel Cladding Localised Corrosion*', Proc. ANS Int'l Topical Meeting LWR Fuel Perform., Orlando, FL, 3–73–3–90, 21–24 April 1985.

Matthews J. R. and Finnis M. W. 'Irradiation Creep Models – An Overview', *J. Nucl. Mater.*, **159**, pp. 257–285, 1988.

Miyashita T. *et al.*, '*Corrosion and Hydrogen Pick-Up Behavior of Zircaloy Component of BWR High Burnup 9 X 9 Lead Use Assemblies*', IAEA Technical Meeting on High Burnup Fuel Experience and Economics, Sofia, Bulgaria, 2006.

Miyashita T. *et al.*, '*Corrosion and Hydrogen Pick-Up Behaviors of Cladding and Structural Components in BWR High Burnup 9x9 Lead Use Assemblies*', Proceedings of the 2007 International LWR Fuel Performance Meeting, Paper 1015, San Francisco, California, 2007.

Möckel A, Cremer I., Kratzer A. and Walter D., '*Controlled Beta Quenching of Fuel Channels Using Inert Gas*', KTG Jahrestagung Kerntechnik, Hamburg, 2008.

Morize P., Baicry J. and Mardon J. P., '*Effect of Irradiation at 588°K on Mechanical Properties and Deformation Behavior of Zirconium Alloy Strip*', Zirconium in the Nuclear Industry: Seventh Int'l Symposium, ASTM STP 939, R. B. Adamson and L. F. P. Van Swam, Eds, American Society for Testing and Materials, Philadelphia, pp. 101–119, 1987.

Motta A., 'Amorphization of Intermetallic Compounds Under Irradiation – A Review', *J. Nucl. Mater.*, **244**, pp. 227–250, 1997.

Motta A. and Lemaignan C., 'A Ballistic Mixing Model for the Amorphization of Precipitates in Zircaloy under Neutron Irradiation', *J. Nucl. Mater.*, **195**, pp. 277–285, 1992.

Münch C., Blavius D. and Hoffmann P., *Ursachanalyse zu Erhöhten Biegung von Brennelementen*, KTG Jahrestagung Kerntechnik, Hamburg, 2008.

Nagase F. and Fuketa T., 'Investigation of Hydride Rim Effect on Failure of Zircaloy-4 Cladding with Tube Burst Test', *J. Nucl. Sci. Technol.*, **42**, pp. 58–65, 2005, Taylor and Francis, www.tandfonline.com.

Nikulina A. V., Markelov V. A., Peregud M. M., Bibilashvili Yu. K., Kotrekhov V. A., Lositsky A. F., Kuzmenko N. V., Shevnin Yu. P., Shamardin V. K., Kobylyansky G. P. and Novoselov A. E., '*Zirconium Alloy E635 as a Material for Fuel Rod Cladding and Other Components of VVER and RBMK Reactors*', Zirconium in the Nuclear Industry: Eleventh International Symposium, ASTM STP 1295, E. R. Bradley and G. P. Sabol, Eds, American Society for Testing and Materials,Garmisch-Parkenkirchen, Germany, pp. 785–804, 1996.

Nikulina A., Shishov S., Cox B., Garzarolli F. and Rudling P., '*Manufacturing of Zr-Nb Alloys*', ZIRAT11 Special Topics Report, ANT International, Mölnlycke, Sweden, 2006.

Nystrand A-C. and Bergquist H., '*Investigation of <c> Component Dislocation Density in Irradiated ZIRLO and Zry-4*', Zirconium in the Nuclear Industry: Twelfth International Symposium, ASTM, Toronto, Canada, Poster Presentation, 2000.

Onimus F., Bechade J-L., Doriot S., Monnet I., Gilbon D., Verhaegue B., Prioul C., Pilvin P. H., Robert L., Legras L. and Mardon J-P., 'Plastic Deformation of Irradiated Zirconium Alloys: TEM Investigation and Micro-Mechanical Modeling', *J. ASTM Int.*, **2**(8), Paper JAI12424, September, 2005.

Onimus F., Monnet I., Bechade J-L., Prioul C. and Pilvin P. H., 'A Statistical TEM Investigation of Dislocation Channeling Mechanism in Neutron Irradiated Zirconium Alloys', *J. Nucl. Mater.*, **328**, 165–179, 2004.

Pierron O. N., Koss D. A., Motta A. T. and Chan K. S., 'The Influence of Hydride Blister on Fracture of Zircaloy-4', *J. Nucl. Mater.*, **322**, pp. 21–35, 2003.

Potts G. A., '*Recent GE Fuel Experience*', International Topical Meeting on Light Water Fuel Performance, American Nuclear Society, Park City, Utah, April 10–13, 2000.

Rebeyrolle V., Mardon J.-P., Charquet D., Soniak A., Robert L. and Bouffioux P., '*Influence of Sulfur Addition on Irradiation Creep of M5 Alloy*', Poster,14th International Symposium on Zirconium in the Nuclear Industry, Stockholm, Sweden, June, 2004.

Regnard C., Verhaeghe B., Lefebvre- Joud F. and Lemaignan, C., '*Activated Slip Systems and Localized Straining of Irradiated Zr Alloys in Circumferential Loadings*', Zirconium in the Nuclear Industry: Thirteenth International Symposium, ASTM 1423, Annecy, France, pp. 384–399, 11–14 June 2001.

Ribis J., Onimus F., Bechade J-L., Doriot S., Cappelaere C., Lemaignan C., Barbu A. and Rabouille O., 'Experimental and Modelling Approach of Irradiation Defects Recovery in Zirconium Alloys Impact of Applied Stress', (ASTM Zr meeting, Sunriver, June, 2007), *J. ASTM Int.*, June, 2007.

Rickover, H. G., '*The Decision to Use Zirconium in Nuclear Reactors*', Zirconium Production and Technology: The Kroll Medal Papers, ASTM RPS2, R. B. Adamson, Ed., ASTM International, West Conshohocken,PA, USA, p. 9, 2010.

Rogerson A., 'Irradiation Growth of Annealed and 25% Cold Worked Zircaloy-2 Between 353 and 673K', *J. Nucl. Mater.*, **154**, pp. 276–285, 1988.

Rudling P., Adamson R., Bairiot H., Cox B., Ford P. Garzarolli F., Riess R. and Strasser A., '*Fuel Material Technology Report, Vol. II*', ANT International, Mölnlycke, Sweden, 2007.

Sabol G. P., Comstock R. J., Weiner R. A., Larouere P. and Stanutz, R. N., '*In-Reactor Corrosion Performance of ZIRLO and Zircaloy-4*', Proceedings of the 10th International Symposium on Zirconium in the Nuclear Industry, ASTM STP 1245, Baltimore, MD, pp. 724–744, 1994.

Schemel J. H., '*Zirconium Alloy Fuel Clad Tubing, Engineering Guide*', 1st Edition, Sandvik Special Metals Corporation, December 1989.

Sedano P. J. G., Ayuela J.I. and Albendea M., '*Shadow Corrosion Behavior of BWR Channels in Cofrentes NPP*', 2010 LWR Fuel Performance, Paper 002, Orlando, FL, 26–29 September 2010.

Seibold A. and Garzarolli F., '*Influence of Composition and Condition on Behaviour of Zr-Sn-FeCrV Alloys*', Zirconium in the Nuclear Industry: 13th International Symposium, 10–14 June 2001, Annecy, France, ASTM STP 1423, Eds, American Society for Testing and Materials, West Conshohocken, PA, pp. 743–757, 2002.

Seibold A., Garzarolli F. and Manzel R., '*Material Development for Siemens Fuel elements*', Proceedings: International Topical Meeting on Light Water Reactor Fuel Performance, American Nuclear Society, 10–13 April, Park City, Utah, 2000.

Shannon D. W., '*Electrical Properties of ZrO_2 Corrosion Films; Electrical Resistance Measurements*', Proceeding of USAEC Symposium on Zirconium Alloy Development, Castlewood, CA, Vol. II, Paper 18, 1962.

Sharp J. V., *Radiat. Eff.* **14**, pp. 71, 1972.

Shi S-Q. and Puls M. P., 'Fracture Strength of Hydride Precipitates in Zr-2.5 Nb Alloys', *J. Nucl. Mater.*, **276**, pp. 312–317, 1999.

Shimada S., Asahi K., Sasaki M., Kogai T., Hayashi H., Kitamura M., Sakamoto M. and Yamawaki M., '*Evaluation of Zircaloy-2 Cladding Corrosion Characteristics by Simulated BWR corrosion Loop Test*', Zirconium in the Nuclear Industry – 12th International Symposium, ASTM STP 1354, pp. 735–755, 2002.

Shimada S., Ito K., Lin C. C., Cheng B., Ikeda T., Oguma M., Takei K., Vitanza C., and Karlsen T. M., '*Parametric Tests of the Effects of Water Chemistry Impurities on Corrosion of Zr-Alloys under Simulated BWR Condition*', Proc. TCM 'Influence of Water Chemistry on Fuel Cladding Behaviour', Rez. Czech Republic, 4–8 October 1993, IAEA-TECDOC-927, pp. 73–89, February 1997.

Shishov V. N., '*The Evolution of Microstructure and Deformation Stability in Zr-Nb-(Sn, Fe) Alloys under Neutron Irradiation*', 16th International Symposium on Zirconium in the Nuclear Industry, ASTM, Chendu, China, 9–13 May, STP 1529, M. Limback and P. Barberis, Eds, ASTM I, West Conshohocen, PA, p. 3766, 2011.

Shishov V. N., Nikulina A. V., Markelov V. A., Peregud M. M., Kozlov A. V., Averin S. A., Kolbenkov S. A. and Novoselov, A. E., 'Influence of Neutron Irradiation on Dislocation Structure and Phase Composition of Zr-Base Alloys', Zirconium in the Nuclear Industry: 11th International Symposium, ASTM STP 1295, E. R. Bradley and G. P. Sabol, Eds, ASTM, pp. 603–622, 1996.

Shishov V. N., Nikulina A. V., Peregud M.M., Kon' kov B. F., Novikov V. V., Markelov V.A., Khokbunova T. N., Kobylyansky G. P., Novoselov A. E., Ostrovsky Z. E. and Obukhov A. V., '*Structure – Phase State, Corrosion and Irradiation Properties of Zr-Nb-Fe-Sn System Alloys*', 15th Symposium 'Zirconium in Nuclear Industry', Sunriver, OR, USA, 24–26 June 2007.

Shishov V. N., Peregud M. M., Nikulina A. V., Kobylyansky G. P., Novoselov A. E., Ostrovsky Z. E., Obukhov A. V. and Pimenov Y. V., '*Influence of Structure and Phase State on Irradiation Induced Growth of Niobium Containing Alloys of Zirconium*', *J. ASTM Int.*, **2**(8), Paper JAI12431, September, 2005.

Shishov V. N., Peregud M. M., Nikulina A. V., Shebaldov P. V., Tselischev A. V., Novoselov A. E., Kobylyansky G. P., Ostrovsky Z. E., and Shamardin V. K., '*Influence of Zirconium Alloy Chemical Composition on Microstructure Formation and Irradiation Induced Growth*', Zirconium in the Nuclear Industry: 13th International Symposium, ASTM STP 1423, G. D. Moan and P. Rudling, Eds, ASTM, pp. 758–779, 2002.

Simpson L. A. and Cann C. D., 'Fracture Toughness of Zirconium Hydride and Its Influence on the Crack Resistance of Zirconium Alloys', *J. Nucl. Mat.*, **87**, pp. 303–316, 1979.

Smirnov, A. V., Kusmin V. I., Smirnov V. P., Dubrovin K. P., Bibilasjvili Yu. K. and Saletnih B. A., '*WWER-1000 and WWER-440 Fuel Operation Experience*', Proceedings of the ANS International Topical Meeting on LWR Fuel Performance, West Palm Beach, Florida, pp. 31–44, 1994.

Soniak A., L' Hullier N., Mardon J.-P., Rebeyrolle V., Bouffioux P. and Bernaudat, C., '*Irradiation Creep Behavior of Zr-base Alloys*', Zirconium in the Nuclear Industry: 13th International Symposium STP 1423, ASTM, pp. 837–862, 2002.

Strasser A. and Rudling P., *Fuel Fabrication Process Handbook*, ANT International, Mölnlycke, Sweden, 2004.

Strasser A., Rudling P., Cox B. and Garzarolli, '*The Effect of Hydrogen on Zirconium Alloy Performance*', ZIRAT13 Special Topical Report Vol. II, ANT International, Mölnlycke, Sweden, 2008.

Tägtström P., Limbäck M., Dahlbäck M., Andersson T. and Petterson H., '*Effects of Hydrogen Pickup and Second Phase Dissolution on the In-Reactor Corrosion Performance of BWR Claddings*', 13th International Symposium on Zirconium in the Nuclear Industry, ASTM STP 1423, 86, Annecy, France, 2002.

Takagawa Y., Ishimoto S., Etoh Y., Kubo T., Ogata K. and Kubota O., '*The Correlation Between Microstructures and in-BWR Corrosion Behavior of Highly Irradiated Zr-Based Alloys*', 14th International Symposium on Zirconium in the Nuclear Industry, Stockholm, Sweden, ASTM STP 1467, pp. 386–401, 2004.

Taylor D. F., Peters H. R. and Yang W. J. S., '*Amorphization of Laves-Phase Precipitates in Zircaloy-4 by Neutron Irradiation*', Proceedings of the 9th International Conference on Environmental Degradation of Materials in Nuclear Power Systems-Water Rectors (TMS), August, 1999.

Tenckhoff E., '*Deformation Mechanism, Texture and Anisotropy in Zirconium and Zircaloy*', ASTM STP 966, ASTM, Philadelphia, PA, 1988.

Tenckhoff E., 'Review of Deformation Mechanisms, Texture and Mechanical Anisotropy in Zirconium and Zirconium Base Alloys', *J. ASTM Int.*, **2**(4), Paper ID JAI 12945, April 2005. (Also ASTM STP 1467, 2005.)

Toffolon C., Brachet J. C., Servant C. *et al*, '*Experimental Study and Preliminary Thermodynamic Calculations of the Pseudo Ternary Zr-Nb-Fe-(O,Sn) System*', Zirconium in the Nuclear Industry, ASTM STP 1423, pp. 361–383, 2002.

Tomalin D. S., '*Localized Ductility of Irradiated Zircaloy-2 Cladding in Air and Iodine Environments*', Zirconium in the Nuclear Industry, ASTM STP 633, A. L. Lowe, Jr. and G. W. Parry, Eds, ASTM, pp. 557–572, 1977.

Trowse F. W., Sumerling R. and Garlick A., '*Nodular Corrosion of Zircaloy-2 and Some Other Zirconium Alloys in Steam Generating Heavy Water Reactors and Related Environments*', Zirconium in the Nuclear Industry, ASTM STP 633; A. L. Lowe, Jr., and G. W. Parry, Eds, ASTM STP 633, pp. 236–257, 1977.

Tucker, R. P. and Adamson, R. B., '*Irradiation Hardening as Measured by Microhardness*', General Electric Co. Report, GEAP-25163–10, 1984.

Tucker R. P., Fidleris V. and Adamson R. B., 'High-Fluence Irradiation Growth of Zirconium Alloys at 644 to 725', *ASTM Spec. Techn. Publ.*, **824**, pp. 427–451, 1984.

Urbanic V. F. and Griffith M., '*Microstructural Aspects of Corrosion and Hydrogen Ingress in Zr2.5Nb*', Zirconium in the Nuclear Industry: Twelfth International Symposium, ASTM STP 1354, G. Sabol and G. Moan, Eds, ASTM, West Conshohocken, PA, pp. 641–657, 2000.

Vaidyanathan S. *et al*, '*Fuel Design and Improved Materials for High Burnup Service*', Proceedings of Jahrestagung Kerntechnik, Bonn, pp. 325–336, 2000.

Vizcaino P. *et al.*, '*Hydrogen Solubility and Microstructural Changes in Zircaloy-4 due to Neutron Irradiation*', 16th International Symposium on Zirconium in the Nuclear Industry, ASTM, Chengdu, China, May, 2010.

Wikmark G. and Cox, B., '*Water Chemistry and CRUD Influence on Cladding Corrosion*', ZIRAT6 Special Topics Report, ANT International, Mölnlycke, Sweden, 2001.

Williams C. D., Adamson R. B. and Olshausen K. D., '*Effects of Boiling Water Reactor Irradiation on Tensile Properties of Zircaloy*', Proceedings, European Conference on Irradiation Behavior of Fuel Cladding and Core Component Materials, Karlsruhe, Germany, December, pp. 189–192, 1974.

Wisner S. and Adamson R., 'Combined Effects of Radiation Damage and Hydrides on the Ductility of Zircaloy-2', *Nucl. Eng. Design*, **185**, pp. 33–49, 1998.

Wisner S. B., Reynolds M. B. and Adamson R. B., '*Fatigue Behavior of Irradiated and Unirradiated Zircaloy and Zirconium*', Zirconium in the Nuclear Industry: Tenth International Symposium, ASTM STP 1245, A. M. Garde and E. R. Bradley, Eds, ASTM, Philadelphia, PA, pp. 499–520, 1994.

Woo C. H., 'Theory of Irradiation Deformation in Non-Cubic Metals: Effects of Anisotropic Diffusion', *J. Nucl. Mater.*, **159**, pp. 237–256, 1988.

Yagnik S., Kuo R. C., Rashid Y. R., Machiels A. J and Yang R. L., '*Effect of Hydrides on the Mechanical Properties of Zircaloy-2*', ANS Topical Meeting on LWR Fuel Performance, Orlando, FL, 2004.

Yang W. J. S., '*Microstructural Development in Neutron Irradiated Zircaloy-4*', Effects of Radiation on Materials: 14th International Symposium, Vol. 1, ASTM STP 1046, N. H. Packan, R. E. Stoller, and A. S. Kumar, Eds, ASTM, Philadelphia, PA, pp. 442–456, 1989.

Yang W. J. S. and Adamson R. B., '*Beta-Quenched Zircaloy-4: Effects of Thermal Aging and Neutron Irradiation*" Zirconium in the Nuclear Industry, Eighth Int'l Symposium, ASTM STP 1023, L. F. P. Van Swam and CM Eucken, Eds, American Society for Testing and Materials, Philadelphia, pp. 541–477, 1989.

Zwicky H-U. *et al.*, '*Enhanced Spacer Shadow Corrosion on SVEA Fuel Assemblies in the Leibstadt Nuclear Power Plant,*' International Topical Meeting on Light Water Fuel Performance, American Nuclear Society, Park City, UT, 10–13 April 2000.

5
Performance and inspection of zirconium alloy fuel bundle components in light water reactors (LWRs)

P. RUDLING, ANT International, Sweden
and R. B. ADAMSON, Zircology Plus, USA

DOI: 10.1533/9780857097453.2.246

Abstract: This chapter highlights integral performance of zirconium alloy fuel bundle components used in nuclear power light water reactors (LWRs). In particular we focus on those behaviours which result in performance issues, and in experimental techniques which are used to quantify the performance. Details in this chapter complement those in the previous chapter on the properties of in-reactor zirconium alloy materials.

Key words: zirconium alloys, nuclear reactors, accidents, dimensional stability, irradiation, mechanical properties, corrosion, inspection, high burnup.

5.1 Introduction

The previous chapter described material properties of zirconium alloys in light water reactors (LWRs). The performance of fuel bundle components is often driven by a combination of singular material properties; for example, mechanical strength and irradiation creep. This chapter extends material behaviour to include integral performance of fuel bundle components such as fuel rods, channels and guide tube assemblies under both normal operational and accident conditions. In all cases, component inspection is needed to verify expected or explore abnormal performance. A thorough understanding of both material and component behaviour is needed to assure safe and efficient reactor operation.

5.2 Materials performance during normal operational conditions

We proceed with sections covering ways in which materials perform during normal operating conditions in the main reactor types.

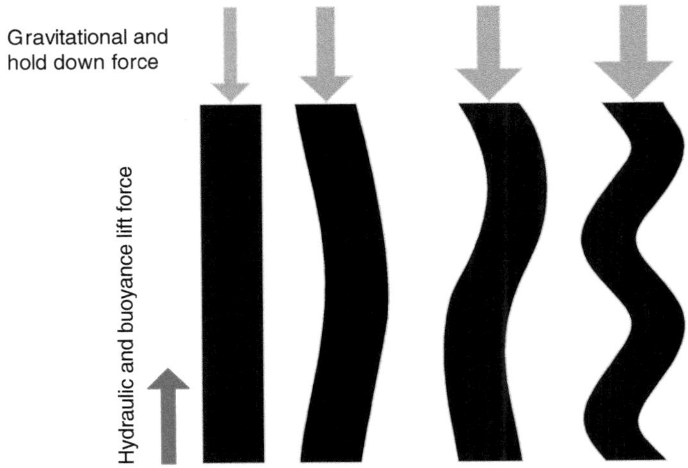

5.1 Schematics showing FA bowing. Increased GT growth may result in larger holding down forces (figures going from left to right). The first mode of bowing is B-shape (the second drawing), while the second and third mode of bowing are S-shape and W-shape, respectively (Strasser *et al.*, 2010a).

5.2.1 Bowing

PWR/VVER fuel assembly bowing may occur due to excessive guide tube (GT) growth that will result in larger holding down forces (Fig. 5.1) (Strasser *et al.*, 2010a). The bowing is caused by the complex interaction of a variety of parameters that include the bowing of the skeleton assembly. The parameters include:

- GT irradiation growth as a function of fluence and temperature, see Section 4.6.1 on irradiation growth.
- GT creep as a function of fluence and temperature, see Section 4.6.2 on irradiation creep.
- GT stiffness and buckling strength as a function of temperature.
- The effect of hydrogen pickup and irradiation on the GT properties, see Section 4.5 on corrosion of zirconium alloys.
- Hold-down force.
- Thermal expansion of the skeleton components, the core plate spacing and their interaction.
- Fuel rod/grid friction force and relaxation over time.

BWR fuel channel bowing was studied by Cantonwine *et al.* (2009). According to them, channel–control blade interference had been a challenging issue over the previous 8 years for operating BWR plants where ~2-year cycles are normal and Zircaloy-2 is the standard channel material. The primary reason for this was the unaccounted channel distortion caused by differential hydrogen across the channel resulting from shadow corrosion on the blade side (known as shadow corrosion-induced bow). Zircaloy-2 is particularly susceptible to this distortion mechanism because it has a high hydrogen pickup fraction (HPUF) that increases with exposure.

Several strategies have been developed to combat bow. As an intermediate resolution to this issue Zircaloy-4 has been reintroduced because it is effectively resistant to shadow corrosion-induced bow and has similar irradiation growth and creep performance to Zircaloy-2. The one disadvantage of Zircaloy-4 is that it has less corrosion resistance than Zircaloy-2. However, based on the extensive experience with Zircaloy-4 channels both in the United States and Japan (plus processing improvements have been made specifically to enhance corrosion resistance), the corrosion performance of Zircaloy-4 is claimed to be adequate for channel applications. Other examples of global nuclear fuel (GNF) publications on channel bow are described by Mahmood *et al.* (2007) and Cantonwine *et al.* (2009).

Other reasons for BWR fuel channel bowing are (Strasser *et al.*, 2010a):

- Fast neutron flux gradients from a variety of causes including the flux gradient at the core periphery (see Fig. 5.2).
- Non-uniform metallurgical structure (e.g. texture difference between the two opposing channel sides leading to difference in irradiation growth rate) or composition.
- Non-uniform wall thickness.
- As-fabricated bow.

The bowing may result in difficulties in inserting the control rods (a safety issue) and/or in a decrease in thermal margins, the latter from two possible causes. First a departure from nucleate boiling (DNB) value: if the fuel rod surface heat flux becomes large enough, the water film adjacent to the fuel rod will convert into a steam film with a much lower thermal conductivity resulting in a rapid large increase in the fuel cladding temperature which, in turn, will accelerate the oxidation and embrittlement of the fuel cladding. The maximum heat flux at which the water is converted into a steam film is referred to as the DNB value. Second, a loss of coolant accident (LOCA) could, for example be caused by a coolant pipe break in the primary circulation system since larger water gaps between assemblies may exist in the core than is accounted for in the core nuclear design. To ensure that the LOCA licensing criteria are met, the fuel rod surface heat flux must be limited.

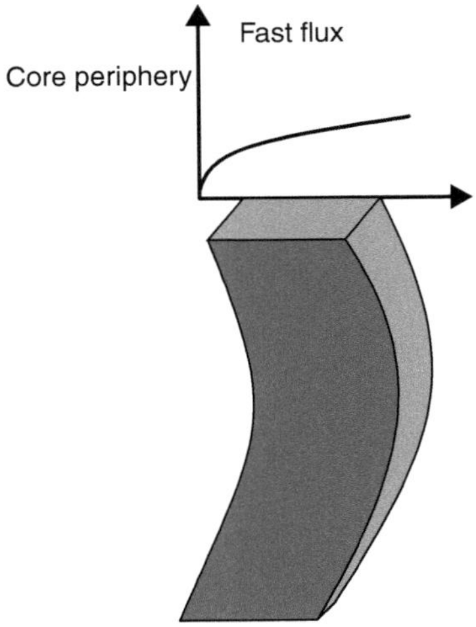

5.2 Schematics showing fuel outer channel bowing at core periphery due to large fast neutron fluence. Largest degree of bowing in *BWR*s occurs at the core periphery due to the flux profile. Also the type of *FA* bow seems to be very dependent on core location (Strasser *et al.*, 2010a).

5.2.2 Pellet-cladding interaction

Stresses which induce both PCI (usually denotes combined mechanical and chemical pellet-cladding interaction) and PCMI (usually denotes pellet cladding mechanical interaction) are caused by expansion of the fuel pellet against the cladding during power increases (Adamson *et al.*, 2006/7; Strasser *et al.*, 2010a). PCI failures are driven by a stress corrosion cracking (SCC) assisted component resulting from fission product release from the fuel, while PCMI failures are generally due to purely mechanical cracking, often enhanced by a reduction in cladding ductility due to formation of local hydrides at the clad outer surface. At the micro level, the PCI crack always starts at the cladding inner surface and propagates towards the outer cladding surface while the PCMI crack propagates from the outer to inner surface.

PCI is associated with local power ramps during reactor start-up or power manoeuvring (e.g. rod adjustments/swaps, load following) as shown schematically in Fig. 5.3, and is caused by the combination of cladding stress due to the power increase and the influence of iodine, caesium and cadmium released during the power increase in a susceptible material (Adamson *et al.*, 2006/2007; Strasser *et al.*, 2010a). This combination of stress, embrittling

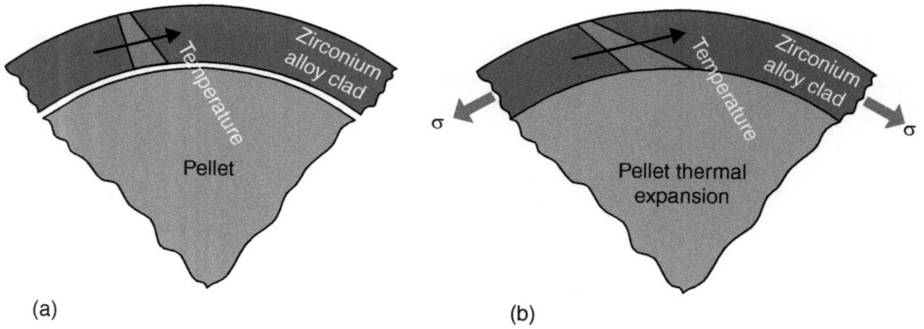

5.3 Schematics showing the fuel rod condition (a) before the ramp and (b) during the ramp (Strasser *et al.*, 2010a).

fission products and susceptible material may result in SCC of the fuel cladding, as shown in Fig. 5.4.

PCI failures may occur in PWRs/VVERs and BWRs (Strasser *et al.*, 2010a). The failure mechanism is much more prevalent in BWRs, since reactor power is controlled in part by control rod movements that subject the fuel to rapid power level changes. (The reactor power in both BWRs and PWRs is also regulated by flow control.) In PWRs and VVERs, reactor power is not normally controlled by insertion and extraction of the control rods in the core; rather, reactor power is controlled by the boron concentration that is continuously decreased during operation to compensate for the decrease in reactivity. This type of reactor power control is much smoother than in the BWR case and, consequently, PCI failures are less common in PWRs. However, during reactor power increases, and specifically during a class II transient (anticipated operational occurrences, AOO), PCI failures may occur in a PWR.

To prevent PCI failures, it is necessary to remove at least one of the fundamental conditions (tensile stress, sensitive material, aggressive environment) which cause SCC. There are two principal types of remedy (Strasser *et al.*, 2010a):

1. One is to develop reactor operation restrictions that will ensure cladding stresses are always below the PCI threshold stress during power increases. This is the main measure in avoiding PCI defects and the only measure used in PWRs. Operating rules (also called management recommendations, or pellet-cladding interaction operating management restrictions (PCIOMRs)) to limit local power increases and 'condition' fuel for power ramping were implemented in BWRs during the late 1970s to mitigate the PCI issue. The rules are usually a function of exposure and were developed by the different fuel vendors, so they differ between various fuel types. To establish and validate these rules, extensive power ramp tests were performed by the fuel vendors in experimental reactors.

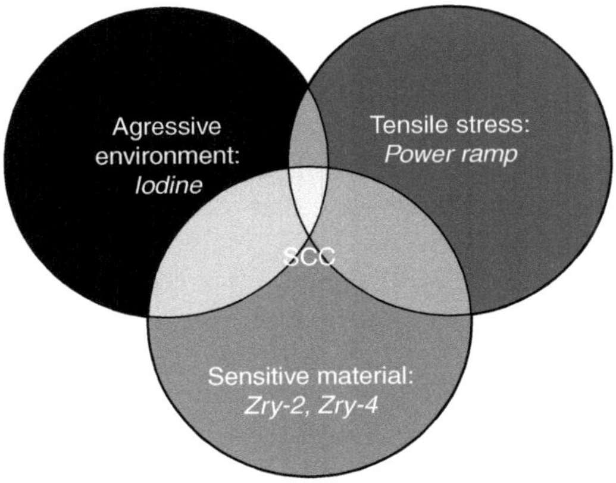

5.4 Schematic showing the three components involved in SCC (Strasser *et al.*, 2010a).

2. The second remedy – design improvement – consists of two approaches:
   2a. Cladding design
      2a1. Development of radial cladding texture and small grain size that may increase cladding PCI resistance.
      2a2. Development of the barrier/liner concept, initially with a 'pure' zirconium (Zr) metal barrier at the cladding inner diameter (ID). The barrier is soft and serves to reduce the local stress, hence giving the cladding resistance to SCC. Later, fuel vendors realized that the Zr could be alloyed with Fe to improve the secondary degradation resistance in case of rod failure. The Fe in the Zr will dramatically improve the corrosion resistance of the liner/barrier but may reduce the PCI performance. Although this remedy has so far only been used in BWRs, it should be equally applicable to PWRs.
   2b. Pellet design
      2b1. Reducing the cladding local strains (and stresses) by shortening the pellet, chamfering the corners and eliminating the dishing.
      2b2. Pellets with additives are being developed both for BWRs and PWRs that will increase the margins towards PCI failures (Adamson *et al.*, 2006/7; Patterson, 2010). The additives of interest fall into two general categories, the first category involves materials that are essentially insoluble in the fluorite lattice and exists as a separate, grain boundary phase, for example, mixtures of alumina and silica (aluminosilicates or Al-Si-O). The second category involves materials that are soluble in the cation sub-lattice, such as chromia, or involve a mixture of soluble and insoluble materials, such as chromia and alumina. Although

many other additives fall into both categories, attention is directed to the aluminosilicate additives and chromia-base dopants as they appear to be the closest to large-scale application.

2b3. Aluminosilicate additives consist of a mixture of SiO_2 and Al_2O_3 and is offered by GNF. During the pellet sintering process, the additive forms a glassy phase that collects on the grain boundaries. It appears that the Al-Si-O additive at the pellet grain boundaries will chemically react with I, Cs and Cd, thus preventing these SCC-promoting elements from accessing the fuel clad inner surface (Matsunaga *et al.*, 2009, 2010). Additive fuel has been irradiated in commercial and test reactors in the US and in Europe. Ramp tests under BWR conditions in the R-2 and Halden reactors of segmented additives rods from commercial reactors show excellent resistance to the PCI failure mechanism (Davies *et al.*, 1999).

2b4. Chromia (Cr_2O_3) is the dopant of greatest commercial significance in this class of additives. Two types of chromia-based additives are being offered. The first consists of Cr_2O_3 in UO_2 as offered by AREVA (Delafoy *et al.*, 2003). The second consists of Cr_2O_3 and Al_2O_3 in UO_2 as offered by Westinghouse (Arborelius *et al.*, 2005). Alumina is reported to be used in the second form to minimize the effects of chromium on the fission cross-section of doped pellets while enhancing grain growth. In both cases, chromia is expected to reside largely within grains as interstitial Cr^{3+} and as insoluble Cr_2O_3 depending on the concentration and temperature. The alumina in the mixed Cr-Al-O dopant should exist as a grain-boundary phase as in the Al-Si-O additive. The cation dopants were developed to increase grain size to reduce fission gas release (FGR) at extended burnup(see for example Delafoy *et al.*, 2007). In addition to improved FGR, chromia-based dopants are reported to improve PCI resistance. Information available in this area is less extensive for the chromia-based dopants than for the aluminosilicate additives. However, ramp tests indicate that the resistance to PCI failures of fuel with chromia-based dopants are improved relative to standard fuel in cladding without PCI-resistant liners (Delafoy *et al.*, 2007).

5.2.3 Cladding liftoff

If the rod internal pressure becomes larger than the reactor system pressure, the fuel cladding may start to creep outwards (Fig. 5.5) (Strasser *et al.*, 2010a). If the fuel cladding outward creep rate exceeds the fuel swelling rate (due to fission product production during irradiation), the pellet–cladding gap may increase. This phenomenon is denoted cladding liftoff. Since this gap constitutes a significant heat flux barrier, such a gap increase may result in increased fuel pellet temperature. This higher temperature will in turn increase the gaseous fission product release rate, further increasing the fuel rod overpressure and leading to an even higher outward cladding creep rate. Such a thermal feedback condition may lead to fuel failure.

A larger fuel rod free volume, lower FGR rate and increased clad creep strength increases the margins towards liftoff (i.e. a larger rod internal pressure can be accepted without getting liftoff) (Strasser *et al.*, 2010a). Free volume refers to the void volume bounded by the inner surfaces of the cladding and end plugs and the outer pellet surface minus the volume of plenum springs and other internal hardware. Note that closed pellet porosity is within the pellet volume, while open porosity, dishes, chips and other surface irregularities with finite, open volume are in the free volume.

5.2.4 Degradation of failed fuel rods

Degradation of failed fuel rod is a situation where the leakage path(s) through the damaged cladding increases to the point where the fuel itself is dispersed into the primary system (Strasser *et al.*, 2008). This may occur if the rods degrade to such a point that the water contacts the fuel pellet, particularly if the contact also involves active flow of the water over exposed fuel pellets, one example being a large axial cladding crack. Steam will not be able to cause fuel washout while water can by oxidising the fuel grain boundaries thereby causing disintegration f the fuel grains. Normally, utilities are much more concerned about fuel washout than high iodine and noble gas release. This is because it may take up to ten years to clean the core from the tramp uranium resulting from the fuel dissolution, while the high iodine and noble gas activities released from the failed rod will be eliminated when the failed rod is extracted from the core.

Degradation has historically been more of an issue in BWRs than in PWRs (Strasser *et al.*, 2008). Failed rods in PWRs may degrade, but the amount of dispersed fuel is lower than in a BWR. The rationale may be that the coolant chemistry in a PWR is more reducing than in BWRs. During the period 1992–93, six plants in the United States and Europe were forced into unscheduled outages because of concerns about failed Zr-sponge liner fuel (IAEA, no. 388, 1998). This is a liner produced from Zr sponge material to which no alloying elements have been added; its major impurities are oxygen (about 600–900 wt.ppm) and iron (about

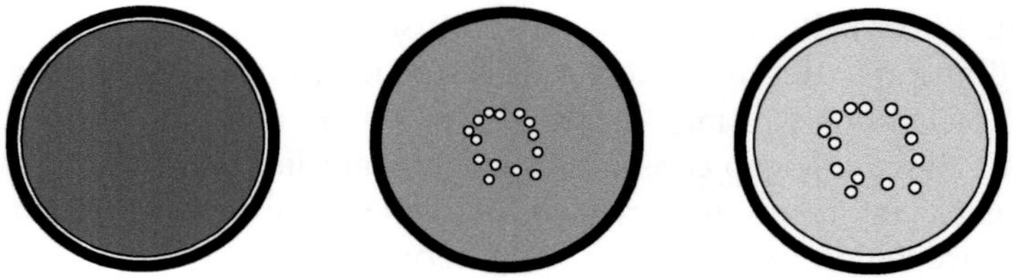

5.5 Schematics showing how the pellet-cladding gap may change over burnup. (a) low burnup – a significant pellet-cladding gap exists; (b) intermediate burnup – no pellet-cladding gap; (c) high burnup in a high power rod with significant fission gas release – reopening of the pellet-cladding gap (Strasser *et al.*, 2010a).

150–500 wt.ppm). In all these cases, the very high off-gas activities and significant loss of fuel pellet material resulted from only one or two failed rods. Other plants in the United States and Europe also elected to shut down during and slightly after this interval to remove failed fuel assemblies and avoid the risk of large residual contamination from tramp uranium. More recently, the risk of degradation and residual contamination has been reduced by the use of corrosion-resistant liners in BWR fuel to the extent that forced and voluntary outages are less common.

Two different types of degradation scenarios have been identified, namely the development of two different types of cracks (Strasser *et al.*, 2008):

1. Transversal breaks (also called guillotine cuts or circumferential break) occurring in BWRs, PWRs and VVERs.
2. Long axial cracks (axial splits), which can occur in BWRs due to the movement of control blades but may also occur in PWRs that are subjected to significant control rod movements during operation. Axial split is a term introduced by GE and represents a failed rod that either has an off-gas level larger than 5000 μCi/s (185 MBq/s) or a total crack length that is larger than 152 mm (6 inches).

Transversal breaks in BWRs – normally occur in low to intermediate burnup rods in the bottom part of the rod with a primary failure in the upper part of the rod (see Fig. 5.6) (Strasser *et al.*, 2008). The primary defect will allow water/steam to gain access to the rod interior (1 in Fig. 5.6) where the steam will oxidize the fuel clad inner surface forming a zirconium oxide the thickness of which will decrease with distance from the primary defect (2 in Fig. 5.6). At the same time a hydrogen partial pressure is being built up in the pellet-cladding gap. At a critical distance from the primary defect, the steam partial pressure will be insufficient to protect the clad inner surface

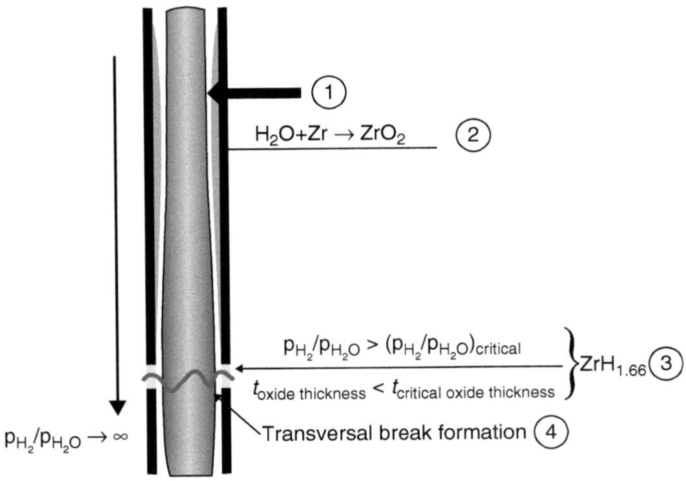

5.6 Schematic showing the events resulting in transversal break formation. The numbers in the figure relate to the sequence of events that may lead to a transversal break as described in the text (Strasser *et al.*, 2008).

from hydrogen ingress thus causing secondary hydriding (3 in Fig. 5.6) (e.g. Olander *et al.*, 1997). If the hydride precipitates along the whole fuel clad circumference, the fuel rod may fracture transversally due to the hydride embrittlement effect (4 in Fig. 5.6).

Transversal break in PWRs/VVERs – are caused by a mechanistic development similar to that of BWRs (Strasser *et al.*, 2008). However, the secondary hydride defects tend to form in the upper part of a PWR/VVER rod. The processes involved in developing a transversal break in a PWR rod are shown in Fig. 5.7.

1. *Axial cracks in BWRs* – Formation of long axial cracks has three prerequisites, (Strasser *et al.*, 2008):
   1a. A sharp primary defect such as a PCI crack or cracks in hydride blisters formed due to a primary defect. However, in this case the hydride blister is very local and does not exist along the whole fuel clad circumference, as seen in formation of transversal breaks.
   1b. A fuel cladding hydrogen content larger than the hydrogen solid solubility.
   1c. A stress intensity (K_I) at the crack tip above the critical value for crack extension. K_I will increase with clad tensile stress level which in turn depends on:

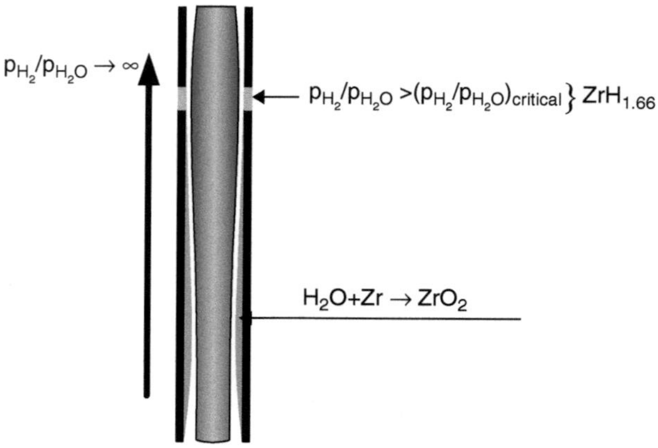

5.7 Schematic description of the events resulting in transversal break formation (Strasser *et al.*, 2008).

1c1. The initial pellet-cladding gap prior to the power ramp, which depends on:
   1c1a. Burnup since the gap is decreasing with increased burnup due to fuel swelling and fuel clad creep-down. This is the reason that axial cracks do not form in low burnup fuel since the fuel pellet-clad gap is so large.
   1c1b. The corrosion properties of the cladding inner surface (Edsinger, 2000). The pellet-cladding gap decreases if the corrosion properties of the cladding inner surface are poor, resulting in formation of a thick porous oxide layer in the failed rod. The decrease in gap is related to the zirconium oxide having a larger specific volume than that of the zirconium metal. It also turns out that, if the corrosion resistance of the cladding inner surface is poor, then formed oxide is less dense due to the many cracks and pores which will decrease the pellet-cladding gap further. The first type of Zr-liner materials used in the nuclear industry were non-alloyed with very poor corrosion properties. Once it was realized that the corrosion properties of the Zr liner have a large impact on the tendency to form axial cracks in failed fuel, all fuel vendors did alloy their liners to improve the corrosion resistance. However, it is important to ensure that the alloying additions will not degrade the PCI performance of the fuel cladding.
1c2. The magnitude of the rod power increase.

The axial split formation is schematically shown in Fig. 5.8 (Strasser *et al.*, 2008). Initially, the control rod is inserted during the time when the primary defect occurs (1 in Fig. 5.8). The same scenario as for transversal breaks in BWRs occurs, but the secondary hydrides are distributed to several fuel clad locations which means that each hydride becomes too small to encompass the whole fuel clad circumference (2 in Fig. 5.8). The tensile stresses in the cladding which are necessary for crack propagation result from a power increase in the failed rod, for example, when a control rod adjacent to the failed rod is pulled out of the core. This will increase the temperature in the fuel stack resulting in a thermal increase of the pellet diameter. If these stresses become large enough the sharp defect may propagate if the resulting K_I exceeds the critical value for crack propagation (3 in Fig. 5.8). It is proposed that the mechanism for crack propagation forming an axial split is a delayed hydrogen cracking (DHC) type failure process (see e.g. Efsing & Pettersson, 1998; Edsinger, 2000; Lysell *et al.*, 2000 for more details). The lower bounds of the crack velocities are in the range 4×10^{-8}–5×10^{-7} ms^{-1} based on assumed constant growth rates in the time between first detection of the defect and removal of the fuel (Strasser *et al.*, 2008).

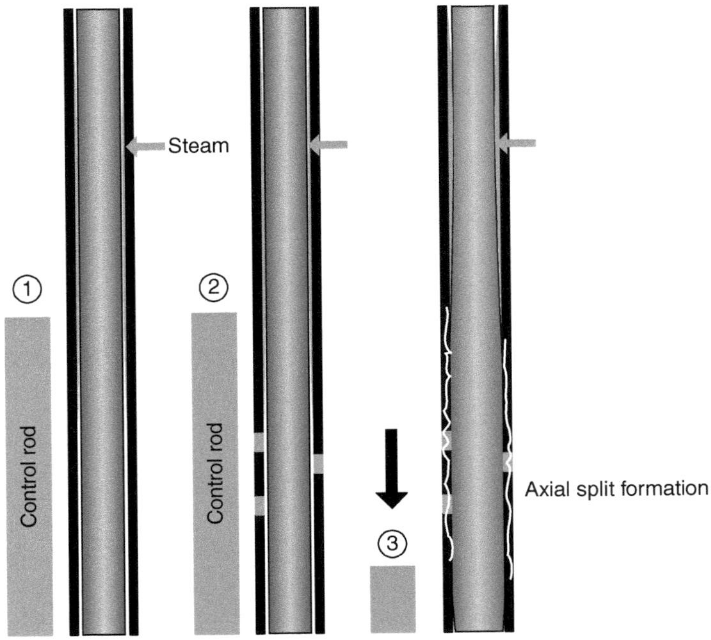

5.8 Schematic showing the events resulting in axial split formation. The numbers in the figure relate to the sequence of the different events that may lead to an axial crack as described in the text (Strasser *et al.*, 2008).

Axial cracks in PWRs/VVERs – Long axial cracks do not form in PWRs as readily as in BWRs (Strasser *et al.*, 2008). The reason for the difference is that in PWRs, the power regulation is done slowly and without pronounced increases in local power by decreasing the boron coolant concentration, while power regulation in BWRs is done by a combination of control rod movements and variations in coolant flow, with the control blade movements leading to rapid increases in local power. However, axial cracks may form in PWRs/VVERs by essentially the same mechanism as formation of long axial cracks in BWRs due to (Strasser *et al.*, 2008):

- A class II transient and/or
- Due to control rod movements in load-following plants.

5.3 Materials performance during accidents

Having considered normal operating conditions, we now move on to cover accident scenarios.

5.3.1 Materials performance during loss of coolant accidents (LOCA)

The LOCA event starts with a decrease and then the loss of coolant flow due to a break in the coolant pipe; at the same time the reactor is depressurized, scrammed and shut down (Strasser *et al.*, 2010b). The fuel starts heating up due to its decay heat until the emergency core cooling systems (ECCSs) are activated and fuel cooling commences. Hypothetical LOCA events are analyzed for each reactor to ensure that the safety criteria, as defined by the regulators for the reactor system and the fuel, are met. The design basis accidents (DBAs) which are analyzed fall into two general categories. The large break, or large break loss of coolant accident (LBLOCA), assumes a double ended break of a primary coolant cold leg of a PWR or a break in the recirculation pump intake line of a BWR, either of which could cause the loss of all the coolant from the core. The small break, or small break LOCA (SBLOCA), assumes a break in one of the smaller primary circuit lines that will cause less coolant loss than the LBLOCA.

The effect of a LOCA cycle on the fuel is shown schematically in Fig. 5.9, plotting the fuel and cladding temperatures as a function of time in the accident (Strasser *et al.*, 2010b). The loss of coolant flow and reactor pressure at the initiation of the accident will decrease heat transfer and allow the fuel and cladding to heat up until the reactor scrams. The fuel will then cool

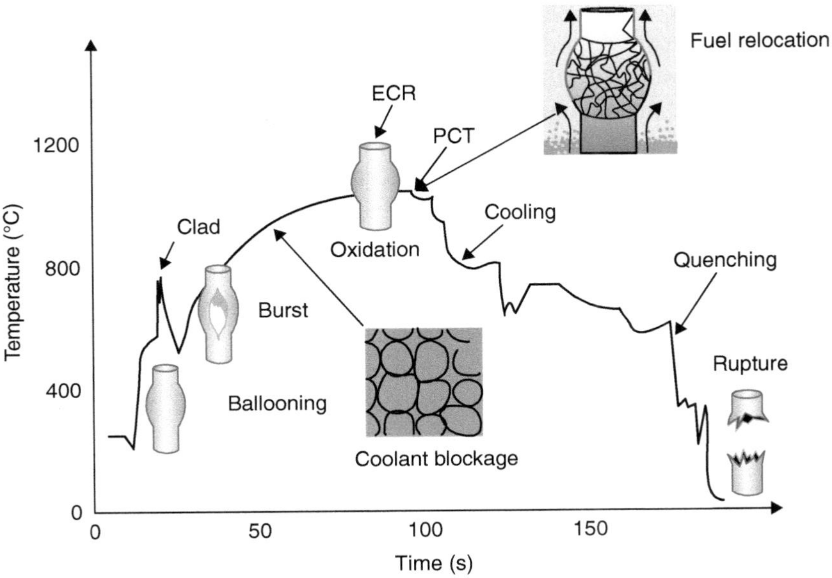

5.9 Typical LOCA in a PWR (Strasser *et al.*, 2010b).

down somewhat, partly due to cooling by the steam-water mixture that is formed, but the cladding temperature will continue to rise.

During and after the LOCA it must be ensured that (Strasser *et al.*, 2010b):

- The core remains coolable (which means that the maximum allowable coolant blockage is limited)
- No fuel dispersal occurs (which means that cladding rupture is not allowed; it is assumed that the cladding burst is so small that only fission gases are released)
- Less than 10% of the fuel rods in the core fail through burst (but without fuel dispersal) (a requirement in Germany only).

Ballooning of the cladding

The loss of coolant flow decreases heat transfer from the fuel, increases the fuel temperature and causes a significant temperature rise of the cladding (Strasser *et al.*, 2010b). The decrease in system pressure causes a pressure drop across and a hoop stress in the cladding. The result is the creep deformation or ballooning of the cladding. Depending on the temperature, the cladding ductility and the rod internal pressure, the cladding will either stay intact or may burst. Ballooning of the fuel rods may result in a blockage of the coolant sub-channel that, in turn, may impact the fuel coolability. If large fuel clad burst strains occur at the same axial elevation, co-planar deformation in the fuel assembly can result and the coolability may be significantly degraded. The extent of the ballooning is also dependent on the fuel clad hydrogen content (picked up during the water-zirconium alloy corrosion reaction during reactor operation prior to LOCA). Hydrogen decreases the $\alpha/\alpha+\beta$ phase transformation temperature, which means that increasing the hydrogen content in the fuel cladding will lower its ductility and result in more fuel rod bursts during a LOCA.

Oxidation of the cladding

The increasing temperatures and presence of steam will cause the intact cladding to oxidize on the outer diameter (OD) and the burst cladding to oxidize on both the OD and ID (two-sided oxidation) (Strasser *et al.*, 2010b). The oxidation process at the high LOCA temperatures will increase the oxygen and hydrogen content in the cladding, reducing its ductility and resistance to rupture. Two sided oxidation can have significant effects on the post-quench ductility (PQD) of the cladding as a result of high but localized hydrogen pickup in addition to the oxidation (Strasser *et al.*, 2010b). The cladding continues to oxidize until the ECCS becomes effective and a peak-cladding temperature (PCT) is reached. The maximum PCT is regulated to be a maximum of 2200°F by the USNRC and 1200°C internationally. The length of time the system may remain at the PCT is determined by the reactor system and regulated by the equivalent cladding reacted (ECR) limit, defined as the total thickness of cladding that would be converted to stoichiometric ZrO_2 from all of the oxygen contained in the fuel cladding as ZrO_2 and oxygen in solid solution in the remaining metal phase.

Embrittlement of the cladding

ECCS activation will stop the temperature rise and start to cool the core by injection from the bottom of the core in a PWR and from the top of the core in a BWR (Strasser *et al.*, 2010b). The 'cooling' process as shown in Fig. 5.9 is relatively slow until the emergency coolant contacts the fuel that has been at the PCT. At that point, in the range of 400–800°C and identified as 'quenching' in Fig. 5.9, the water from the ECCS will reduce the cladding temperature at a rapid rate (1–5°C/s) by re-wetting the cladding heat transfer surface. The process will collapse the vapour film on the cladding OD and cooling will be by nucleate boiling. Thermal shock due to the sudden change in heat transfer conditions can fracture the cladding at this stage and the ability of the cladding to withstand the thermal stresses will depend on the extent of oxidation and degree of cladding embrittlement that occurred during the LOCA transient.

The oxidation embrittlement process and final structure of the cladding after completion of the LOCA cycle is as follows (Strasser *et al.*, 2010b):

- First, the increasing water and steam temperatures during heat-up increase the reaction rates with the cladding and increase the conversion of the cladding surface into thicker ZrO_2 films.
- As the LOCA temperature passes the levels where $\alpha \rightarrow \beta$ transformations start and finish, the resulting structure consists of:
  - The growing ZrO_2 layer.
  - A zirconium alloy layer with a very high oxygen content which stabilizes the α phase.
  - The bulk cladding which is now in the β phase.
- The ECCS initiated quenching phase cools the cladding back down through the $\beta \rightarrow \alpha$ transformation temperature and the bulk cladding is now re-transformed from the β into the α phase and referred to as the 'prior or former β phase.'

Oxygen and hydrogen affect the formation of the structure as follows during the oxidation (Strasser *et al.*, 2010b):

- Oxygen diffuses from the ZrO_2 to the bulk cladding which is in the β phase at the high temperature (HT); however, the β phase has a low solubility for oxygen.
- Increased hydrogen levels from the oxidation reactions prior to and during the LOCA increase the diffusion rate and solubility of oxygen in the β phase >1000°C.
- Wherever the solubility limit of oxygen in the β phase is exceeded, the excess oxygen stabilizes the α phase.
- The oxygen stabilized α phase forms next to the ZrO_2 layer and grows, as does the ZrO_2 layer, at the expense of the bulk cladding in the α phase and as a result after quenching in the 'prior β phase.'

The final integrity of the cladding is based on the properties of the prior β phase, since the ZrO_2 and oxygen stabilized α zones are too brittle to sustain a load (Strasser *et al.*, 2010b). 'Oxygen is the major source of cladding embrittlement as noted above and hydrogen is less likely to contribute to the embrittlement except to the extent that its presence increases the oxygen solubility' (Strasser *et al.*, 2010b).

5.3.2 Materials performance during reactivity-initiated accidents (RIA)

The design basis RIA in a PWR is the control refection rod accident (REA) and in a BWR the control rod drop accident (RDA) (Strasser *et al.*, 2010b). The REA is based on the assumption of a mechanical failure of the control rod drive mechanism located on the reactor vessel top, followed by the ejection of the mechanism and the control rod by the internal reactor pressure. The resulting significant power surge is limited partly by Doppler feedback and finally terminated by the reactor trip. The BWR RDA is assumed to occur if a control rod is detached from its drive mechanism in the core bottom, stays stuck while inserted in the core and then, if loosened, drops out of the core by gravity, without involvement of a change in reactor pressure as in the REA. As a result the BWR power pulses are slower and the pulse widths wider than for a PWR. The pulse widths for PWRs are in the range of 10–30 ms and for BWRs in the range of 20–60 ms.

The reactivity transient during a RIA results in a rapid increase in fuel rod power leading to a nearly adiabatic heating of the fuel pellets (Strasser *et al.*, 2010b). In a fresh fuel rod, the fissile material consists predominantly of U-235, which is usually uniformly distributed in the fuel pellets. Hence, both power and fission products are generated with a relatively small variation along the fuel pellet radius. However, with increasing burnup, there is a non-uniform build-up of fissile plutonium isotopes through neutron capture by U-238 and formation of Pu-239 and heavier fissile isotopes of plutonium. Since the neutron capture takes place mainly at the pellet surface, the distributions of fissile material, fission rate and fission products will develop marked peaks at the pellet surface as fuel burnup increases. The highest temperatures are occurring at the fuel pellet periphery.

The RIA-simulation experiments conducted in the 1960s and 1970s using zero or low burnup test rods showed that cladding failure occurred primarily by either (Strasser *et al.*, 2010b):

- Post-DNB brittle fracture of the clad material occurring during the re-wetting phase of the overheated heavily oxidized (and thereby embrittled) clad due to the abrupt quenching resulting in large thermal clad stresses. This failure mode is imminent if the cladding is severely oxidized due to the RIA fuel clad temperature excursion.
- Cladding contact with molten fuel.

Contrary to low burnup rods, the failure mechanism for BWR/PWR high burnup rods not subjected to DNB is PCMI and potentially creep burst (for rods with a rod internal overpressure and subjected to DNB) (Strasser *et al.*,

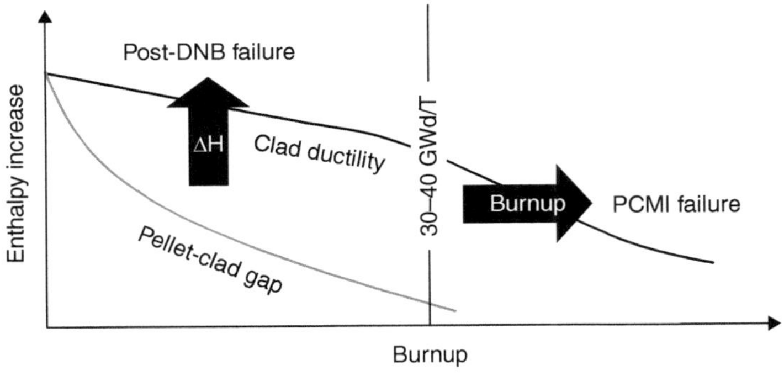

5.10 Clad failure mechanisms (Strasser *et al.*, 2010b).

2010b) while VVER high burnup rods only fail through creep burst (due to very low hydrogen contents in the fuel cladding).

PCMI – The change in failure mechanism is due to the decrease in pellet-cladding gap and the embrittlement of the cladding (due to corrosion induced hydriding) with increased burnup (Fig. 5.10). The rapid increase in power leads to nearly adiabatic heating of the fuel pellets, which expand thermally and may cause fast straining of the surrounding cladding through PCMI. At this early heat-up stage of the RIA, the cladding material is still at a fairly low temperature (<650K), and the fast straining imposed by the expanding fuel pellets may cause a rapid and partially brittle mode of clad failure (Chung & Kassner, 1998). The survival of a high burnup fuel rod in a RIA is dependent on the ability of the cladding to resist PCMI, which depends primarily on the imposed stress and the cladding ductility. The ductility is dependent on the temperature to a large degree, which in turn is dependent on the pulse width and enthalpy increase of the transient. The condition of the cladding has a significant effect on the ductility, specifically the alloy composition, microstructure and texture. *In addition, the cladding hydrogen content – most importantly the hydrogen distribution – has a significant impact on the PCMI response. More specifically, hydride rims/blister*s and/or radial hydrides *at the clad outer surface may result in a significant embrittlement effect.* The degree of embrittlement due to precipitation of hydrides in the cladding is dependent on the amount of hydrogen in excess of the solubility limit, as well as on size, orientation and distribution of the hydrides. Hydride-induced embrittlement is a complex matter, and several mechanisms contribute to the loss of clad strength and ductility (Northwood & Kosasih, 1983).

Hydride blisters can only be formed once the hydrogen content of the cladding has significantly exceeded the solubility limit and a certain temperature gradient is introduced by local oxide spallations (Strasser *et al.*, 2010b). Thus the probability of significant hydride blisters forming depends on the average hydrogen content of the cladding, the thickness of the oxide pieces flaked off (and the difference in oxide layer thickness adjacent to the spalled region and at the position of oxide spallation) and on the heat flux.

Hydrides precipitate in the form of thin platelets on planes that depend on cladding microstructure, heat flux and stress (Strasser *et al.*, 2010b). The orientation and continuity of these platelets with respect to residual or applied tensile stresses strongly influences the embrittlement. The orientation of hydrides in clad tubes is affected by the thermo-mechanical treatment of the tubes under manufacturing, and by the stress state prevailing under precipitation. In the presence of a radial heat flux, hydride platelets are generally oriented with their surface normals preferentially aligned to the clad tube radial direction, and the width of the platelets along the tube axial direction is significantly larger than in the circumferential direction. These hydrides, usually termed 'circumferential' hydrides, have only a moderate embrittling effect, since there is no tensile stress in the clad tube radial direction, that is in the direction perpendicular to the hydride platelets (Northwood & Kosasih, 1983).

However, there are also hydride platelets oriented with their surface normals more or less aligned to the clad circumferential direction (Strasser *et al.*, 2010b). These hydrides, which are usually termed 'radial' hydrides, are much more deleterious, since they are perpendicular to the dominating tensile stress in clad tubes of high-burnup fuel rods. The fraction of these detrimental radial hydrides is larger in recrystallization annealed (RXA) clad materials than in SRA cladding (Northwood & Kosasih, 1983). The former heat treatment results in a larger fraction of grain boundaries in the radial direction, and since hydrides tend to precipitate along grain boundaries, this could to some extent explain the differences in hydride orientation between RXA and SRA materials. However, there are also other causes to these differences, such as difference in hydride size.

It is noteworthy that VVER fuel cladding will not fail due to PCMI because of the very low hydrogen clad contents in VVER fuel claddings (Strasser *et al.*, 2010b). Instead VVER fuel rods with rod internal overpressures may fail due to creep burst.

If the cladding fails, fragmented fuel may disperse into the coolant (Lespiaux *et al.*, 1997). This expulsion of hot fuel material into water has the potential to cause rapid steam generation and pressure pulses, which could damage nearby fuel assemblies and possibly also the reactor pressure vessel and internal components. Hence, 'the potential consequences of fuel dispersal are of primary concern with respect to core and plant safety' (Strasser *et al.*, 2010b). The fuel dispersal tendency for high burnup fuel is very much dependant on the pulse width. For energy deposition with narrow pulse widths, heat conduction from the rim region is low and leads to higher local temperatures in the rim region due to the radial power peaking. Energy deposition with wider pulses allows for heat conduction from the pellet to the cladding, thus minimizing the temperature peaking in the pellet rim and maximising the fuel clad ductility. Lowering the pellet rim temperature decreases the potential for fuel particle dispersal.

5.4 Materials performance during interim dry storage

Most countries that generate nuclear power are in the process of developing criteria, designs and sites for the permanent disposal of spent nuclear fuel, but they have yet to become licensed realities (Adamson *et al.*, 2010). The most significant fuel related criteria for dry storage are compared in Table 5.1. Meanwhile the pools at the nuclear plant sites are filling up with spent fuel and the utilities are transferring the spent fuel from the pools to dry cask storage sites that are mostly located at the plant sites. Exceptions are the central, large intermediate pool facilities that serve all the plants in Sweden (CLAB facility) and all the plants in Finland (KPA-STORE). The lack of a licensed permanent fuel repository in any country has placed total reliance on intermediate storage. As a result the dry storage technology has become a major activity and business component of today's back-end fuel strategies.

The key differences between dry storage and in-reactor performance of fuel are (Adamson *et al.*, 2010):

- Long storage time – 40 years or more.
- Inert gas, helium (He) storage atmosphere instead of pressurized or boiling water (decreased heat transfer, but no corrosion).
- Decay heat that can raise the cladding temperature to 400°C or higher, then decreases over time.
- Atmospheric storage pressure that, combined with high fuel rod temperature and internal gas pressure, results in a high clad ΔP and clad stresses that decrease with time as well.
- No external radiation (no additional radiation damage).
- Dry cask storage containers dissipate the fuel decay heat by natural convection of the cask He atmosphere and conduction through the cask container walls; there are no moving parts or forced cooling in this system. As a result, the cladding can reach temperatures of several hundred degrees Celsius. The pressure differential across the cladding can be significant since the fuel rod internal gas pressure is made up of (Adamson *et al.*, 2010):

Table 5.1 Comparison of international dry storage regulations (Adamson *et al.*, 2011)

Dry storage criteria	USA	Germany	Hungary	S. Korea (CANDU only)	Spain (follows USNRC)	Switzerland
Pellet clad °C, storage, drying, other	400° 570°	370° 370°, 100 hr	410° 410°	None but maintain clad integrity	400° 570°	None 570°
Temperature cycling Max. no. of cycles and Max. ΔT (°C) per cycle	10/65°	None	None	None	10/65°	10/65°
BU, GWD/MT	None	65 assembly avg.	49–50	None	None	None
Clad hoop stress, MPa	90 (if T_{clad} > 400 °C)	120@370°	None	None	90	Depends on fuel supplier*; typical values are 90 and 120
Clad strain limit	None	1% circumferential	None	None	1% for HB	1% during storage
Creep rupture limits	None	None	None	None	Limited by °T and stress limits	
Cladding condition limits	Oxide thickness for stress calculation	Oxide thickness for stress calculation	None	None	Limited by °T and stress limits	None

Failed fuel definition	NUREG-1536 Rev 1a (ISG1, rev. 2)	Any assembly with defected rods	Failed assembly identified by sipping	Clad penetration that emits fission products	NRC ISG1, rev. 1	None
Failed fuel placement in storage cask	NUREG-1536 Rev 1a (ISG1, rev. 2)	Yes in special container (ZIRAT11/IZNA6, Sect. 11.3.2)	Not approved	Not approved	Yes, case by case to meet safety criteria	Not approved
Reactivity requirement (max, K_{eff})	0.95	0.95 normal 0.97 accidents	$K_{eff}+\Delta K_{eff}$ 0.95	0.95	0.95	<1
BU credit	Yes, with actinides & fission products	Yes, with actinides & fission products	No; storage is based on enrichment of fresh fuel	"none"	Yes, with actinides, only for transportation license	Yes, acceptable for transportation license
Analysing with flooding	Yes	Yes	Yes	"none"	Yes	Yes
Moderator exclusion claim	No, likely	No	No	"none"	No	No
Worst accident assumed	9 m cask drop	9 m cask drop	PSA approach for max. radioactive exposure	Coincident failure of "cylinder" and 600 CANDU assemblies	No criticality in transport accident with flooding	Airplane crash + kerosene fire, earthquake for storage + 9m drop etc. for transport

(*Continued*)

Table 5.1 (continued)

Dry storage criteria	USA	Germany	Hungary	S. Korea (CANDU only)	Spain (follows USNRC)	Switzerland
Fuel failure allowed in worst case accident	Yes	Yes	Yes	No failure during cooling blockage	Yes, within off-site dose limits	Yes
Acceptable failure conditions	10CFR72 & source term limit	10% failed rods 1% fuel release	Retrievability Criticality Coolability	–	% fuel failures 1% – normal 10% – off normal 100% – accident + maintain max cask internal pressure	Retrievability
Modelling codes	ABACUS LS DYNA FALCON	MCNP	MCP4, SCALE 5.1, MONK 6B, RISK SPECTRUM, TRANSURANUS, ENIGMA + others	–	MCNP, SCALE, ANSYS, FLUENT ABACUS, LS DYNA 3D	CSAS (AREVA) for normal conditions

ANT International, 2011

* Differences in the allowable cladding stress among fuel vendors and plants was identified through private communication subsequent to the survey (Hellwig, 2011).

Table 5.2 Maximum BUs achieved vs Regulatory limits (excludes LUAs), A. Strasser in Adamson *et al.* (2010, ZIRAT 15 Annual Report)

Country	BU (GWD/MT)				
	Batch	Assembly	Rod	Pellet	Regulatory limit
USA	54	58	62	73	62.5 peak rod
Belgium		50–55			55 UO_2 assy, 50 MOX assy
Czech Republic	51	56	61		60 peak rod
Finland	45.6	46.5	53		45 assy
France	47	51 UO_2 42 MOX			52 assy
Germany	58	62	68		65 assy
Hungary		50	62		
Japan	50	55	62		55 UO_2 assy, 45 MOX assy
Korean Republic	46				60 rod
Netherlands	51.5	58	64.5		60 rod
Russia	45	56	60	68	
Spain	50.4	57.4	61.7	69	
Sweden	47	57.2	63.6		60 assy, 64 rod
Switzerland	58	60	65	71	75 pellet
Taiwan					60 rod (P), 54 assy (B)
UK	44.3	46.5	50		55 pellet
Ukraine		50			

Source: A.N.T. International, 2011

- He gas pre-pressurization during fabrication
- fission gases
- He from transmutation of B in burnable absorbers
- alpha decay of the Pu isotopes in the fuel during storage

and their pressure is further raised by the fuel decay heat. All of these pressure sources, except the pre-pressurization level, are burnup dependent.

The pressure outside the cladding in the cask is only slightly above atmospheric. Creep deformation of the cladding will occur at a relatively constant rate early in life, when the internal gas pressure, the cladding stresses and the cladding temperature are at their highest. As the decay heat decreases with time, the gas pressure and the cladding temperature both decrease. In addition, the internal free volume of the fuel rod increases as the cladding creeps outward, decreasing the gas pressure and cladding stresses further. All three of these factors eventually reduce the creep rate to a negligible value (Adamson *et al.*, 2010).

Creep-rupture is the most likely cladding failure mode during dry storage and there is general consensus on this mechanism among the industry and the regulators (Adamson *et al.*, 2010). The parameters that determine the potential for creep rupture are the cladding stress level, the cladding temperature and the rate of decay heat decrease, all parameters that are burnup dependent.

The cladding temperature, currently one of the primary USNRC licensing criteria, is determined by the decay heat generated in the fuel, the heat transfer capability of the cask and the surface temperature of the cask in its storage environment (Adamson *et al.*, 2010). The decay heat is generated primarily by absorption of the alpha decay either directly or indirectly from the plutonium (Pu) isotopes. Increasing burnup will increase the level of Pu isotopes formed by transformation of the ^{238}U, and increase the cladding temperature in dry storage conditions. In comparison, MOX fuel will have significantly higher temperatures under the same conditions.

Several other potential failure mechanisms were considered, but eliminated as highly unlikely (Rashid, 2006). They are summarized below (Adamson *et al.*, 2010).

Stress corrosion cracking (*SCC*) is not a credible failure mechanism in dry storage because:

- There is insufficient elemental iodine present to cause SCC.
- At the stress and strain rates in dry storage, initiation of intergranular cracking is nearly impossible; the 180–200 MPa stresses needed for SCC are well above those for high burnup fuel rods.
- Hydrides, including radial hydrides, will not affect iodine induced SCC.
- The occurrence of all the conditions that cause DHC is highly unlikely, but cannot be ruled out. The initial conclusions are based on the following evaluation:
  - Analyses indicated that at a hoop stress of 250 MPa (well above dry storage conditions) in a cladding wall thickness reduced by 100 μm oxide with an 83 μm crack size, the stress intensity factor is below that needed to initiate the DHC process,
  - Hydride re-orientation that might assist crack propagation is intended to be minimized or prevented by current regulations and industry practices, but cannot be ruled out.
  - In addition, propagation of a crack assisted by radial hydrides may not occur for many of the hydride morphologies.

Also under accident conditions during storage or subsequent transportation, the fuel must remain subcritical and should be recoverable by normal methods (Adamson *et al.*, 2010). The hypothetical accident conditions that these criteria have to meet, as defined by the USNRC, are specified in 10CFR71.73 (NRC, Rules and Regulations, Title 10 Code of Federal Regulation, Chapter 71). Of all the accident scenarios the most limiting scenario is a free drop of the cask for a distance of 9 m (30 ft) onto a flat, unyielding horizontal surface, striking the surface in a position that would cause the maximum fuel damage.

Radial hydrides in zirconium alloy cladding are undesirable because they reduce the critical stress intensity required to propagate a radial crack through the wall of the cladding during handling or transportation (Adamson *et al.*, 2010). The objectives of the dry storage regulations are to limit the conditions that could result in hydride re-orientation.

A certain fraction of the hydrogen (H) picked up during the oxidation reaction is soluble in the zirconium matrix and the remainder forms zirconium hydrides (Adamson *et al.*, 2010). The solubility of the H is a function of temperature, alloy composition and microstructure. Solubility is also a function of irradiation history, heating or cooling rates during service. The orientation of the hydrides formed during normal reactor operation are generally circumferential near the cooler cladding OD and remain so during wet storage of the spent fuel.

The hydrides can reorient in the radial direction if they are precipitated from solid solution by cooling the alloy from a higher temperature under a tensile or hoop stress (Adamson *et al.*, 2010). The hydrides will align themselves in the direction perpendicular to the tensile stress. This can occur during reactor operation although it is generally unlikely. It could occur during dry storage if the internally pressurized cladding is at a high temperature, holds sufficient hydrogen in solution and is then cooled while under the hoop stress. The hydrides in solution will precipitate in the radial orientation (provided the hoop stresses are large enough), while the hydrides that did not dissolve will remain in their original circumferential orientation. This is most likely to occur during rapid cool-down from high temperatures after cask drying or evacuation procedures rather than during storage when the rate of temperature and pressure reduction that control the stress levels are extremely slow.

In summary, the factors that affect hydride re-orientation in irradiated cladding are (Adamson *et al.*, 2010):

- Hoop stress.
- Maximum temperature.
- Cool-down rate and final temperature.
- Solubility of H in the specific alloy at its specific burnup that will determine the amount of H in solution at the maximum temperature and the amount of circumferential hydrides.

- Microstructure features such as grain size and shape, amount of CW, and perhaps others.
- Texture.

The radial hydrides can be present in a wide variety of sizes and distributions as well as fractions of the total hydrides present and each type of structure can have a different effect on mechanical properties. This emphasizes the importance of characterizing the structures when they are related to the mechanical properties measured.

5.5 Inspection methods

There are several reasons why poolside and/or hot cell examinations are undertaken on fuel assembly (FA) components (Rudling & Patterson, 2009). These are:

1 Root cause investigations of failed FA components
   (a) A 'failed' FA component has a wider meaning in this respect. It not only means that the component has physically failed but it could also mean that the component does not behave satisfactorily, for example FA bowing that is so large that control rods cannot be inserted.
2 Maintaining good fuel reliability by:
   (a) Providing baseline data before a change in operational environment of the fuel.
   (b) Getting early warnings of potential issues.
3 Fuel vendor design and licensing data such as:
   (a) Providing data to material models and fuel performance codes
   (b) Verification of the good performance of a new fuel design
   (c) Assessment of the effects of changes in the operating environment; for example, water chemistry improvements or higher exposures

The in-pool examinations are usually non-destructive, but can also involve destructive operations such as breaking a flow tab off a spacer or cutting coupons from a channel for measurement of hydrogen concentration. Hot cell examinations normally start with non-destructive followed by destructive examinations. The costs for hot cell examinations are much more expensive than those carried out in pool. However, certain material characteristics can only be assessed in a hot cell. In the following subsections examples of different examination techniques and results obtained are discussed. The interested reader is referred to Rudling & Patterson (2009).

5.5.1 Poolside examinations

Sipping

If a core is suspected or known to contain leaking fuel, the fuel is almost universally checked for leaks during the subsequent refuelling (or sometimes forced) outage. Various techniques are used to identify leaking fuel assemblies and are collectively referred to as 'sipping' methods. They rely on changes in the concentration of fission products that are released from the FA or rod being tested to signify the presence of a leak; such as, changes in gamma activity, beta activity, isotopic composition of fission products or combinations of such measurements. The technique used to identify a leaking assembly can vary depending on the size of the leak, the background activity from tramp uranium and on the time of sipping relative to shut down. Changes in the gross (total) gamma activity of water or noble gas samples representative of the fuel being tested are sometimes used to identify a leaker, particularly in cases of low background activity. In general, however, changes in the gamma or beta activity of nuclides with moderate-to-long half lives are typically used to minimize the effects of background activity and decay time; examples include Xe-133 (5.25 d), I-131 (8.04 d) or Kr-85 (10.72 y). In addition, many of the current sipping methods also involve the collection and measurement of noble gases to enable the detection of leaks that are too small to allow the release of soluble fission products in quantities clearly detectable in the sipping process.

Historically, sipping methods have been classified as wet, dry or vacuum, based on the manner in which fission products are collected for measurement (Lin, 1996). In practice, however, sipping techniques can better be classified as 'open' or 'closed' methods based on the manner in which the fuel being tested is isolated from other assemblies during the sipping process. The wet, in-reactor methods represented by the TELESCOPE, INMAST and various hood systems are open methods. They have become the primary means of leak detection because of the small amount of time needed to inspect a BWR, PWR or VVER core; for example, ~16 h for a large BWR with an in-core sipping hood versus close to a week for vacuum sipping (Knecht *et al.*, 2001). Wet sipping is also used in storage pools to test individual assemblies or fuel rods. The dry method makes use of a canister to isolate the fuel being tested and is a closed method. Dry sipping is not used today in power reactors because of issues related to handling and test time and to decay heat and cladding temperature. Vacuum sipping also makes use of a canister to isolate individual fuel assemblies and is frequently used to supplement the in-reactor methods because of its higher resolution and detection capabilities.

The wet, in-core or in-mast sipping methods are used almost exclusively in power reactors because of their speed and the need for short refuelling outages. Vacuum or similar wet canister sipping is now used to confirm or resolve the results of the in-reactor methods.

Ultrasonic testing

Ultrasonic testing (UT) is being used to detect leaking BWR and PWR fuel rods. The testing method makes use of differences in scattering by water and gas of ultrasonic (pressure) waves as they are reflected between the inner and outer surfaces of fuel cladding. The UT process makes use of two probes, which move laterally across a FA. The probes are arranged in a "pitch-catch" configuration in which one probe transmits a signal, which is picked up by the cladding, reflected between the inner and outer surfaces and ultimately received by a second probe. The scattering by water at the inner surface of a failed rod is greater than the scattering by gas at the inner surface of a sound rod. Such differences are used to identify the specific rod or rods that are leaking.

Eddy current testing

The zirconium oxide thickness can be measured by the eddy current (EC) technique. However, before measuring the fuel rod oxide thickness it is important to remove as much crud from the rods as possible. This may be done by brushing the rod before the oxide is measured by the EC probe. If crud is not removed, the oxide thickness value normally obtained is too large. In many cases the crud deposited onto the fuel rods may be ferromagnetic and therefore the EC technique may fail to give reliable results unless the EC equipment has been designed to compensate for ferromagnetic crud. Eddy Current technique can also be used to detect fuel claddings defects, for example non-penetrating cracks.

Visual examinations

Visual examination of the fuel assembly is used to:

- Get a general impression of the condition of the fuel assembly.
- Assess zirconium oxide type and whether oxide spallation has occurred or not.
- Identify the primary failure cause and characterize degradation of failed fuel.
- Measure component dimensions by means of translation stages and position encoders.

Fuel assemblies are sometimes inspected while supported on the fuel handling mast. More commonly, visual inspections are performed in fuel preparation machines located in fuel storage pools. The fuel preparation machines are usually equipped with a rotating fixture to enable inspections of all four assembly faces without the use of the bridge crane and with a sensor to record axial position.

Visual inspection equipment, typically consists of:

- Television camera and spotlights.
- In-pool support system for camera and lights.
- Data acquisition, processing and storage equipment.

Linear variable differential transformers (LVDTs)

Poolside equipment equipped with linear variable differential transformers (LVDTs) can be used to determine the dimensions of fuel assembly components. As an example, LVDTs may be used to measure the dimensions of the outer flow channel of BWR fuel assemblies. In this case the position of a channel relative to a reference surface is measured. Each of the four channel sides is measured by three transducers and consequently 12 axial traces are obtained simultaneously over the circumference of the channel. From these measurements, the bulge, bow, and twist may be calculated over the total length of the channel. Bulge is an outward deformation of the channel faces which results from the difference in pressure between the inside and the outside of the flow channel and from the effects of fast neutrons on channel creep. Twist is an angular deformation over the length of an assembly.

The length of the fuel channel is also of interest and can be measured with in-pool devices that range from tape measures to caliper-like gauges equipped with LVDTs or similar sensors. Length data are used to assess the effects of differential growth in the axial direction on the fit and remaining growth margin of the channel relative to its fuel bundle. Length data can also be used to estimate lateral bow in cases where time or equipment availability prevents more accurate, explicit measurements.

5.5.2 Hot cell examinations

Shielded containments are commonly referred to as hot cells, the word 'hot' being used as a synonym for radioactive. Hot cells are used in both the nuclear power and the nuclear medicines industries. They are required to protect individuals from radioactive isotopes by providing a safe containment box in which they can control and manipulate the equipment required.

Hot cells are used to inspect spent nuclear fuel rods and to work with other items, which are high-energy gamma ray emitters.

The methods utilized in a hot cell examination vary with the objective of the program, but usually begin with a visual inspection. Visual inspections are used to characterize the external surfaces of a fuel rod, including the crud and oxide layer, and to identify features of interest for subsequent examinations. High resolution diametral profilometry and EC lift-off measurements are also used to characterize the thickness of the crud and oxide layers relative to axial and azimuthal position, particularly if such measurements were not performed before transporting the rod to the hot cell.

EC measurement with encircling, pancake or pencil coils can detect the axial and tangential location of cracks and other damage that might not be readily discernible in visual inspections; for example, small or incipient cracks due to PCI.

Neutron radiography – If a test reactor is adjacent to a hot cell, neutron radiography can be performed on the fuel rod to non-destructively locate regions with different hydrogen levels in the cladding. Depending on the method used for capturing and displaying the neutrographic image, accumulations of hydrogen are indicated by light or dark areas. Neutron radiography can also reveal fuel washout in the case of degraded failed fuel. In addition, neutron radiography provides geometrical information of the fuel column for the selection of cutting positions of samples for ceramographic examinations.

Destructive examinations – The fuel and cladding can be examined by means of a number of destructive methods including:

- Fission gas collection and analysis.
- Cladding oxide thickness (ID and OD) measurement.
- Hydride concentration and morphology.
- Second phase particle size and distribution.
- Microstructure characterization.
- Ceramographic characterization of fuel pellets.
- Micro-gamma scanning across the diameter of fuel pellets.
- Electron microprobe analysis of the fuel pellets and the pellet-cladding interface.

The fission gas that was released from the fuel to the free volume within a rod is normally collected and analyzed as the first step in the destructive examinations of sound fuel rods. The collection process involves puncturing the cladding and collecting the free gas in one or more containers of known volume, pressure and temperature. This process not only captures the gas from the open volume within a fuel rod, but also provides data for the calculation of the net internal void volume and the total quantity of gas in the open volume. The gas sample is analyzed for composition by techniques such as gas chromatography to identify the concentration of Xe, Kr, CO-CO_2, CH_4 and other constituents. The results are adjusted for decay and combined with calculations or measurements of exposure to establish the fraction of the generated fission gas that was released to the free volume within the fuel rod.

Metallographic mounts of the cladding and fuel may be prepared, polished and photographed. The mounts may be examined for clad oxide layer (external and internal), fuel/clad interactions, fuel restructuring, rim effect and agglomerate behaviour. The mounts may also be etched to reveal grain boundaries for an analysis of the grain size.

Hydrogen/hydride analysis – A hot vacuum extraction process may be used to determine clad hydrogen content such as that employed by the LECO method. (LECO is a corporation which provides instrumentation for elemental determination in organic and inorganic materials.) When such analyses are done, it is important to separate the total hydrogen evolved from the sample into concentrations from the surface layer(s) and from the metal itself (which is of interest). Specifically for samples with thick oxides, water molecules adsorbed at the cracked and porous zirconium oxide may contribute to the recorded hydrogen content and give a metal hydrogen content that is too large. It is therefore crucial that the adsorbed water molecules are removed before the hydrogen content in the sample is measured. This can be done by grinding of the oxide. but in this case care is needed so as not to remove massive hydrides at the outer metal surface of a fuel rod. For components without a surface heat flux there is no concentration of hydrides at the outer metal surface and there is lower risk of removing hydrides from the sample by grinding.

In addition to gas-extraction methods, detailed destructive post-irradiation examination (PIE) can be used to characterize the local oxide thickness and hydrogen concentration. Backscattered electron imaging (BEI) of polished cladding in the SEM may be used together with image analysis to determine the local hydrogen concentration and radial hydrogen profile through the cladding wall. This is a relatively new application of BEI which allows the relative positions of hydrides and oxides to be located more precisely than the gas-extraction methods.

Alternatively, the hydride distribution and orientation in the cladding material is determined by optical examination after etching the sample to reveal the hydride locations.

Crud analysis – The morphology of the *crud* may be examined by electron probe microanalysis (EPMA).

Mechanical tests – Mechanical tests of differently shaped samples from the fuel cladding can be carried out in the hot cell. Some of the tests that can be done with irradiated cladding are:

- Room and elevated temperature axial tensile test yield strength (YS), ultimate tensile strength (UTS), uniform elongation (UE) and total elongation (TE).
- Cladding room and elevated temperature ring tensile test (YS, UTS, UE and TE).
- Burst testing.
- Cladding thermal creep rate test.
- Hardness testing.

5.6 Future trends and research needs

Improved fuel reliability and operating economics are the driving forces for changing operating conditions, while at the same time maintaining acceptable margins to operating and regulatory safety limits. Table 5.1 gives the trends for burnup (BU) achieved compared to regulatory limits in various countries. An approximate conversion of BU to fluence is 50 GWd/MT, which is equivalent to about 1×10^{22} n/cm^2, $E > 1$ MeV (or about 17 dpa), but this depends on many nuclear parameters such as enrichment, extent of moderation and neutron energy spectrum. In general PWRs operate to higher discharge burnups compared to BWRs because of higher PWR power densities and neutron fluxes, but the differences are decreasing with time. There are some incentives to reach burnups of 60–70 GWd/MT batch average, but the economic values of doing so are decreasing. The majority of US plants and many in Europe have undergone power uprates from a few per cent to up to 20%. This increases the number of fuel assemblies in a core that operate at high power, thereby decreasing the margin to established limits. In cooperation with utilities, fuel suppliers have operated lead test assemblies (LTA) or lead use assemblies (LUA) to very high burnup, in some cases approaching 100 GWd/MT peak rod exposure (Strasser in Adamson *et al.*, 2010).

As discussed in earlier sections, as burnup and fluence become higher so material properties and microstructure evolve. Examples include:

- In PWRs it is found that Zircaloy-4 no longer meets corrosion and hydriding needs therefore virtually all current PWR cladding uses a zirconium alloy containing Nb.

- Although not a new phenomenon, observed second phase precipitate dissolution and re-precipitation have required a new perspective on alloy development and hydrogen pickup.
- BWR channel bow at high burnup has required a new understanding of the relationships between hydrogen pickup, shadow corrosion and irradiation growth.

A broader listing of issues needing resolution includes:

- Corrosion related to oxide thickness and H pickup
  - BWRs and PWRs:
    - Mechanism of solid hydrides on corrosion mechanisms.
    - Effect of Nb.
  - BWRs:
    - Shadow corrosion mechanisms and their relation to channel bow.
    - Late increased corrosion and HPU of Zircaloy-2 at high burnups.
    - Crud-chemistry-corrosion interaction.
    - Effect of water chemistry impurities, as well as specific effects of NMCA, with or without Zn-injection.
  - PWRs:
    - Effects of surface contaminations and/or boiling on Zr-Nb alloys.
    - Welding of the new alloys may need improved processes (Zr-Nb alloys).
    - Effect of increased Li together with increased duty (subcooled boiling) with and without Zn-injection.
    - Effects of increased hydrogen coolant content (to mitigate PWSCC).
    - Axial offset anomaly (AOA) mechanisms.
- Mechanical properties related to irradiation and H pickup
  - Decreased ductility and fracture toughness as consequence of the increased HPU and formation of radial hydrides during any situation (e.g. RIA, PCMI, LOCA and post-LOCA events, seismic event, transport container drop-accident conditions).
  - Quantification of the effect of irradiation on hydrogen solubility and mechanism(s) by which the phenomenon occurs.
  - Details of deformation mechanisms in zirconium alloys, including being able to predict the dislocation channelling system.
  - Development of micromechanical models applicable to deformation at appropriate component conditions.

  - DHC mechanism (degradation of failed fuel, outside-in cracking and dry storage).
  - Role and kinetics of Fe, Cr, Ni from dissolving SPPs in Zircaloy and Zr-Nb alloys for corrosion, mechanical properties and dimensional stability.
- Dimensional stability
  - Effect of hydrogen on irradiation growth mechanisms.
  - PWR fuel assembly bowing mechanism.
  - BWR fuel channel bowing mechanism and parameters affecting them such as: texture, residual stress, flux gradient and hydrogen gradient.
  - Mechanism of loop formation in zirconium alloys.
  - Mechanisms of both irradiation and post-irradiation creep.
    - Role of Nb in decreasing irradiation growth.
  - The effects of texture and hydrogen pickup of Zr-Nb alloys as related to growth of PWR guide tubes.
  - Effects of thermal and radiation induced relaxation of Zr- and Ni-alloys, particularly relative to spacer grids.
- PCI and PCMI
  - PCMI failure mechanism during out-side in cracking, and possible relevance to failure mechanism for high burnup fuel?
  - PCI failure mechanisms due to Missing Pellet Surfaces (MPS).
  - PCI mechanism and performance of liner/barrier and pellets with additives at high burnup.
- LOCA
  - Verification of coolant blockage with real fuel rods in lattice design, related to maintenance of coolable geometry.
  - Mechanism of runaway oxidation in Russian E110 alloys.
  - Conditions when alpha-Zr layers are formed due to fuel clad bonding.
- RIA
  - Effects of hydride orientation, hydrogen distribution and hydrogen content on PCMI fuel clad failure mechanism.
- Severe accidents
  - Performance and phenomena when coolable geometry cannot be sustained.
- Intermediate dry storage
  - Effects of the projected longer dry storage times before final disposal on hydride re-orientation and its consequences during a cask drop accident
- Effects of hydride orientation, hydrogen distribution and hydrogen content as well as temperature during a cask drop accident
  - Effects of irradiation on hydrogen solubility of various Zr alloys.

– The effects of hydrogen, temperature, stress and time on DHC in relation to extended storage of current, high burnup fuel.

5.7 Sources of further information and advice

Major sources are many, including:

- ZIRAT Annual Reports and ZIRAT Special Topic Reports, A N T International, Mölnlycke, Sweden (www.antinternational.com).
- The series of *Zirconium in the Nuclear Industry, International Symposiums*, ASTM International, West Conshohocken, PA, USA, held every 2–3 years.
- *Zirconium Production and Technology: The Kroll Medal Papers 1985–2010*, editor, R. B. Adamson, ASTM International RPS2, ASTM I, West Conshohocken, PA, USA, 2010.
- Proceedings of the LWR Fuel Performance Meeting/Top Fuel/WRFPM, held annually in the US, Europe or Asia.
- References given in Section 5.9.
- The previous chapter of this book, 'Properties and Performance of Zirconium Alloy Components for Nuclear Power Light Water Reactors', R. B. Adamson, Zircology Plus, Fremont, CA, USA and P. Rudling, ANT International, Mölnlycke, Sweden.

5.8 Acknowledgements

The authors sincerely thank our colleagues in the expert network staff of ANT International: Alfred Strasser, Friedrich Garzarolli, Brian Cox and Charles Patterson. Their discussions, their expertise, their comments and their contributions to the ZIRAT programme reports have made Chapter 5 a possibility.

5.9 References

Adamson R., Cox B., Davies J., Garzarolli F., Rudling P. and Vaidyanathan S., '*Pellet-Cladding Interaction (PCI and PCMI)*', ZIRAT11/IZNA6, Special Topics Report, ANT International, Mölnlycke, Sweden, 2006/2007.

Adamson R. B., Garzarolli F., Patterson C., Rudling P., Strasser A. and Coleman K., '*ZIRAT15 Annual Report*', ANT International, Mölnlycke, Sweden, 2010.

Adamson R. B., Garzarolli F., Patterson C., Rudling P., Strasser A., Coleman K. and Lemaignan C., '*ZIRAT16 Annual Report*', ANT International, Mölnlycke, Sweden, 2011.

Arborelius J., Backman K., Hallstadius L., Limbäck M., Nilsson J., Rebensdorff B., Zhou G., Kitano K., Löfström R. and Rönnberg G., '*Advanced Doped UO_2*

Pellets in LWR Applications', Proc. Water Reactor Fuel Performance Meeting, pp. 35–46, Kyoto, Japan, 2–6 October 2005.

Cantonwine P., Crawford D., Downs M., Joe B., Bahensky T., Reimer J., del la Hoz C., Petersen K., Reitmeyer M., Morris J. and Zbib A., '*Channel – Control Blade Interference Management at LaSalle 1 and 2 during 2007 and 2008*', Proceedings of Top Fuel 2009, Paper 2154, pp. 6–15, Paris, France, 6–10 September 2009.

Chung H. M. and Kassner T. F., '*Cladding Metallurgy and Fracture Behavior during Reactivity-Initiated Accidents at High Burnup*', Nucl. Eng. Design, **186**, pp. 411–427, 1998.

Davies J., Vaidyanathan S., and Rand R., '*Modified UO_2 Fuel for High Burnups*', TopFuel 1999, Avignon, France, pp. 385–395, 1999.

Delafoy Ch., Blanpain P., Maury C., Dehaudt Ph., Nonon Ch. and Valin S., '*Advanced UO_2 Fuel with Improved PCI Resistance and Fission Gas Retention Capability*', TOPFUEL, Proc. Int. Conf. Würzburg, Germany, 16–19 March 2003.

Delafoy C., Dewes P. and Miles T., '*AREVA NP Cr_2O_3-Doped Fuel Development for BWRs*', Proceedings of the 2007 International LWR Fuel Performance Meeting, pp. 1–8, San Francisco, California, USA, paper 1071, 2007.

Edsinger K., '*A Review of Fuel Degradation in BWRs*', Proc. Int. Topical Meet. on Light Water Reactor Fuel Performance, Park City, UT, 10–13 April, Vol. 1, pp. 523–40, 2000.

Efsing P. and Pettersson K., '*Delayed Hydride Cracking in Irradiated Zircaloy Cladding*', Proc. 12th Int. Symp. on Zr in the Nuclear Ind., Toronto, ON, 15–18 June, ASTM-STP-1354, pp. 340–355, 1998.

Hellwig C., *Comments on ZIRAT15 Annual Report*, 2011 (private communication).

IAEA, '*Review of Fuel Failures in Water Cooled Reactors*', Technical report series no. 388, IAEA, Vienna, 1998.

Knecht K., Stark R. and Habeck K., '*In-Core Sipping at BWR Plants in Only 16 Hours*', Top Fuel 2001, paper 2–24, 2001, pp. 1–2.

Lespiaux D., Noirot J. and Menut P., '*Post-test Examinations of High Burnup PWR Fuels Submitted to RIA Transients in the CABRI Facility*', Proc. ANS Topical Meeting on Light Water Reactor Fuel Performance, Portland, Oregon, 2–6 March 1997, pp. 650–658, 1997.

Lin C. C., '*Radiochemistry in Nuclear Power Reactors, National Research Council*', Nuclear Science Series, NAS-NS-3119, National Academy Press, Washington, D.C., 1996.

Lysell G., Grigoriev V. and Efsing P., '*Axial Splits in Failed BWR Rods*', Proc. Int. Topical Meet. on Light Water Reactor Fuel Performance, Park City, UT, 10–13 April, Vol. 1, pp. 541–555, 2000.

Mahmood S.T., Lin Y-P, Dubecky M. A., Edsinger K. and Mader E. V., '*Channel Bow in Boiling Water Reactors – Hot Cell Examination Results and Correlation to Measured Bow*', Proc. of the 2007 International LWR Fuel Performance Meeting, Paper 1061, pp. 124–133, American Nuclear Society, San Francisco, CA, USA, 2007.

Matsunaga J., Takagawa Y., Kusagaya K., Une K., Yuda R., Hirai M., Makovicka M. D. and Hogan P. K., '*Fundamentals of GNF Al-Si-O Additive Fuel*', Proceedings of Top Fuel 2009, paper 2003, Paris, France, 6–10 September 2009, pp. 767–772.

Matsunaga J., Une K. and Kusagaya K., '*Chemical Trap Effect of Aluminosilicate Additive Fuel*', 2010 LWR Fuel Performance Meeting, Orlando, Florida, USA, 26–29 September 2010.

Northwood D. D. and Kosasih U., '*Hydrides and Delayed Hydrogen Cracking in Zirconium and Its Alloys*', Int. Metals Reviews, **28**, pp. 92–121, 1983.

Olander D. R., Wang W., Kim Y. S., Li C. and Lim K., '*Chemistry of Defective Light Water Reactor Fuel*', EPRI report No. EPRI TR-107074, 3564–02, 1997.

Patterson C., '*Processes Going on in Nonfailed Rod during Normal Operation – Volume I*', ZIRAT15/IZNA10 Special Topical Report, ANT International, Mölnlycke, Sweden, 2010.

Rashid J., '*Spent-Fuel Transportation Applications: Modeling of Spent-Fuel Rod Transverse Tearing and Rod Breakage Resulting from Transportation Accidents*', EPRI Report #1013447, October, 2006.

Rudling P. and Patterson C., '*Fuel Material Technology Report, Vol. IV*', ANT International, Mölnlycke, Sweden, 2009.

Strasser A., Rudling P., Cox B. and Garzarolli, '*The Effect of Hydrogen on Zirconium Alloy Performance*', ZIRAT13 Special Topical Report Vol. II, ANT International, Mölnlycke, Sweden, 2008.

Strasser A., Epperson K., Holm Jerald, Rudling P., Lundberg S., '*Fuel Design Review Handbook*', ANT International, Mölnlycke, Sweden, 2010a

Strasser A., Garzarolli F. and, Rudling P., '*Processes Going on in Nonfailed Rod during Accident Conditions (LOCA and RIA) Volume II*', ANT International, Mölnlycke, Sweden, 2010b.

6

Ageing of electric cables in light water reactors (LWRs)

H. M. HASHEMIAN, Analysis and Measurement Services Corp., USA

DOI: 10.1533/9780857097453.2.284

Abstract: This chapter will address the ageing of nuclear power plant cables and test methods for these cables to manage ageing and verify reliability. The focus will be on instrumentation and control (I&C) cables, low-voltage cables and medium-voltage cables. Ageing due to long-term exposure to temperature, radiation, humidity, and other environments can cause the cable insulation material to deteriorate, allowing moisture into the cable. This can in turn cause cable failure and jeopardize plant safety. Various techniques are available to assess cable condition and health, including electrical and mechanical measurements, and chemical tests. Of these, electrical measurements are preferred as they allow *in-situ* cable testing in operating plants. Prognostic techniques estimate residual life of cables using data from periodic tests. To guard against ageing, nuclear power plants are implementing ageing management programs and regulators are writing new requirements for acceptable programs and techniques for cable ageing management.

Key words: insulation resistance, high-potential (Hi-Pot), partial discharge, quality factor, dissipation factor, AgeAlert™, LCR (inductance, capacitance, and resistance) tests, time domain reflectometry, frequency domain reflectometry, reverse time domain reflectometry.

6.1 Introduction

The thousands of miles of electrical cable and wire in light water reactors deliver the power and the signals enabling safety- and non-safety-related equipment to operate in normal and in post-accident conditions (U.S. NRC, 2010a; Hashemian, 2010; AMS Corp., 2011). All plant instrumentation and control (I&C) systems depend on reliable plant wiring (AMS Corp., 2010). They bring the necessary signals to the operators, control equipment, and safety systems, as well as delivering commands to activate relays, pumps, valves and motors. Reliable instrumentation signals are often essential to maintaining redundancy or containing an accident, and the loss of a cable can result in the loss of crucial performance and operational data. Similarly, dependable power cables enable pumps and valves to be activated or deactivated as the plant recovers (Hashemian, 2010; U.S. NRC, 2001).

In normal but particularly in accident situations, cables and their coatings can be exposed to a range of stressors, from heat and humidity to thermal and mechanical shock and radiation. When instrumentation cables and power cables become exposed to or submerged in water, chemicals, high pressure steam or other environments during an accident, accident recovery and mitigation can be severely endangered (Hashemian, 2010).

Because cables have been regarded as passive, durable components that have proven their ability to evade the need for replacement; they have traditionally received insufficient attention (U.S. NRC, 2010b). This neglect was partly because the wholesale replacement of cables is almost impossible; cables are only readily replaceable in short sections (AMS Corp., 2011). However, the issue of cable ageing and degradation has intensified as the trend toward license renewal has enabled existing plants to operate for up to 60 or even 80 years – twice their original planned life. The accident at the Fukushima nuclear power plants – the oldest of which is more than 40 years old – has raised questions about the advisability or practicality of extending the life of old plants. But whether plants are operated for 40, 60, 80 years or more, the Fukushima tragedy has underscored the importance of safety, in which cables play a ubiquitous and important role (AMS Corp., 2011).

Given that wholesale replacement of cables is neither prudent nor practical, cost-effective ageing management strategies and techniques that objectively assess cable condition and remaining life are essential (AMS Corp., 2010). This chapter addresses the ageing of nuclear power plant cables – especially I&C, low-voltage, and medium-voltage cables – and methods for testing these cables to manage their ageing and verify their reliability.

6.1.1 Cable component types and properties

Several hundred different cable types and sizes are used throughout a typical light water plant. Plant conditions determine which type of cable is used; for example, cables for control rod drive mechanisms must withstand higher temperatures and have additional shielding capacity (IAEA, 2011). I&C cables are by far the most common cable type (Hashemian, 2010). Instrumentation cable, which includes thermocouple (T/C) extension wires, is a low-voltage (< 1 kV), low-ampacity cable used to transmit digital or analog measurement signals from transducers such as resistance temperature detectors (RTDs) and pressure transmitters. Control cable, also low-voltage and low-ampacity, is used in the circuits of control (rather than monitoring)

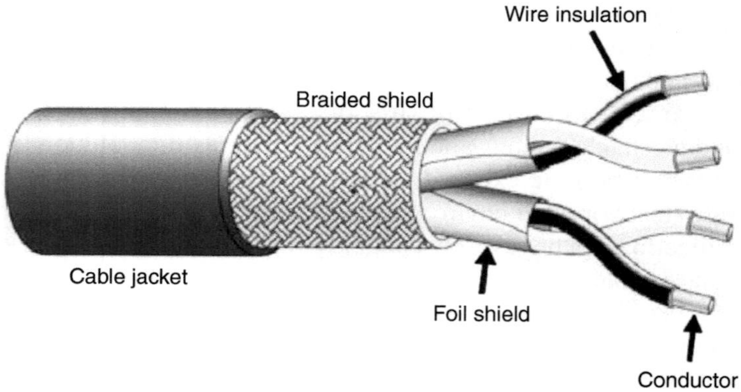

6.1 Cable components.

components such as control switches, valve operators, relays, and contactors (Hashemian, 2010).

A complete cabling system (see Fig. 6.1) may include any or all of the following components: conductor, insulation, shield, jacket, terminations, penetrations, splices, connectors, and/or end devices (sensor, transmitter, detector, motor, etc.) (AMS Corp., 2010). However, the main components of an I&C or low-voltage power cable are conductors, electrical insulation or dielectric, shielding, and the outer jacket.

Power cables and I&C cables both operate by providing a conductive route for an electric circuit by using metallic conductors – typically copper or aluminum that are insulated with a polymer and have different configurations such as coaxial, triaxial, twisted pair, or multi-conductor arrangements of single-strand or bundled wires (AMS Corp., 2011; U.S. NRC, 2001). The cable insulation and jacket are made of different polymers, including polyethylene (PE), cross-linked polyethylene (XLPE), polyvinyl chloride (PVC), ethylene propylene diene-monomer (EPDM) rubber, ethylene propylene rubber (EPR), Hyplon, Lipalon, and others (AMS Corp., 2010; 2011). More than three-quarters of cable insulation and jacket used in nuclear plants is constructed from such polymers (U.S. NRC, 2010a). Another type of cable, fiber-optic, is used to transmit signals based on optical fiber technology. Though its outer jacket is similar to copper and aluminum cable, fiber-optic cables have unique ageing, degradation, and failure characteristics (U.S. NRC, 2010a). As such, they are not covered here.

Connectors are also part of the conductor in a cable circuit. A multitude of connectors, terminations, terminals, splices, etc., join the conductor to other cables or electronic equipment. The failure of these components may appear as a problem with the conductor in test data and may be the result of corrosion, loose terminations, and other faults (AMS Corp., 2010).

Cable manufacturers qualify cables for a specific service life (e.g. 40 years for nuclear power plant cables) and specific voltage class at a given maximum ambient temperature (U.S. NRC, 2010a). Service life is affected by everything from voltage and temperature rating of the cable and the material and thickness of its insulation and conductor jacket, to the conductor size and construction (e.g. solid or stranded), the type of metal and coatings used in the conductor, the cable configuration (e.g. single or multiple), and the presence of ground conductors, shields, braids, or binding and filler material (U.S. NRC, 2001).

Because I&C cables are used at low current, their typical operating temperatures are between about 40°C and 65°C (IAEA, 2011). In contrast, power cables can operate at 80–90°C because of continuous current flow, which generates ohmic self-heating, and the higher voltages and currents used to power medium- and high-voltage equipment such as pump motors (IAEA, 2011).

Because of their typical length, cables can experience multiple operating environments as they travel through different areas of the plant, including harsh temperature, radiation, humidity, and moisture conditions, which may include submersion in water (IAEA, 2011; U.S. NRC, 2001).

6.2 Cable degradation issues

As long as cables are installed properly and not exposed to environmental conditions beyond their design basis, they are generally durable and relatively long lived, typically lasting 40–50 years (Hashemian, 2010; IAEA, 2011). In fact, compared to other I&C components, cables have historically experienced few problems. A Japanese study, for example, found that most nuclear power plant I&C cables will maintain their electrical function capabilities over 60 years of operation (Hashemian, 2010).

The IAEA defines a 'mild' operating environment as one that 'would at no time be significantly more severe than the environment that would occur during normal plant operation, including anticipated operational events.' In contrast, a 'harsh' environment is one that results from a design basis accident (DBA) involving, for example, a loss-of-coolant accident (LOCA) or the failure of a high-energy line or main steam line (IAEA, 2011). Mild and harsh operating environments can be distinguished from unanticipated operating conditions such as those caused by poor installation, operation or movement of the cable. All of these can accelerate cable ageing and degradation (IAEA, 2011).

Cable ageing is a subset of cable degradation and primarily consists of cracking, embrittlement, or other changes to the cable jacket or insulation material. In most cases, these changes are produced by a combination of physical age and environmental stressors such as temperature or radiation exposure (AMS Corp., 2010). Cable circuits can be subjected to any or a combination of the following stressors: oxidation, water intrusion, contamination, vibration, thermal variations, electrical transient, voltage variations, temperature, installation damage, and handling and physical contact (AMS Corp., 2010). However, the three principle ageing factors for cables are (1) elevated ambient temperature or humidity; (2) cyclic mechanical stress; and (3) exposure to radiation (Hashemian, 2010). Cable degradation is mainly dependent on environmental factors such as temperature, radiation, humidity, or contaminants (IAEA, 2011).

6.2.1 Individual cable stressors: temperature, humidity, mechanical stress, and radiation

Elevated temperatures cause the polymers in the cable insulation to degrade through loss of elongation, embrittlement, and cracking (U.S. NRC, 2001). Cable polymers are primarily degraded by thermal oxidation in the presence of oxygen, accelerating with increases in temperature as defined by the modified Arrhenius equation (IAEA, 2011):

$$k = A \exp(-\mathrm{EA}/RT) \qquad [6.1]$$

where EA is the activation energy, A is the frequency factor, and R is a constant. Temperature is the most important ageing stressor for most cables in a light water reactor (IAEA, 2011).

As a result of internal ohmic self-heating, power cables age uniquely, depending on how long the cable carries electric current, which current it carries, and the specific configuration of the cable installation itself. Treeing (the appearance of small tree-shaped cracks in the insulation caused by electrochemical reactions) and the loss of the dielectric properties of cable insulation are characteristic results of power cable ageing (IAEA, 2011; U.S. NRC, 2001).

Exposure to moisture can also degrade cables that have been installed directly in the ground or in ducts or conduits where water has access. 'Wetting' describes conditions in which a cable is exposed to moisture or high humidity for extended periods of time, including limited periods of complete submergence. Submersion describes conditions when the cable is completely submerged in water for extended periods. So long as the insulation and outer jacket are not damaged, intermittent wetting will not damage most cables, but extended submersion is beyond the qualified operating conditions for most cables (U.S. NRC, 2010a).

Moisture can cause water treeing where voids or contamination are present in the cable. This combination of water and electrical stress degrades the insulation's dielectric properties (U.S. NRC, 2001). In fact, the U.S. NRC (NUREG 6704) identified wetting as the primary ageing-related cause of failure (specifically, short circuit) for medium-voltage cables, in particular the insulation (U.S. NRC, 2001). Such failure could allow currents and voltages to spread into the adjacent power distribution system, potentially causing other degraded power cables to fail too (U.S. NRC, 2010b). For this reason, cables in hard-to-access underground ducts and conduits, covered trenches, bunkers, and manhole vaults are the subject of special concern (U.S. NRC, 2010b).

Both power cables and I&C cables are directly affected by mechanical stress including bending, abrasion, cutting, contact, deformation, and perforation, as a result of installation and maintenance, for example. Cables connected to vibrating machines are also subjected to stress, leading to chafing, cutting, or cracking of the cable insulation material (AMS Corp., 2011). Cable jacket and insulation material as well as cable conductors can be damaged by electromechanical forces caused by high levels of short circuit current passing through a power cable (U.S. NRC, 2010a).

Radiation is another significant cause of cable degradation. During normal operation, gamma and neutron radiation cause oxidative degradation in increasing (nonlinear) relation to the radiation dosage absorbed by the cable. During accidents, beta radiation may also affect cables unprotected by a conduit (IAEA, 2011).

6.2.2 Cable failure modes and consequences

The basic cable failure modes resulting from exposure to stressors include: short circuits between cable conductors, short circuits between one or more cable conductors or the shield and ground (ground fault), open circuits in the cable conductors, and breakdown of the cable insulation (AMS Corp., 2010). The most common failure mode is ground fault, in which the cable faults to ground from one or multiple conductors (U.S. NRC, 2001). Ninety-five per cent of cable problems occur at the cable connector where age factors are combined with mechanical damage and wear (AMS Corp., 2010).

For power cables, such failure modes can cause circuit protection devices to trip or partially discharge, resulting in excessive heating and degradation of the cable insulation and ionization of the air around the discharge. This failure mode – degraded insulation resistance – can lead to a conductor short-circuit to ground failure, conductor to conductor short circuit, or potentially both. Power cable failures have resulted in reactor trips, weakened engineered safety features, loss of redundancy, and reduced power operation (U.S. NRC, 2001).

For I&C cables, conductor short-circuit to ground failure and conductor-to-conductor-short-circuit failure interrupt the transmission of control signals through the cable.

Degraded insulation resistance failure can impair the functioning of I&C cable and/or increase the rate of error (U.S. NRC, 2010a). The transmitted signal may become erratic, causing errors in measurement, spikes, noise, and other problems. When cables become bare, shunting and short circuits can occur, and if the cable insulator is degraded, the insulation material can become brittle and flammable (Hashemian, 2010). I&C cables are the most susceptible to ageing degradation (U.S. NRC, 2001).

In addition to signal anomalies and problems with plant control and safety systems, cable ageing has resulted in loss of critical functions and fire (AMS Corp., 2011). In light water reactors, the most severe cable failure scenario is loss of normal function during a LOCA when hot steam under pressure can cause cables to malfunction if insulation ageing, cracks, or other damage allow moisture to enter the cable. Hot steam combined with high pressure is the primary cause for cable malfunction in a LOCA, because steam penetrates smaller cracks more easily than water. Such consequences explain why the Hungarian Paks Nuclear Power Plant has described cable ageing as 'the most significant I&C ageing issue' in its plant (Hashemian, 2010).

A 2007 U.S. NRC report found that 93% of reported cable failures occurred in normally energized power cables: 'More than 46% of the failures were reported to have occurred while the cable was in service and more than 42% were identified as 'testing failures' in which cables failed to meet testing or inspection acceptance criteria' (U.S. NRC, 2010a). The majority of these cable failures occurred between 11 and 30 years of service – less than the typical 40-year licensing period of a plant (U.S. NRC, 2010a).

While many cases of cable failure are identified through routine cable testing, some occur before a failure is identified (e.g. on cables that are not normally tested or powered). This fact underscores the importance of implementing a cable condition monitoring program (U.S. NRC, 2010b).

6.3 Analysis and assessment methods

Cable components such as the conductor wires, insulation, shielding, and jacket material can all be tested to reveal signs of degradation. By applying the right testing method or combination of methods effectively faults that typically occur at cable connections can be confirmed; these include terminations, penetrations, and/or splices that have been exposed to mechanical stress, oxidation, or corrosion. Other faults include end-device failure such as motor windings, sensors, transmitters, and detectors that may also be detected using appropriate testing methods (AMS Corp., 2010).

Cable analysis and assessment methods require observing, measuring, and trending indicators of cable condition that correlate to the physical condition of the cable or its functional performance (U.S. NRC, 2010b). According to the NRC, an 'ideal' condition monitoring technique should have the following desired attributes: 'nondestructive and nonintrusive, capable of measuring property changes or indicators that are trendable and can be consistently correlated to functional performance during normal service, applicable to cable types and materials commonly used in nuclear power plants, provides reproducible results that are not affected by the test environment or, if they are so affected, the results can be corrected for those effects, able to identify the location of any defects in the cable, allows the establishment of a well-defined end condition, and provides sufficient time before incipient failure to allow corrective actions' (U.S. NRC, 2010b). However, because the nuclear industry relies primarily on manufacturer qualification data, it does insufficient testing to confirm that cables can operate dependably in the long term (IAEA, 2011).

Cable testing methods can be characterized in multiple ways (see Table 6.1). In the broadest terms, two cable ageing methods are available: laboratory tests (involving microsampling, e.g. conducted in non-operational conditions in a lab) and *in-situ* tests (conducted on cables as installed in a plant) (U.S. NRC, 2010a; IAEA, 2011). However, another way of categorizing cable ageing methods is life testing versus electrical testing. Life-testing techniques involve testing – visually, physically, or chemically – the physical properties (e.g. hardness) of spare cable samples of the same cables actually installed and in operation at the plant. When such 'real-time' testing is not possible or desirable, accelerated life testing can compress the time required to test the ageing processes by 'pre-ageing' cable samples and monitoring their performance when installed in the same environment as actual in-service cables.

Electrical testing of cables involves the testing of electrical properties such as insulation resistance/polarization index, voltage withstand, dielectric loss/dissipation factor, time and/or frequency domain reflectometry, and partial discharge. These electrical testing methods can be further categorized according to whether the inspection or test is performed *in-situ* on electric cables in the plant or whether it is a laboratory-type test performed on representative material specimens in a controlled laboratory setting.

Which of these testing techniques is used will depend on the type of insulation material in the cable and type of environmental stressors to which the cable is subjected (Hashemian, 2010). For example, historically, visual and tactile inspection techniques have been the most commonly used methods for cables that are accessible. Some, such as the gel content and other

Table 6.1 Cable testing and diagnostic techniques

Cable testing technique	Part of cable evaluated	In-situ*	Remote testing	Non-destructive	Applicable to low- and medium-voltage cables
Visual/tactile (visual screening test)	I,CN,P,S,T	✓	No	✓	✓
Hi-Pot test (high voltage test may damage insulation)	I	No	✓	No	✓
AgeAlert™ (wireless sensors attached to cables)	I	✓	✓	✓	✓
TDR (time domain reflectometry)	I,C,CN,P,S,T	✓	✓	✓	✓
RTDR (reverse TDR for coaxial shielded cables)	C,CN,P,S,T	✓	✓	✓	✓
Impedance measurements (inductance, capacitance, and AC resistance)	C,T,CN,P,S,T	✓	✓	✓	✓
Partial discharge	I,P,CN,S,T	✓	No	✓	✓**
Insulation resistance (dielectric absorption ratio, polarization index) (sustained higher DC voltage resistance)	I,CN,P,S,T	✓	✓	✓	✓
Dissipation factor (Tan Delta)	I,CN,P,S,T	✓	✓	✓	✓**
AC voltage withstand (low frequency, high voltage)	I,CN,P,S,T	✓	✓	✓	✓**
DC step voltage	I,C,CN,P,S,T	✓	✓	✓	✓**

FDR (frequency domain reflectometry). A version of FDR technique is referred to as LIRA, which stands for Line Resonance Analysis. There is also a technique which is referred to as Joint Time and Frequency Domain Reflectometry or JTFDR	I,C,CN,P,S,T	✓	✓	✓	✓
Elongation at break (tensile strength)	I	No	No	No	✓
Indenter test (compressive modulus)	I	✓	No	✓	✓
Oxidation induction time/temperature (classical oxidation measurement)	I	No	No	No	✓
Fourier transform infrared measurement	I	No	No	No	✓
Gel content test	I	No	No	No	✓

*In-situ: Test that can be performed without disconnecting the end device while the cable is installed in its normal environment.
Legend: Insulation and Jacket (I), Conductor (C), Connections (CN), Penetrations (P), Splices (S), Terminations (T). **Applicable to medium voltage cables only

chemical and mechanical tests like the cable indenter, were developed specifically for evaluating the condition of the protective jacket or insulation on a cable (AMS Corp., 2010).

6.3.1 Visual and mechanical measurements

Visual inspection involves examining the cable throughout its length during a formal plant walkdown, a useful practice when, as is often the case, degradation is visible to the naked, well-trained eye (IAEA, 2011). Visual inspection can identify changes in physical/visual appearance, surface texture, and damage as a result of manufacturing or operation (U.S. NRC, 2010b). More sophisticated techniques can then be used to determine the degree of ageing more accurately.

The advantages of visual inspection are that it is low cost and easy to perform, requires no specialized equipment, does not require that samples be removed from the cable, and can be performed on operating equipment *in-situ*. Its disadvantages include the requirement that the cable be accessible and visible, inspectors must be trained to evaluate what they are looking at (subjectivity), it generally only provides information on the cable jacket, and it does not provide quantifiable results (no trending possible) (AMS Corp., 2010; IAEA, 2011).

Mechanical testing is a subset of life-testing techniques that involves inspecting cables for cracks or changes in color, texture or hardness, mass loss, visco-elasticity properties, or size (swelling, shrinkage, deformation). Among the most conventional and popular means of mechanical cable testing are measuring the elongation-at-break of the cable and its tensile strength when pulled apart. The elongation-at-break test measures the strain on the cable when it breaks and is a recognized standard for assessing the health, integrity, and functionality of a cable insulation material (IAEA, 2011). This test is performed by stretching a 'dog bone'-shaped cable sample until it breaks. The elongation-at-break test yields information on the tensile strength and modulus of elasticity of the cable, but the percentage of elongation is the most important criterion in evaluating cable health. When the percentage elongation-at-break is less than 50%, the cable is considered to be unhealthy – potentially unable to survive DBA conditions (AMS Corp., 2010; IAEA, 2011).

The tensile test measures the stress needed to break the cable. For polymeric materials like thermoplastics, tensile strength only begins to fall after substantial ageing has already occurred. Both the elongation-at-break and tensile strength tests can be performed using a tensile testing machine.

A third mechanical test, measuring compressive modulus, involves checking the ductility of the cable insulation or jacket material to determine if the cable has become dry, brittle, or prone to crack. Developed in the mid-1980s by the Electric Power Research Institute (EPRI), this test is performed with a device known as a cable indenter, which uses a small probe to press against the cable jacket or insulation. A PC-based system analyzes cable hardness by measuring the probe force and polymer deformation, thus providing diagnostic insights (Hashemian, 2010).

The difficulty with these classic life-testing techniques is that they can check for problems only at the locations on the cable where the cable is tested. Such passive maintenance methods can thus fail to detect problems or hot spots in other areas. Similarly, the elongation-at-break and tensile strength test are also destructive to the tested material and require that the cable be removed from operation for testing (IAEA, 2011). For these reasons, mechanical life-testing techniques should be combined with other measurements, such as electrical or chemical functionality.

6.3.2 Electrical measurements

The IAEA has stated that electrical properties – voltage withstand, insulation resistance, capacitance, attenuation, and/or signal propagation – are the 'most important functional properties' of cables (IAEA, 2011). Such properties provide a direct measure of the loss of cable resistance or dielectric parameters and therefore its loss of functionality (IAEA, 2011). Electrical measurements are primarily suitable for cable conductors, connectors, splices and penetrations. Most electrical measurement techniques are less sensitive to problems with cable insulation material though they can reveal them (IAEA, 2011).

Two types of electrical tests are available: destructive methods, which identify cable failure locations before the cable is installed, and non-destructive methods, which are better suited to identifying cable degradation (U.S. NRC, 2001). Non-destructive tests can be categorized by direct current (DC) and alternating current (AC) methods. DC tests generally require the least expensive test equipment, but may be less appropriate for some power cables and cables used in AC applications (U.S. NRC, 2001). The primary advantage of electrical techniques is that they can be used *in-situ* on installed and less accessible cables, providing information on the entire length of a cable, not just those points which are tested (IAEA, 2011). Because some methods also enable trending based on baseline measurements, electrical techniques can be used to note changes over time for ageing management purposes (IAEA, 2011). The most important electrical parameters in cables are insulation resistance, leakage current, loss factor, permittivity, and breakdown voltage. Provided that one or both ends of the cable are accessible to measurement and the cable can be de-energized, these electrical parameters can usually be measured on any cable (Hashemian, 2010).

There are two general types of *in-situ* electrical cable tests:

1. Insulation quality tests, which include insulation resistance (IR), high-potential (Hi-Pot), partial discharge, quality factor, dissipation factor, and AgeAlert™.
2. Impedance tests, which include LCR (inductance, capacitance, and resistance), time domain reflectometry (TDR), and frequency domain reflectometry (FDR).

Many of these electrical tests are simple and have been in use for decades. In recent years, LCR measurements have been added to the TDR test to improve cable diagnostics, help identify the nature of a fault, and pinpoint its location along a cable. Used together, electrical methods like the TDR and LCR tests provide an overall picture of cable health as well as information for expediting any repairs that may be needed (Hashemian, 2010).

Many conventional electrical test methods such as IR and LCR are only used to give a snapshot of the current condition of a cable. Others, such as TDR, FDR and RTDR, can identify the fault location within the length of cable, but may not differentiate whether the problems are in the connection or the end device. Additional tests are normally required to help distinguish whether the fault is in the cable or to diagnose the cause of the end device problem (AMS Corp., 2011). In recent years, the TDR and FDR techniques have been either combined or packaged and introduced under such names as LIRA (line impedance resonance analysis) a method which seems to be essentially the same as the FDR technique and JTFDR (joint time and frequency domain reflectometry) which combines TDR and FDR in a single test.

Insulation quality tests

Insulation resistance, 'Hi-Pot,' partial discharge, AgeAlert™, and quality/dissipation factor are electrical measurement tests for the entire cable circuit (cable, connections, and end device) to identify cable insulation degradation, failed end devices, and moisture intrusion on the cable.

Insulation resistance (IR) – the simplest and most common test for monitoring cable ageing – quantifies the quality of cable insulation by energizing the cable conductor and measuring for leakage current through degraded insulation (AMS Corp., 2011; U.S. NRC, 1990). One of two fundamental wire insulator properties, insulation resistance is the resistance to current leakage through and over the surface of the cable material. Insulation can also be impacted by cable length; humidity or moisture in the cable and insulation as well as dirt, oil, and other surface contaminants (U.S. NRC, 2010a). IR changes in a progressive, ongoing basis as a cable is exposed to these environmental stressors (IAEA, 2011).

In an IR test, a high voltage (e.g. 100 V DC), is applied between each cable lead and the cable shield, and also between the cable shield and ground. The principle of the IR test is that the application of a DC voltage to an insulated conductor induces a small current in the insulation to ground (U.S. NRC, 2010a). As the high voltage is applied, the leakage current through the insulating material is measured to establish the quality of the cable insulation and to determine if there are any contaminants (moisture, grease, dirt, etc.) in the cable (Hashemian, 2010). The IR test is easy to perform with inexpensive equipment, but it is a simple pass/fail test for cable dielectric and results are too inconsistent for trending purposes (U.S. NRC, 2010a; U.S. NRC, 2010b).

Two other insulation quality tests, polarization index (PI) and dielectric absorption ratio (DAR), provide the trendability that the IR test lacks. This is essential because if insulation is badly deteriorated, wet, or contaminated, the leakage current will exceed acceptable levels and could continue to increase over time. Insulation resistance is therefore typically measured at different time intervals, with the ratio of these two measurements constituting the polarization index (PI) or polarization ratio (PR) (U.S. NRC, 2010a). The PI test detects cracking induced by heat, radiation, moisture, and surface contamination. Although the PI test is trendable, easy to perform and does not require access to the entire cable, it requires that the end terminations be disconnected and is insensitive to insulation degradation (U.S. NRC, 2010b).

DAR is another index of the quality of cable insulation over time. To determine DAR, the IR is measured 60 s after applying the test voltage, and that result is then divided by the IR measurement after 30 s. Depending on how fast the system polarizes, the IR will increase and then start to plateau over time. DAR is somewhat subjective and should be considered in the context of IR; it is not an absolute indicator of insulation quality (Hashemian, 2010). DAR, PR, and PI values of less than 1.0 usually indicate degradation in the insulating material, which may be due to dirt, moisture, cracking, ageing, or other problems.

Another insulation quality test, the direct current (DC) high-potential test (Hi-Pot), is a pass/fail test used on medium-voltage power cables and all insulation and jacket materials to detect embrittlement and cracking caused by heat and radiation, mechanical damage, water treeing, moisture intrusion, and surface contamination. The principle of the test is that if a cable contains defects, the high test voltage will force the defects to fail. In the DC Hi-Pot test, a high-voltage potential is applied to the insulation to see if it can withstand higher DC potential than it normally experiences in operation. Because cable insulation can normally endure sustained DC potential without damage, the Hi-Pot test is typically used to repetitively test insulation at sufficient voltage to indicate whether insulation is weak enough to begin failing in service (breakdown voltage) but without damaging sound insulation (withstand voltage). The Hi-Pot test is easy to perform, provides trendable data, and does not require access to the entire length of a cable. However, the cable must be disconnected in order to perform the test, and the high voltages applied may damage the cable insulation (U.S. NRC, 2010b).

The partial discharge (PD) test is another insulation quality assessment method. Partial discharges are small electrical sparks that occur at voids, gaps, and similar defects within the insulation in medium and high voltage cables. Over time, these partial discharges will erode the insulation and ultimately break down the cable completely, resulting in embrittlement and cracking, mechanical damage, and water treeing (U.S. NRC, 2010b). The lower the PD inception voltage, the greater the degradation of the insulation material (IAEA, 2011).

Measurements of partial discharge are performed in both time and frequency domain by a monitor connected to a cable circuit. If a sufficiently high voltage (called the inception voltage) is applied across a cable insulation, an electrical discharge (partial discharge or corona) can occur in small voids or air gaps in the insulation or between insulation and a ground plane or shield (U.S. NRC, 2001). The monitor or oscilloscope measures the peak magnitude of the partial discharge pulse, phase angle, and pulse shapes of the partial discharge signals acquired. PD test equipment can determine the location of the voids or gaps by measuring the time lag between direct and reflected pulses from the discharge site or by using acoustic emission monitoring techniques (U.S. NRC, 2001). These measurements can be made continuously or intermittently and identified on- or off-line (Hashemian, 2010).

The PD test does not require access to the full length of the cable and enables both the quantification of the severity of insulation defects and identification of their location in the cable. However, the end terminations of the cable must be disconnected to perform the test, the test itself requires highly skilled personnel, and the high testing voltage can weaken or damage cable insulation (U.S. NRC, 2010b).

Another insulation quality test – AgeAlert™ – is a wireless microsensor that measures ageing or degradation of electrical insulation. Constructed out of cable insulation and nano-size conductive particles, it is installed in multiple locations along cables or embedded in motors and allowed to age together with the insulation material being monitored (IAEA, 2011). The parameter that is measured and correlated to cable condition is the resistivity of the microsensor as a function of its age. Because it is constructed out of the same material as the cable it measures, it responds to temperature, humidity and radiation environments much as the insulation does (IAEA, 2011). Under thermal-oxidative conditions, the polymer material becomes denser and loses volume. This change results in a change in resistance of the AgeAlert™ microsensor. The resistance data can be transmitted wirelessly using radio frequency identification (RFID) technology for short distances and can be acquired by a handheld RFID reader or a personal digital assistant (PDA) type device. If needed for long distance transmission, active wireless transmission technology could also be used. The AgeAlert™ has no internal power source, receiving power from the radio frequency (RF) signal that is used to interrogate it and read the resistivity data from the sensor (AMS Corp., 2010). The AgeAlert™ sensors can be installed by wire/cable manufacturers during manufacture of the cable or bonded to the cable after installation (IAEA, 2011).

Dissipation factor (DF) and quality factor (QF) represent a final insulation quality test. The ratio of the energy loss in a dielectric to the total energy transmitted through the dielectric, DF represents the departure of a cable from an 'ideal' capacitor. If the cable is free of defects or contaminants, its dielectric properties are similar to a perfect capacitor. If the cable dielectric contains impurities, the resistance of the insulation decreases, and it no longer acts as a perfect capacitor (Hashemian, 2010). Similarly, QF represents the departure of a cable from an 'ideal' inductor. Quality factor applies to electrical circuits that contain resistance, inductance, and capacitance and is the ratio of energy stored to energy dissipated in a system at a specific frequency.

Impedance tests

LCR, TDR, and FDR are cable testing techniques that measure impedance in cables in order to detect anomalies. There are two basic types of impedance tests – lumped data and distributed measurement – based on the ability of the test to localize its measurement. Lumped data tests like LCR typically identify anomalies in cables with greater accuracy than distributed measurement methods. But once the fault is detected, the distributed measurement methods (TDR and FDR) can determine the distance to the fault.

The LCR test uses an LCR instrument or meter at specific frequencies to make impedance measurements along the cable at specific frequencies to verify the characteristics of the cable conductor, insulating material, and the end device. The results are evaluated to determine if they are as expected for the type of circuit being tested. Imbalances, mismatches, or unexpectedly high or low impedances between the cable leads indicate problems caused by cable degradation and ageing, faulty connections and splices, or physical damage. For example, abnormal capacitance measurements indicate a change in cable dielectric or insulation. In addition to providing information about cables, connectors, and end devices, LCR measurements can identify circuit problems such as moisture or loose connections (Hashemian, 2010; IAEA, 2011).

The most popular and effective cable testing technique today, TDR, is used to locate problems along a cable, in a connector, or at passive devices at the cable end by sending a test signal through the conductors in the cable and measuring its reflection. It works on the same principle as radar. A pulsed or swept DC signal is sent through the cable, and its reflection is measured to identify the location of any impedance discontinuity or change in the cable and the end device (load). It measures the time taken for the signal to travel down the cable to where the impedance change is located, and return. This propagation time for a known distance is then converted, and depending on the type of display used, the information can be presented as a waveform and/or a distance reading (IAEA, 2011).

Any significant change in impedance along the cable will cause a reflection that will appear on the TDR signature as a peak or valley whose amplitude depends on the characteristics of the cable impedance. Depending on the impedance of the load, the TDR trace representing the end of the cable may step up or step down. That is, reflected voltage waves occur when the transmitted signal encounters an impedance mismatch or discontinuity (fault) in the cable, connector, or end device. Any such change in impedance along the cable due to a short, open, shunt, or other electrical effect can thus be identified and located using the TDR test. A rise in the reflected wave is indicative of an increase in impedance, and a decrease in the reflected wave is indicative of a decrease in impedance. Thus, the peaks and dips in a TDR plot are used to identify the location of normal and abnormal electrical effects throughout the cable (Hashemian, 2010).

The TDR test is typically performed using a pulse generator, which produces a step pulse, and a recorder, oscilloscope, or automated computer-controlled data acquisition system, which captures the reflected wave. The test signal is applied between pairs of lead wires, a cable shield, and a ground plane, and the results are displayed as a plot of the reflected wave versus time or distance (Hashemian, 2010).

Yielding diagnostic information about the cable conductor and any connector or connection, the TDR method relies on comparisons with a baseline TDR. Its success therefore typically shows significant improvement if there is a baseline TDR for comparison (IAEA, 2011). In light water reactors, the TDR method is useful for testing instrumentation circuits, motor and transformer windings, pressurizer heater coils, thermocouples, RTDs, motor-operated valve cables, neutron detector cables, and other components that are normally inaccessible, such as in high temperature and high radiation zones. The simplest and perhaps most important application of TDR is to locate an open or short lead, moisture, or problems such as erratic behavior along a cable or in an end device (e.g. resistance temperature detector).

The TDR test non-destructively identifies and locates cable defects and discontinuities on an installed cable *in-situ*, providing trendable measurements. However, end terminations of the cable must be disconnected in order to perform the test (U.S. NRC, 2001; 2010b). In addition, the size of the wave that the TDR method sends down a tested cable is limited by the bandwidth of the pulse and sampling circuitry. Because it sends only very broad DC pulses, the TDR method can locate only DC open- or short-circuit conditions.

Like TDR, the frequency domain reflectometry (FDR) technique can measure the distance to and severity of a fault in a cable conductor, connectors, and end device. However, because the FDR technique uses a selected set of much smaller or narrower bandwidth frequencies, it is also able to locate RF faults in cables, unlike the TDR test. The FDR technique can also help identify degradation in cable insulation material. There are three types of FDR which calculate distance based on the sine wave property they measure – namely, frequency, magnitude, and phase.

In FDR, a stepped (or variable) frequency sine wave generator sends stepped-frequency sine waves down the cable. These waves are reflected back from the cable end as well as from any faults encountered along the cable, and are sensed by either a frequency counter, received signal strength indicator, or another technology for measuring high or intermediate frequency voltage magnitudes. Using pulses of discrete frequencies can identify and locate small faults in connectors or cables, making possible a more realistic picture of cable condition than TDR provides.

FDR measures reflection responses in the frequency domain and then converts the data into the time domain using an inverse Fourier transform. Similarly, FDR data can be acquired by using a TDR to measure the reflected wave over the large bandwidth and then using Fourier transform to convert from time to frequency domains. Decreasing the time required for a signal to change from a specified low value to a specified high value (rise time) of a TDR test will increase its accuracy, as will increasing the bandwidth of an FDR test. Similarly, increasing the number of frequency samples in an FDR test increases its maximum range, as does increasing the period between the rise and fall of the pulse in a TDR pulse (Hashemian, 2010).

Like TDR, FDR is a non-destructive technique that can send a swept signal through miles of cable without attenuation as long as the cable under test is shorter than the signal wavelength (IAEA, 2011).

Reverse time domain reflectometry (RTDR) is a technique developed by CHAR Services Inc., a division of Analysis Measurement Services Corporation (AMS). It tests the quality of the shielding around the conductor of a coaxial or triaxial electrical cable. The RTDR method estimates the distance to a fault in a cable by coupling a repetitive DC pulse to the shield and allowing the pulse to travel the length of it. The time delay between the DC pulse and when the signal is received can be measured by simultaneously monitoring the cable signal path (Hashemian, 2010). These time delays make it possible to identify the point at which the electromagnetic interference (EMI) couples into the cable system. This reveals the location of degraded connectors or cable shields because such interference usually couples at cable connections or terminations that tend to degrade through ageing or damage (IAEA, 2011).

6.3.3 Chemical measurements

Chemical measurement techniques determine cable condition by measuring a chemical property of the cable insulation and then correlating the results with a known measure of electrical performance (U.S. NRC, 2001). In chemical cable testing, a small piece (a few milligrams) of cable insulation or jacket material is shaved off for chemical analysis in a laboratory using one of the following techniques:

- oxidation induction time/temperature (OIT/OITP) test;
- Fourier transform infrared (FTIR) spectroscopy;
- gel content or gel fraction tests;
- density tests.

Chemical tests are not considered *in-situ* since they require a small sample, but the sample is so small that they are sometimes considered non-destructive.

The polymers used in cable insulation respond to radiation and thermal degradation through oxidation processes. The greater the radiation or thermal ageing conditions imposed on the cable insulation, the more antioxidants (manufactured into the insulation to slow degradation) in the polymer are consumed. Differential scanning calorimetry (DSC) instruments can be used to measure the rate of oxidation induction time (OIT) and oxidation induction temperature (OITP) in polymers. The OIT and OITP values correlate with the degree of cable insulation degradation. OIT – a measure of the remaining antioxidant in the insulation polymer – decreases with age (U.S. NRC, 2001).

The DSC instruments measure the difference in heat flow between a polymer sample oxidizing under heat and an identical empty sample pan (acting as a control) also being heated. The levels of antioxidant in the sample will determine how long it takes for the heated polymer sample to begin oxidizing. The complete depletion of antioxidants would typically simulate polymer degradation after a 20-year operational life. A sample that takes a long time to begin oxidizing has substantial antioxidant levels, and therefore minimal degradation (U.S. NRC, 2001).

Fourier transform infrared (FTIR) spectroscopy, a laboratory technique for studying the molecular structure of materials,, can identify operating conditions where heat may cause cable insulation to become brittle or crack. FTIR involves applying infrared radiation to a small piece of cable insulation using a spectroscope. The spectroscope measures the ability of the material to absorb or transmit the radiation. When the chemical bonds in the sample absorb the radiation they begin to vibrate at specific wavelengths. The FTIR technique compares the actual measured maximum vibrations of these chemical bonds to their known maximum vibrations to ascertain to what extent the bonds have already oxidized or degraded over time. The more the cable surface has been exposed to heat over time, the more likely the measured vibrations of the sample chemical bonds will differ from the original values.

FTIR is extremely accurate – measuring to one tenth of a degree Fahrenheit – and enables relatively easy, trendable, non-destructive degradation monitoring. However, it requires expensive equipment and a small sample, which may be difficult to obtain from remote sections of a cable circuit (U.S. NRC, 2001; 2010a; 2010b).

6.3.4 Limitations of individual analysis and assessment methods

The techniques outlined here are often applied individually to assess cable problems. However, none of these techniques can characterize the ageing condition of a cable with confidence, and most have never been evaluated comprehensively to determine if the changes they identify in a cable are correlated to cable age. Moreover, performing all these destructive/non-destructive mechanical, electrical, and chemical tests on the 9.1 million feet of cable and wiring in an LWR would be daunting and time consuming (AMS Corp., 2010). To more accurately determine or model the residual life in the cable network of a plant, a system and program is needed that combines and integrates these methods in such a way as to provide a more objective assessment of the health and ageing condition of low- and medium-voltage cables (AMS Corp., 2010).

6.4 Residual life modeling

The servicing and maintenance of the miles of I&C, low- and medium-voltage cables in each light water plant has historically been reactive in nature. Such reactive efforts have successfully resolved connector problems, corrected signal-to-noise ratios, and improved grounding and shielding. However, they have done little to identify the condition, age, or remaining useful life of cables, especially the insulation material (AMS Corp., 2010). Not enough research has been completed to identify a useful, practical method, procedure or technique for accurately evaluating the ageing condition of plant wiring or correlating the condition of cables to measurable electrical, mechanical, or chemical properties (AMS Corp., 2010). In 2010, the U.S. NRC (NRC DG1240) stated that 'research and experience have shown that no single, nonintrusive, currently available condition monitoring method can be used alone to predict the survivability of electric cables under accident conditions' (U.S. NRC, 2010b).

Because of the safety-related importance of I&C cables functioning effectively on an ongoing basis, efforts to use prognostic techniques to predict residual life in cables continue.

Such techniques attempt to establish relationships between condition indicators and ageing stressors (IAEA, 2011). To predict future performance, a trendable indicator and a well-defined end point are essential. From them, a trend curve can be used to estimate the time remaining before the end point is reached (U.S. NRC, 2001). Used with appropriate material ageing models and knowledge of environmental conditions, such trend data can be used to estimate residual cable lifetimes, but only when sufficient data has been generated to validate predictive ageing models (IAEA, 2011). Currently, both the NRC and DOE are sponsoring research at AMS, national laboratories such as Sandia National Laboratories (SNL), Oak Ridge National Laboratory (ORNL), Idaho National Laboratory (INL), and elsewhere to address cable aging and cable qualification issues.

In recent years researchers have developed analytical ageing models based on experimental data from cable samples that have been subjected to accelerated ageing. For example, the power law extrapolation model extrapolates radiation ageing data obtained under isothermal conditions at several dose rates. Similarly, the superposition of time-dependent data model combines data from both thermal and radiation ageing to account for both dose rate effects and the synergistic relationship between radiation and thermal ageing. The superposition of end-point dose data model also uses a superposition approach to radiation and thermal ageing data, but can be used in materials where a single dominant degradation mechanism is lacking (AMS Corp., 2011).

There have been recent efforts toward integrated cable residual life analysis systems that combine existing methods to provide cable testing, ageing assessment and cable management as part of a plant-wide cable ageing assessment program (AMS Corp., 2010) (see Table 6.2). For example,

Table 6.2 Benefits of a cable ageing management solution to LWR nuclear plants

Current cable maintenance	Cable ageing maintenance program
• Reactive	• Periodic, proactive
• Manual testing	• Manual and automated testing
• Requires access to the cable	• Some tests may be performed remotely
• Typically tests for the cause of problems after they have occurred	• Detects cable ageing problems early to allow for scheduled maintenance
• Problems may lead to plant shutdowns	• Early detection may prevent shutdowns

Analysis and Measurement Services (AMS) Corp. of the United States is developing methods to 'calibrate' results from classical testing methods so they can be categorized, evaluated consistently, and if necessary improved. Correlations between measurable parameters and the health and condition of the cable using classical ageing tests such as the elongation at break (EAB) test would be identified. The classical tests would then be integrated with promising new cable testing technologies, such as the wireless AgeAlert™ micro-sensors (AMS Corp., 2010) (see Table 6.3). Testing methods are then categorized according to their capability to show a particular fault, faults or developing cable conditions that indicate degraded performance (see Fig. 6.2). These tests are performed using laboratory and plant-aged cables of the types found in nuclear power plants.

The correlations between the changes measured by the various methods and condition or age of the cable will form the foundation of a database that will be the core of the integrated cable testing and analysis system (AMS Corp., 2010). This database would contain the information to provide default configuration settings for the various devices that could be tested, optimized data acquisition parameters for the equipment under test, control of data acquisition hardware, and the ability to analyze and store the results of the testing. The program for the AMS integrated cable testing practice would incorporate eleven different modules (see Fig. 6.3): user interface; test lead compensation; test data acquisition; data storage; data qualification; data review; statistical analysis; historical data trending; similar equipment data comparison; report generation; default equipment setting (AMS Corp., 2010).

The result will be a user-friendly and technically feasible solution for examining low- and medium-voltage plant cables and wiring to determine their ageing condition and residual life (AMS Corp., 2010).

Table 6.3 Best cable measurement techniques for integrated cable condition monitoring program

Testing method	Part of cable evaluated	*In-situ**	Remote testing	Non-destructive
Visual inspection	I,CN,P	✓	No	✓
Indenter	I	✓	No	✓
AgeAlert™	I	✓	✓	✓
TDR	I,C,CN,P,S,T	✓	✓	✓
RTDR	C,CN,P,S,T	✓	✓	✓
Impedance measurements	I,C,CN,P,S,T	✓	✓	✓
Partial discharge	I,C,CN,P,S,T	✓	✓	✓
Insulation resistance	I,CN,P,S,T	✓	✓	✓
Dissipation factor (Tan Delta)	I,CN,P,S,T	✓	✓	✓
FDR	I,C,CN,P,S,T	✓	✓	✓
Infrared Thermography	I,C,CN,P,S,T	✓	No	✓

*In-situ: Tests that can be performed without disconnecting the cable from its in-service connections or removing the end device.
Legend: Insulation and Jacket (I), Conductor (C), Connections (CN), Penetrations (P), Splices (S), Terminations (T).

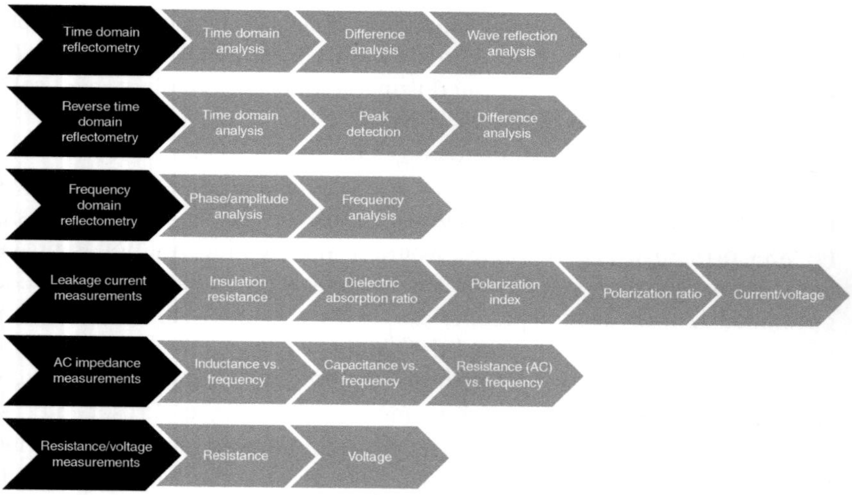

6.2 Testing and analysis techniques in an integrated cable condition monitoring system.

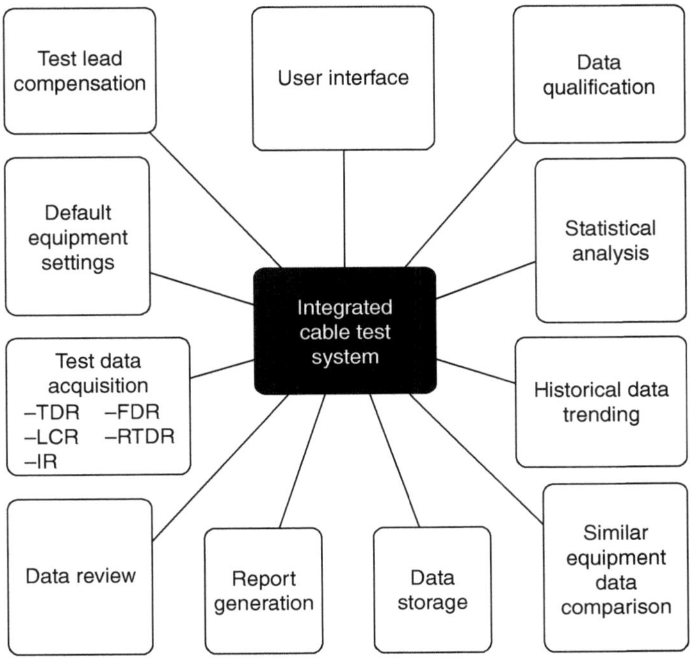

6.3 Conceptual design of the AMS integrated cable condition monitoring program.

6.5 Development and application of cable ageing mitigation routes

Cable testing and diagnostics is not a new field, and many standards have been developed that provide guidance on different methodologies for assessing their performance (AMS Corp., 2010). Over the years, the nuclear industry has suffered from a variety of plant issues resulting from reactive cable ageing management practices and failed cables, including plant trips, damage to plant equipment, radiation exposure to maintenance personnel, increased outage activity, and more (AMS Corp., 2010). Because cables have been the source of serious accidents, national and international organizations such as the U.S. Department of Energy (DOE), U.S. Nuclear Regulatory Commission, International Atomic Energy Agency (IAEA), International Electrotechnical Commission (IEC), American Society for Testing and Material (ASTM), Institute of Electrical and Electronics Engineers (IEEE), and Electric Power Research Institute (EPRI) have sponsored research and development projects, developed cable testing techniques, and written standards and guidelines to preserve the integrity, health, and reliability of the cables used in nuclear and other applications (Hashemian, 2010).

As a result, today cable testing is often recommended in nuclear industry standards and guidelines as a method for performing predictive maintenance and managing the ageing of I&C equipment. For example, the IAEA stipulates that 'aged I&C cables are expected to fully function to carry the I&C signals to a control room for normal operation, Design Basis Event (DBE) management and recovery.' Similarly, IEC standard 62465 ('Aging of Electrical Cabling Systems') outlines requirements for *in-situ* testing techniques to detect problems in cable conductors and cable insulation material. IEEE standards for testing fire travel and cables under fire conditions are similar to several of the IEC standards. There are also ANSI and ASTM standards covering general cable testing as well as specific cable tests such as partial discharge testing (AMS Corp., 2010).

Both the regulatory and industry pressure to manage cable ageing in light water reactors has only intensified as plants have been granted license renewal to operate cables for an extended qualified life as part of their efforts to extend the initial design life of a nuclear power plant from 30–40 years to 60 years (Hashemian, 2010). In some countries, plants have been able to replace some of their critical cables as an ageing management strategy. For example, the Beznau nuclear power plant in Switzerland has implemented a comprehensive cable maintenance program and has thereby emerged as a leader in cable ageing management in the worldwide nuclear power industry (AMS Corp., 2011).

Today, the nuclear power industry can obtain guidance for managing and testing plant cables and wiring from an extensive collection of various cable specifications and cable testing standards. This collection presents plant staff with a variety of recommendations for testing, monitoring, and managing the maintenance of plant cables (AMS Corp., 2010). However, because replacing cables is expensive, radiation intensive, and typically impractical, utilities operating nuclear power plants are not adopting wholesale replacement of cables as a strategy. Rather, they are searching for ageing management techniques that can identify cable problems and areas where maintenance or replacement is needed (AMS Corp., 2011).

On a regulatory level, cable ageing has not been an afterthought, and the U.S. NRC, DOE and others have sponsored ongoing research to improve currently available techniques so as to enhance preventative maintenance and proactive management of cable ageing. Others, such as the U.S. Department of Defense (DOD), NASA, the National Institute of Standards and Technology (NIST) and numerous international organizations have also sponsored and performed research and development (R&D) on cable condition monitoring and residual life estimation (AMS Corp., 2010).

As an example of increased regulatory concern over cable ageing, in February 2007 the NRC issued NRC Generic Letter 2007–01; 'Inaccessible or Underground Power Cable Failures that Disable Accident Mitigation Systems or Cause Plant Transients.' This letter required responses from license holders to the issues of undetected ageing problems associated with underground power cables that had resulted in plant shutdowns and unusual transients. After evaluating licensee responses, the NRC summarized the current plant circumstances with respect to cable condition monitoring in its recommendations: 'Plants undergoing license renewal have agreed to a cable testing program for the extended period of plant operation for a limited number of cables that are within the scope of license renewal, but only a few have established a cable testing program for the current operating period. The data... show an increasing trend of cable failures. These cables are failing within the plants' 40-year licensing periods... Licensees have identified failed cables and declining insulation resistance properties through current testing practices; however, licensees have also reported that some failures may have occurred before the failed condition was discovered ... The 10 CFR Part 50 regulations require licensees to assess the condition of their components, to monitor the performance or condition ... in a manner sufficient to provide reasonable assurance that they are capable of fulfilling their intended functions, and to establish a test program to ensure that all testing required to demonstrate that components will perform satisfactorily in service is identified and performed' (AMS Corp., 2010).

Regulators are increasingly urging that cable ageing be taken into account to ensure that plants continue to operate safely throughout the remainder of their original licenses and during any extended operation (AMS Corp., 2011). For example, the U.S. NRC published the Regulatory Guide 1.218 in April 2012 to describe the technique that the NRC staff considers acceptable for monitoring the performance of electrical cables that are important to safety (U.S. NRC, 2012). The title of this regulatory guide is 'Condition Monitoring Techniques for Electric Cables Used in Nuclear Power Plants'.

6.6 Sources of further information

Cable ageing and condition monitoring have been the subject of numerous research and development (R&D) projects, reports, and standards produced by the worldwide nuclear power industry. For example, in the mid-1990s, the International Atomic Energy Agency (IAEA) produced one of the first documents (known as TECDOC 1188) on cable degradation, ageing, and testing techniques. In the meantime, the Electric Power Research Institute (EPRI) conducted a number of research projects on this subject and has published a number of reports that are available to EPRI member utilities. Beginning in 2007, the International Electrotechnical Commission (IEC) began to prepare new standards, technical reports, and guidelines on the subject of cable ageing, condition monitoring, and testing techniques. Recently, the Nuclear Energy Agency (NEA) in Paris has taken up the subject and has already issued a comprehensive report on ageing of nuclear power plant components, systems, and structures (SSCs) including cables. The NRC's office of research has been evaluating the cable ageing issue and has drafted a regulatory guide entitled 'Condition Monitoring Program for

Electric Cables Used in Nuclear Power Plants' (DG-1240). In April 2009, the NRC released the Regulatory Guide 1.211 'Qualification of Safety-Related Cables and Field Splices for Nuclear Power Plants,' which is concerned with acceptable qualification, maintenance, and testing of power cables as well as instrumentation and control (I&C) cables.

Cable Testing Standards
Organization Standard Number Standard Name
IEEE 383–2003 Standard for Qualifying Class 1E Electric Cables and Field Splices for Nuclear Power Generating Stations
IEEE 400–2001 Guide for Field Testing and Evaluation of the Insulation of Shielded Power Cable Systems
IEEE 400.2–2004 Guide for Field Testing of Shielded Power Cable Systems
Using Very Low Frequency
IEEE 400.3–2006 Guide for Partial Discharge Testing of Shielded Power Cable Systems in a Field Environment
IEEE 576–2000 Recommended Practice for Installation, Termination, and Testing of Insulated Power Cable as Used in Industrial and Commercial Applications
IEEE 1017–2004 Recommended Practice for Field Testing Electric Submersible Pump Cable
IEEE 1407–2007 Guide for Accelerated Aging Tests for Medium-Voltage (5 kV–35 kV) Extruded Electric Power Cable Using Water-Filled Tanks
IEC 60811 Common Test Methods for Insulating and Sheathing Materials of Electric Cables
IEC 60840 Power Cables With Extruded Insulation and Their Accessories for Rated Voltages Above 30 kV up to 150 kV – Test Methods And Requirements
IEC 60332 (10 in total) Test on Electric and Optical Fiber Cables Under Fire Conditions
ASTM D257 – 07 Standard Test Methods for DC Resistance or Conductance of Insulating Materials
ASTM DF2765 – 01 Standard Test Methods for Determination of Gel Content and Swell Ratio of Crosslinked Ethylene Plastics

6.7 References

AMS Corp. 2011 Analysis and Measurement Services Corporation (April 2011), *Integrated system for management of cable ageing in nuclear power plants*, U.S. Department of Energy Proposal, Knoxville, TN, Analysis and Measurement Services Corp.

AMS Corp. 2010 Analysis and Measurement Services Corporation (June 2010), *A holistic approach for in-situ cable condition monitoring in nuclear power plants*, U.S. Department of Energy Proposal, Knoxville, TN, Analysis and Measurement Services Corp.

Hashemian 2010 Hashem M. Hashemian, Wendell C. Bean (December 2011), 'Advanced cable-testing techniques for nuclear power plants', *Nuclear technology*, **176**, no. 3, 414–29.

IAEA 2000 International Atomic Energy Agency (2000), *Assessment and management of ageing of major nuclear power plant components important to safety: In-containment instrumentation and control cables*, Volume 1, IAEA-TECDOC-1188, Vienna, Austria, International Atomic Energy Agency.

IAEA 2011 International Atomic Energy Association (2011), *Assessing and managing cable aging in nuclear power plants*, IAEA Nuclear Energy Series Report D-NP-T-3.6, Vienna, Austria, International Atomic Energy Association.

IEC 2010 International Electrotechnical Commission (2010), *Nuclear power plants – instrumentation and control systems important to safety – management of ageing of electrical cabling systems*, IEC 62465, Geneva, Switzerland, International Electrotechnical Commission.

U.S. NRC 1990 U.S. Nuclear Regulatory Commission (July 1990), *Aging of cables, connections, and electrical penetration assemblies used in nuclear power plants*, NUREG/CR-5461 SAND89–2369, Washington, DC, U.S. Nuclear Regulatory Commission.

U.S. NRC 2001 U.S. Nuclear Regulatory Commission (February 2001), *Assessment of environmental qualification practices and condition monitoring techniques for low-voltage electric cables*, NUREG/CR-6704, Vol. 1 BNL-NUREG-52610, Washington, DC, U.S. Nuclear Regulatory Commission.

U.S. NRC 2010a U.S. Nuclear Regulatory Commission (January 2010), *Essential elements of an electric cable condition monitoring program*, NUREG/CR-7000 BNL-NUREG-90318–2009, Washington, DC, U.S. Nuclear Regulatory Commission.

U.S. NRC 2010b U.S. Nuclear Regulatory Commission (June 2010), *Condition monitoring program for electric cables used in nuclear power plants*, Draft Regulatory Guide DG-1240, Washington, DC, U.S. Nuclear Regulatory Commission.

U.S. NRC 2012 U.S. Nuclear Regulatory Commission (April 2012), Condition monitoring techniques for electrical cables in nuclear power plants, Regulatory Guide 1.218, Washington, DC, U.S. Nuclear Regulatory Commission.

Part III

Materials management strategies for light water reactors (LWRs)

7

Materials management strategies for pressurized water reactors (PWRs)*

Y. H. JEONG and S. S. HWANG, Korea Atomic Energy Research Institute, Korea

DOI: 10.1533/9780857097453.3.315

Abstract: This chapter discusses management strategies in terms of mitigation and repair techniques for degradation in pressurized water reactors (PWRs). We begin by introducing PWR materials management strategies followed by details of ageing and life management of the PWR components. International cooperation activities for ageing management are also included. Development and application of mitigation techniques for reactor pressure vessels, reactor internals, steam generators, pressurizer nozzles, control rod drive mechanisms, and secondary piping are described next. Mitigation and repair methods for degradation of PWR components include: material changes, isolation techniques, weld material changes, design changes, weld overlays, stress improvements, environment improvement, mechanical repair and component replacement or removal. Finally, cracking history of components around the world and countermeasures are introduced.

Key words: PWR, life management, reactor pressure vessel, reactor internals, steam generators, pressurizer nozzles, control rod drive mechanisms, primary and secondary pipings.

7.1 Introduction

In the case of PWR power plants, Korea, China and Europe continue to build nuclear power plants which show advanced performance, whereas the

---

* Copyright information: Please note that some material in the following sections has been published previously in the cited articles:

Section 7.2 (IAEA 2011. Chapter 3) Reprinted with permission from the International Atomic Energy Agency.

Sections 7.2.1 and 7.3.3 (Shah and MacDonald 1993. Chapter 3.6 and 3.7) Reprinted with permission from © Elsevier 1993.

Sections 7.2.1, 7.2.2 and 7.2.3 (Morgan and Livingston 1995. Chapter 2.1.2, 3.1, 4.2 and 4.5) Courtesy Pacific Northwest National Laboratory, operated by Battelle Memorial Institute for the U.S. Department of Energy.

United States has built no new power plants in the last 30 years and struggles to operate its existing ones which are over 40 years old. Material degradation management and maintaining the integrity of plants became important in renewal of the operating licenses. Stress corrosion cracking of Alloy 600 steam generator tubing and nozzles, has been a significant mechanism of degradation since the 1970s, and it is still a poorly understood technical problem. This chapter examines strategies of degradation management for PWR reactor pressure vessels, reactor internals, stream generators, pressurizers, control rod drive mechanisms (CRDMs) and primary/secondary piping, and describes some cases of component degradation.

7.2 Materials management strategies

In order to establish measures for managing materials degradation, the degradation mechanisms must first be fully understood. Inspection techniques, mitigation methods and repair technologies depend on knowledge grounded in experimental studies of degradation mechanisms or in field operating experience within power plants (IAEA, 2011). There are three stages to managing materials ageing in nuclear power: preventive action; monitoring and inspection; and repair and replacement. In preventive action, improvement of the materials, reduction of stress and improvement of water chemistry can be used as measures to prevent cracks of Ni alloys (IAEA, 2011). Surveillance of pressure vessels can be carried out through monitoring and inspection to check the soundness of parts. For example, by checking for leakage of primary coolant through wall cracks in J-welds of the upper vessel head penetration (VHP) or lower bottom mounted instrumentation (BMI) nozzles, pressure boundary performance can be maintained. In the case of coolant leakage, boric acid residues on the outside of the pressure vessel or carbon steel corrosion products can be detected through visual inspection. Cracking can occur in operating power plants due to material properties or residual stress, therefore the timing of cracking can differ from the experimental result. Regardless of the cause, it is important to detect cracks as early as possible. Besides visual inspection, methods such as penetrant testing or eddy-current testing (ECT), and ultrasonic testing (straight beam and longitudinal wave angle beam UT) can also be used. In repair and replacement, the damaged parts should be isolated from the corrosive environment, or the tensile stress upon them reduced. In the case that these two measures are inappropriate, the parts should be replaced with others made from more corrosion-resistant materials. In order to systematically manage PWR structural materials, a common objective has been established, and much research has been done as a collaborative effort between many countries. The joint research programme for 2011 is as shown in Table 7.1.

Table 7.1 International research programme for PWR materials ageing management

Organization	Programme name	Objectives
IAEA	IGALL	International Generic Ageing Lessons Learned (IGALL)
	PLIM	To facilitate decisions concerning when and how to repair, replace or modify SSCs in an economically optimized way, while assuring that a high level of safety is maintained.
		To assure a safe and reliable NPP operation, provide a forum for information exchange, provide key elements and good practices related to safety aspects of ageing management and long term operation.
EC	COPRIN, CORTEX	PWSCC, SG tubes 600 & 690 of welds and Ni-base alloys in primary water
	INTERNALS PERFORM	IASCC of the lower core internals, baffle bolts management, IASCC of stainless steels, focus on mechanistic modelling
	RPV Lifetime	Methodologies applied to justify RPV margins and lifetime
	COFAT	Fatigue crack initiation and propagation, environmental effect
	Halden Reactor Project	Clad Corrosion and Water Chemistry Issues (PWR corrosion studies and BWR crud studies) Plant Lifetime Extension (IASCC crack initiation & growth studies, stress relaxation, reactor pressure vessel integrity)
DOE (USA)	IFRAM (International Forum for Reactor Aging Management)	To facilitate the appropriate exchange of information among parties and organizations around the world that are presently, or are planning to, address issues on nuclear power plant (NPP) materials ageing management. Three objectives support this purpose: (i) cooperating to achieve common objectives; (ii) sharing information/data; and (iii) entering into joint research/ demonstration projects.

7.3 Management techniques: development and application

In this section, we discuss in further detail management techniques for both reactor vessels and internals as well as steam generator tubes, pressurizer nozzles and the CDRM and finally, look at applied management practice around the world.

7.3.1 Management techniques for a reactor vessel

In a pressure vessel and its internals, degradation areas are welds of beltline regions, inlet–outlet nozzles, CRDM, instrumentation nozzles and flange closure studs. The degradation mechanisms are largely radiation embrittlement, fatigue, IGSCC and boric acid corrosion.

Embrittlement of pressure vessels is a more significant problem in PWR than in boiling water reactors (BWRs). This is, because in a PWR the layer of coolant around the core is thinner, so the PWR core generates a 20–100 times greater neutron fluence. The current design of RPVs does not feature welded joints in the beltline region, as this is the most radiation-embrittled zone, but in older vessels there are both circumferential and axial welds in this area since vessels were manufactured from plates. Current materials regulations describe the application of low copper materials and low-alloy steel of SA533B-1 for the fabrication of pressure vessels, so that the parent metal in the beltline part of the shell is damage resistant. In older type vessels, the most important issue is radiation embrittlement around the weld zone of the beltline area. The weld zone can easily become more embrittled than the parent metal not only because copper, nickel and phosphorous impurities are present, but also because it is the point of connection of various metals and the heat-affected zone (HAZ). When materials have been embrittled, the nil ductility transition temperature or the ductile-brittle transition temperature increases, and the upper shelf energy (USE) value from the Charpy impact test decreases. As a result, the permissible pressure temperature (PT) of power plants is limited. Damage by fatigue cracking occurs in the beltline weld zone (under normal operating pressure/heat cycle and abnormal events), closure head studs (during loading cycles in normal operation and repair), primary coolant entrance and exit nozzle (under heat cycling) and penetration and CRD housing (under heat cycling). Heat cycling can occur in normal operation during the heat-up or cool-down phases associated with servicing, or may be unexpected (Morgan and Livingston, 1995).

Degradation management strategies can be categorized as: mitigation, inspection and surveillance, or repair, as in Table 7.2.

Table 7.2 Materials ageing management for PWR

RPV components	Ageing mechanisms	Management categories	Management techniques	Description of techniques
Beltline region welds	Radiation embrittlement; fatigue	Mitigation	Thermal stress reduction	Change operating procedures so as to create off-normal sequences; change of plant system (e.g. installing monitoring instrumentation)
			Neutron flux reduction	Low leakage loading pattern (Moylan *et al.* 1987)
			Thermal annealing	Annealing below 343°C (650°F) (Wet anneal) or at 430°C (806°F) (Dry anneal) per ASTM E509 (ASTM 1988c)
		Inspection and Surveillance	Ultrasonic examination	Per ASME Section XI code requirements
			Acoustic emission	Per ASME Section XI code requirements
			Surveillance programme	Surveillance capsule of base metal, weld metal and HAZ metal hung between thermal shield and RPV inner wall
		Repair	Repair of beltline welds by temper bead technique	Automated gas tungsten arc welding process having temper bead effect with no grinding of first layer of the welds
Inlet–outlet nozzles	Fatigue	Mitigation	Reduction of cyclic stress	Change operating procedures so as to create off-normal sequences
CRDM, instrumentation nozzles	IGSCC, fatigue, boric acid corrosion	Mitigation	Reduction of cyclic stress and residual stress	Change operating procedures so as to create off-normal sequences
			Material change	Use of corrosion-resistant materials (Alloy 690)
			Improvement of water chemistry	Optimized hydrogen, temperature reduction operation
Flange closure studs	Fatigue, boric acid corrosion	Inspection and surveillance	UT inspection, replacement	Per ASME Section XI code requirements

The references are from Morgan W.C & Livingston J.V. (1995) except for noted.

For thermal stress reduction, a device that inspects the state should be installed or risk abnormal events. To reduce damage from neutron radiation, the low leakage fuel loading technique can be applied which minimizes the influence of the neutrons on the materials of the pressure vessel through appropriate arrangement of burned and fresh nuclear fuel. The thermal annealing technique which returns the hardened pressure vessel materials to the nature of their raw materials can be carried out near 343°C (650°F) in water or 430°C in air.

There are two kinds of in-service inspections for vessels: ultrasonic testing and acoustic emission testing. The ultrasonic testing is described in ASME Section XI, and it is used to characterize cracks of the HAZ and weld zone. The uncertainties in this method, especially when it is used on cracks under cladding, have resulted in conservative regulatory requirements for use of these flaw estimates to set the permissible PT limits and evaluate pressurized thermal shock (PTS) events. ASME Section XI requires four inspections every ten years, and during this period, it recommends 100% volumetric inspection on repair welds on all shells, heads and flanges in the shell, nozzles in the vessel and beltline parts (Morgan and Livingston, 1995). This enables closer monitoring at the beginning and growth of potential fatigue cracks. The sharp cracks found on the surface of the vessel or in the embrittled beltline are most important to PTS but it is difficult to detect or inspect these cracks. Some studies have developed advanced ultrasonic techniques for this purpose (Shah and MacDonald, 1993). Acoustic emission monitoring can be used in online monitoring the growth of cracks if the surface of a vessel is accessible (Morgan and Livingston, 1995).

7.3.2 Management techniques for reactor internals

Materials used for PWR internals include ferritic steel, wrought austenitic stainless steel, cast stainless steel (CASS) and Ni alloys. The internals maintain the soundness of the geometrical core. The core consists of the upper core structure, core baffle/former/barrel, thermal shield and lower core support structure. The factors which influence degradation of these parts are: thermal plant transient, flow-induced vibration, radiation, high temperature, mechanical and thermal stress, and corrosive coolant. The main degradation mechanisms are: fatigue; radiation and thermal embrittlement; void swelling; and irradiation assisted stress corrosion cracking (IASCC). IASCC is a type of SCC indicated by a large quantity of neutrons in a material. The main objective of degradation management in the case of reactor internals is to ascertain if the internals support the core and can protect the CRDM (Morgan and Livingston, 1995).

In-service inspection and surveillance and changing of the materials are some of the measures used to manage degradation of the internals.

It is difficult to inspect the inside of a nuclear reactor, but it is possible to obtain information on physical damage, leakage and mechanical and structural states through visual inspection of the accessible zone. When an in-service inspection is being conducted, all flange closure studbolts and heads are removed. At this time, damaged equipment can also be removed. Equipment moved to the pool or which remains in the pressure vessel can be inspected using a remote control camera. It is difficult to conduct ultrasonic testing on this equipment or to interpret the results, but eddy-current testing is effective in measuring reduced thickness of pipes. Inspecting the inaccessible zone using the monitoring systems is complicated. Therefore, more effective remote control inspection equipment is needed (Morgan and Livingston, 1995).

The general regulation of in-service inspection of reactor vessel internals in ASME Section XI requires a visual inspection every ten years. Recently the requirement has become a visual inspection (VT-1, VT-3) supplemented with ultrasonic inspection of the baffle former bolts. The baffle former bolts comprise the weakest part of the internals. Supplementary ultrasonic examination is carried out in accordance with ASME Section XI subsection IWB, examination category B-N-3 in the United States and some other countries. Development of the ultrasonic examination equipment used for inspecting these bolts should take into consideration the existence of locking bar style bolts and the accessibility problems.

7.3.3 Management techniques for steam generator tubes

Degradation management in steam generators is possible with the help of research and development or a technical support programme. Strategies can be established by supplementing inspection and repair programmes based on operating experience in power plants. Management of ageing can be divided into the understanding, prevention, detection, monitoring and mitigation of ageing. A measure to systematically combine the management strategies is needed for steam generators which are widely used globally. An effective strategy could also be established and efforts to reduce duplication made through the cooperation of the equipment vendors and energy utility companies as shown in Fig. 7.1.

With the techniques developed to date, such as shot peening, rotopeening and heating and temperature reduction of the hot leg side, it is possible to reduce the tensile stress inside steam generator tubes. These measures markedly postpone PWSCC initiation. Plugging, sleeving or changing the affected pipes is effective in terms of repair. Secondary water chemistry control is the best defence against ageing damage on steam generator tubes. Measures to expand the life of the steam generator include

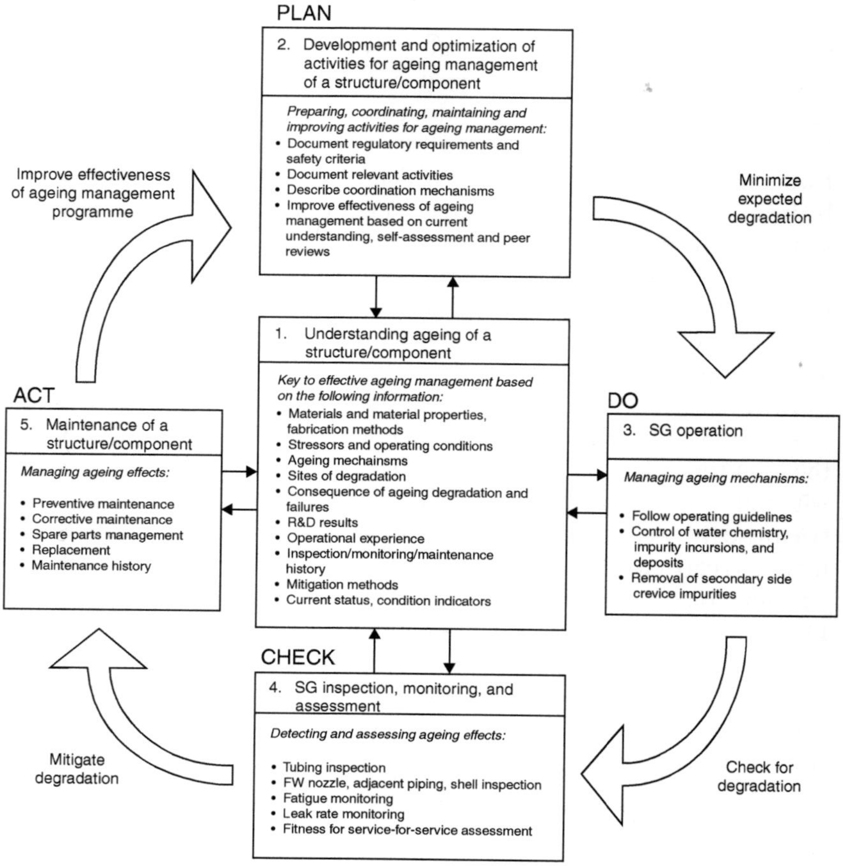

7.1 General structure of a steam generator ageing management strategy. (Reproduced with permission from the Electric Power Research Institute © 2008).

controlling impurities (chloride, iron and copper ions in the primary side) and oxygen (in the secondary side) and to prevent the accumulation of sludge. Ingress of chlorine-containing inorganics through condenser leakage, resin releases from condensate polisher and make-up water, are factors that compromise water chemistry control. It is proven that certain chemical additives (e.g. boric acid or morphine) decrease intergranular attack (IGA), SCC and denting of the tubes. However, it is not known whether such additives influence the equipment of other power plants (Morgan and Livingston, 1995). A continuous monitoring and control programme should also be followed to reduce impurities in the secondary water (Wood, 1990).

Denting of the tubes, fretting wear and erosion–corrosion can be detected through a normal in-service inspection before leakage occurs, whereas it is difficult to detect regional pitting corrosion and cracking (fretting-fatigue, stress corrosion and formation of intergranular) before leakage occurs. The fundamental cause of fretting is related to the design of the stream generator. As a result, the most effective management option is dependent upon the design. In most cases plugging of the affected tubes is an effective solution when damage is found in a particular part of a certain design. The occurrence of erosion–corrosion and corrosion fatigue is limited to once through steam generators (OTSGs), and management options vary depending on the characteristics of particular power plants (Morgan and Livingston, 1995).

To avoid maintenance cost increases, suspension of operation or reduction of output, it has become possible to replace the existing steam generator with one using corrosion resistant alloys (Alloy 690). As of 2011, over 100 steam generators have been replaced around the world. For most of the replaced stream generators, thermally treated Alloy 690TT has been used. The power plant Cook-2 used this alloy for the first time in 1989. With advanced methods and greater experience, it no longer takes much time to replace a steam generator. Developments in design and material allows newer steam generators to have a long service life. Crevices can be removed, allowing a steam generator to have low residual stress. New generator designs also have improved accessibility for secondary lancing and chemical cleaning (Morgan and Livingston, 1995). Improved corrosion-resistant materials for SG tubes include high temperature mill annealed Alloy 600 (Alloy 600 HTMA), mill annealed Alloy 690 (Alloy 690 MA) and Alloy 690TT. Alloy 690TT has only recently been used in new steam generators. Ferritic stainless steel is used for tube support structure.

7.3.4 Management techniques for pressurizer nozzles and the CRDM

To minimize occurrence of damage to pressurizer nozzles and the CRDM, basic management strategies consist of: operation within operating guidelines; inspection and monitoring; assessment of any degradation that is detected; and maintenance. The main degradation mechanisms which can occur in pressurizer nozzles and the CRDM are thermal fatigue, vibratory fatigue, SCC and boric acid corrosion. Because coolant leakage through the heater sheath, instrument penetrations or manway cover gasket can cause corrosion and SCC of other equipment of the pressurizer system, this must be controlled. Molybdenum di-sulphide lubricant should also not be used in a steam exposed environment, because experience suggests that MoS_2 has a pronounced tendency to decompose in the presence of high temperature and moisture conditions releasing sulphide which is a known promoter of SCC. In the United States, to manage ageing of vessel head penetrations and nozzles, the utilities are forced to conduct a regular inspection under ASME Code Case 694. Some plants also conduct supplementary inspections on the PWSCC sensitive zone. Many power plants conduct supplemented inspection and replace their RPV heads with new ones (IAEA, 2007).

7.3.5 Management practices in selected countries

In Japan, utilities inspect in accordance with JEAC-4205 (Japan Electric Association Code for ISI Requirements), and the inspection requirements are similar to those in the United States. For the RPV weld lines, a volumetric examination is conducted on a regular basis. In France, utilities conduct inspection according to their RSE-M (Rules for In-service Inspection of Nuclear Power Plant Components), and also undertake water pressure testing with acoustic emission monitoring, non-destructive inspection during the outages, loose-parts (noise) monitoring during operation, leak detection during operation and fatigue monitoring. The range of inspection covers the beltline region of the shell, all welds, top and bottom heads, nozzles and safe end welds, penetrations, control rod drive housings, studs, threaded holes and supports. In Germany, utilities conduct regular inspection using the non-destructive inspection method in accordance with German Code KTA 3201.4.

7.4 Case studies of management strategies

In this section we look at applied management practice around the world.

7.4.1 Degradation of reactor vessels

There have been no cases of ageing degradation of reactor vessels. Thermal annealing at high temperature (475 ± 15°C) for 100–150 h has been applied to plants operating in Russia since 1987 as a preventative measure (Badanin, 1989; Cole and Friderichs, 1991).

7.4.2 Degradation of reactor internals

There is a report that a guide tube support pin made of Alloy 750 was damaged in Mihama Nuclear plant in Japan in 1978. The damaged pin was moving around as a foreign body and it was discovered in the steam generator chamber. It was determined that the damage was from high stress in primary water at high temperature. After this incident, research studies were conducted over many years to understand the damage mechanisms. It was found that Alloy 750 was a material sensitive to PWSCC, and that sensitivity increased largely depending on heat treatments. Since it was replaced with a new cold worked material (CW316 SS), no additional cracks have been found.

Regarding baffle former bolts, some cracks were found in an old French power plant (Fessenheim and Bugey) in 1988. Some bolts had 10–25 dpa fluence after being in operation for 10–20 years. They were made of a 316 cold worked stainless steel and showed intergranular stress corrosion cracking (IGSCC). The ageing mechanism was assumed to be IASCC. The cracks had spread from the shank zone of the head to the lower part of the head. The material was found to be hardened and radiation-induced segregation was found in the grain boundary. According to the hardness profile measurement, it was approximately 5–10 dpa, and there was no evidence of swelling. The bolts with cracks were detected at rows 2 and 3 from the lower part of the nuclear reactor where a considerable amount of neutron radiation had accumulated. According to the report, until that point, cracks in the baffle former bolts had been found in the 'down-flow' design in which inlet coolant flows downward (Gérard, 2009). From 1989 to 1993, the flow of the coolant in the nuclear reactor of the CP0 (name of French PWR 900 MW pre-series units) plant was changed to up-flow and, between 2000 and 2003, one third of the bolts were replaced. The cracking rate of baffle bolts increases slowly depending on dose. Based on the information on all the baffle bolts researched during in-service inspection (ISI) for all CPO units, the dose threshold is estimated to be approximately 3–4 dpa. The number of cracked bolts increases slightly at higher doses.

In Japan, two kinds of approach have been applied to the PWR plants since 1998 in light of the baffle former bolt issues experienced in France and also in the United States. The first approach was to replace the baffle former bolts. From 2001 to 2002, type 347 stainless steel bolts were replaced with cold worked 316CW stainless steel bolts in Mihama Unit 1 and 2. The second approach was to replace the internal structure of the nuclear reactor. In this case, the lower zone including the baffle former bolts and the upper zone were replaced. Since 2004, the internals of the nuclear reactors in three PWR plants have been replaced entirely. The measures described above were applied to the baffle former bolts in a 2-loop PWR plant which was built in the early 1970s.

A research programme aimed at preventing defects of the internals of nuclear reactors has been ongoing since 2000. For example, there has been research on IASCC of austenitic stainless steel used in PWR internals such as in baffle former bolts. Valuable data was collected through this research, which showed that IASCC initiation is closely related to stress and exposed neutron fluence. In other words, the threshold stress which determines IASCC occurrence is dependent on the amount of neutron fluence. The threshold stress value tends to decrease as neutron fluence increases.

Based on the data and experiments, it was decided that a guideline would be published with detailed interpretation. Japan published a guideline on management activities such as inspection of the baffle former bolts of actual power plants in 2002. This guideline will be revised in the future to reflect up-to-date knowledge obtained through international collaboration.

In Korea in 2007, defects were found in the control rod guide tubes made of Inconel X-750 in the internals of a nuclear reactor. The control rod guide tubes were replaced with CW 316 stainless steel tubes. Since then, no cracks have been found in the baffle former bolts (Hwang *et al.*, 2010).

In the United States, the Westinghouse Owners Group (WOG) has inspected cracks in baffle former bolts from PWRs in other countries. They have also provided information on activities planned for Westinghouse power plants with potential cracking. The WOG has clearly demonstrated that destruction of minor bolts would not have a serious impact on safety because a number of baffle former bolts would still support the structure.

The WOG activities are as follows:

- Development of analytical methods and acceptance criteria for bolt analysis.
- Performance of risk-informed evaluations.

The Nuclear Energy Institute (NEI) which consists of the Materials Technical Advisory Group (MTAG) of the United States is formed of energy company representatives. The MTAG has received support from EPRI to prepare a guideline for In-Service Inspection (ISI) of RVI equipment which has significant impact on continued and safe operation of power plants. In preparing an inspection guideline, the damage to equipment inside the nuclear reactor by inspection, fatigue, abrasion and corrosion were considered. Recently, Westinghouse and AREVA have published a report using screening based on various significant damage mechanisms as a part of EPRI MRP on reactor internals.

7.4.3 Degradation in steam generators

Mechanisms

The first crack found in a steam generator tube was in a hot leg side tube of a steam generator in the Obrigheim plant in 1971. This was the first reported incidence of PWSCC at this site (Shah, 1992). As of 1994, at least 61 power plants had experienced PWSCC in the tubes; 32 plants experienced U-bend PWSCC and 5 plants experienced PWSCC in the denting zone. In most cases, it occurred in hot leg side tubes, but in some it occurred in cold leg side tubes. The cracks were found in 1–10 EFPY (effective full power year) for Alloy 600 LTMA (low temperature mill annealing), but they were detected only after 10 EFPY in Alloy 600 HTMA (high temperature mill annealing). However, some Korean power plants that have installed Alloy 600 HTMA in an explosive expansion method experienced circumferential SCC in three to seven years. PWSCC of the U-bend zone usually causes axial cracks. If there is mechanical expansion, axial cracks are primarily found, but in a French power plant, a circumferential crack was found in kiss roll expanded tubes in the sludge accumulation area.

Denting of Alloy 600 was first reported in the tube support plate (TSP) region in 1975 immediately after the secondary system water treatment changed from phosphate to all volatile treatment (AVT). Denting refers to the state where the cross-section of a tube does not maintain its original form due to the growth of oxides around the pipe. This damage mechanism was a major cause of tube plugging from 1976 to 1980, although more recently it has not been particularly serious. If chloride intrudes into the secondary system by steam condenser leakage and is concentrated between the tube and the support structure, it creates an acidic environment. When the oxygen content is high, corrosion of carbon steel tube support plates increases and a porous magnetite is formed which has double the volume of the base metal. The degree of superheat, presence of chloride and concentration of oxygen in the secondary system influences the corrosion rate of carbon steel. Copper oxides or copper ions can also be factors that supply oxygen to the water. Sulphate, like chlorides, can accelerate corrosion of carbon steel. The most serious cases occurred when seawater was used as the coolant of the steam condenser, such as in the RSG steam generators of Westinghouse and CE (Combustion Engineering in the United States). For a power plant that uses phosphate treatment, the environment remains alkaline so denting hardly occurs. Denting occurs rapidly, not occurring or spreading gradually like PWSCC or outer diameter stress corrosion cracking (ODSCC). When exposed to seawater containing chloride, denting occurs in a very short time (20 ppb acid chloride: 2.5 years), though when exposed to water containing neutral salt, it takes a more considerable length of time (20 ppb neutral chloride: 50 years).

The secondary IGSCC and IGA of Alloy 600 have been considered a serious corrosion problem since they were first reported in the early 1970s. These problems have been found in many freshwater cooling power plants, whereas fewer problems have been found in the power plants that use seawater coolant. The causes of IGSCC corrosion are impurities concentrated by surface boiling due to coolant flow not being smooth in a certain area, as well as stress, materials, temperature, etc. IGA is different because it occurs when there is no stress; however, sometimes stress creates IGA.

Erosion–corrosion describes pieces of corroded metal peeling off the metal surface by the action of solid particles repeatedly colliding with the

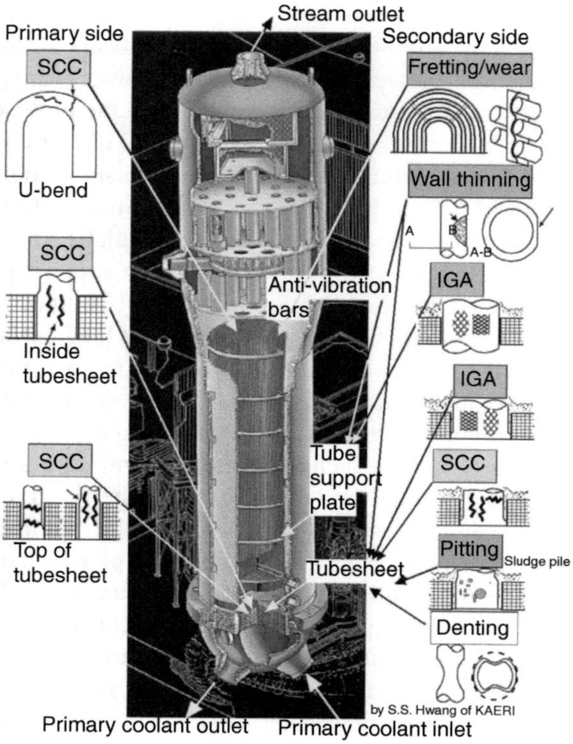

7.2 Various types of corrosion found inside and outside the steam generator tubes.

metal surface when there is stable protection film on the metal surface. Depending on the size, shape and hardness of the particles and the corrosion environment, mechanical damage can be accelerated. If there was no protective film on the surface, only erosion of the metal would occur, but once the protective film is removed, both corrosion and erosion will be accelerated. Such corrosion occurs in OTSGs.

Figure 7.2 shows the diverse types of corrosion found inside and outside steam generator tubes.

Experiences in different countries

Pull tube examinations of 92 tubes from some Korean nuclear power plants have been carried out since 1989 (Hwang *et al.*, 2007). The tubes had different types of failures such as pitting, ODSCC, primary water stress corrosion cracking (PWSCC) and intergranular attack (IGA). A new type of 'PWSCC'crack was found during the ISI carried out after the chemical cleaning in 1990, and 22 tubes in SG A and 26 tubes in SG B had to be sleeved.

The pitting of plant A was related to high copper dissolved from condenser material, chloride and high levels of dissolved oxygen. Transgranular SCC of plant B seemed to be related to lead compounds. ODSCC and IGA in plant A were related to a caustic environment in the crevices. PWSCC in plant A and plant C originated from the inherent characteristics of the materials, which were not properly thermally treated (Hwang, 2003). After failure analysis, the performance of non-destructive testing was evaluated based on destructive metallographic examination, and some counter measures, such as material change, inhibitor injection, molar ratio control and temperature reduction operation, were suggested.

In a typical case of high cycle fatigue, a complete 360-degree break occurred in the cold leg side tube in Row 9 of the North Anna Unit 1 plant in the United States on 15 July 1987. The case was explained as follows (Shah and MacDonald, 1993):

1 An anti-vibration bar (AVB) was not installed around row 9.
2 A small dent was found in the tube. It opened due to mean stress as the fatigue strength of the material had dropped.
3 An uneven AVB was installed around the troubled tube, which caused sectional high speed coolant flow.
4 High amplitude and deteriorated fatigue strength caused fatigue destruction.

After this accident in North Anna Unit 1 plant, the US Nuclear Regulatory Commission (USNRC) ordered an inspection of the power plants that showed potential fatigue destruction due to denting around the TSP and fast sectional flow. Except for the U-bend area, well-installed TSP structures have caused no high cycle fatigue.

It is said that in the mechanical ageing progress, fretting, wear and thinning are caused by the vibration between the tube and tube support structure (TSP and AVB). But thinning occurs where there is no flow-induced vibration, so it is difficult to say that tube vibration is a cause of thinning, Only in certain cases can we can say that thinning derives from pure corrosion wastage. There are many factors influencing fretting, wear and thinning including the distance between the tube and support plate, coolant flow, oxide film formation and corrosion product accumulation. Of these, friction of the same side causes fretting and a large vibration causes wear. When a combination of vibration and corrosion predominates, thinning results.

One example of low temperature PWSCC is the stress corrosion cracking which was detected in the tube of OTSG in the Three Mile Island (TMI-1) plant in 1981. Most of the cracks were circumferential, and were found mostly in the HAZ of the weld or Top of Tubesheet (TTS) of the expansion part. The tube of the lower defective zone was repaired using the explosive expansion method, and the unrepaired tube was plugged. The tube of the OTSG made by B&W was sensitized, heat treated and carbides were created at the grain boundary. The tube was SCC-sensitive if it was exposed to acid because of the tensile stress in the material from the manufacturing process.

In France, cracks have been found in the divider plate of a steam generator. As of the end of 2007, defects were found in the divider plates of ten steam generators. Various inspections have highlighted the fact that these defects are located in the stub of the hot branch, with no signs of significant evolution, either by fatigue or corrosion.

In Japan in 1976, leakage occurred from the U-bend of the steam generator row 1 of the Takahama Unit 1. It was assumed that the crack was caused by the PWSCC due to plastic deformation of the tube. The deformation, which was located between the U-bend and the tube, had been created by passing a ball mandrel through the tube, or had developed by the curving process during tube manufacture. This area usually has high residual stress. PWSCC of the U-bend also occurred in Ohi Unit 1 and Mihama Unit 2. Further, another leakage occurred in a small radius U-bend in Ohi Unit 2 in 1994. In this area, the ovality was larger, relatively speaking.

Since 1982, PWSCC of the tubesheet zone has been detected in many power plants. In Mihama Unit 3 and Ohi Units 1 and 2, PWSCC was found in both hard rolled areas and expansion transitions (made using full depth expansion) in tubesheet. PWSCC was detected at expansion transitions with part depth rolling in tubesheet at Takahama Unit 1 and Mihama Unit 2. PWSCC in the expansion transition occurs due to high residual stress in the zone where materials are mechanically rolled. This is caused by insufficient expansion during mechanical rolling over by uneven tubesheet holes. Such PWSCC of 600 MA pipe can be removed by replacing the steam generator with one with 690 TT tubes.

In a case of PWSCC of 600 TT tubes in the Japanese Kansai power plant, some cracks were found by ECT on Alloy 600 TT in tubesheet region of three power plants since 1999. Inspection was carried out on the damaged tubes and PWSCC was proven as a result. In three power plants (Sendai Unit 1, Takahama Units 3 and 4) the depth was expanded by full depth mechanical rolling after full depth hydraulic expansion. Cracks were found in the upper part of transitions to the hydraulic expansion area, indicating that cracks are located near the area expanded by the mechanical roll. But cracks did not occur where there was hydraulic expansion transition. Inspection of Takahama Unit 4 found the diameter of the tube hole to be sectionally large. It is considered that the oval shape was made by polishing the eccentricity of the tube hole during manufacturing. Cracks occurred in the zone where oval shaped holes were present. From a mock-up experiment, high residual stress was observed at the zone where there were tube holes that were irregular in shape, and mechanical rolling had been carried out. It is thought that mechanical rolling causes PWSCC where there are irregularly shaped tube holes.

7.4.4 Degradation of CRDM and pressurizer nozzles

In September 1991, leakage occurred from the Bugey 3 T54 vessel head penetration in France. After 10 years of monitoring the leak was detected using the acoustic emission method as part of the thermal-hydraulic test and it was estimated to be approximately 1 L/h. Non-destructive inspection using dye penetrant testing, eddy-current testing and ultrasonic testing confirmed a vertical crack penetrating the pipe in the opposite direction to the weld. Through metallographic analysis, it was concluded that the crack was PWSCC. By the end of 1992, Non-Destructive Examination (NDE) programmes using Eddy Current Test (ECT), Ultra sound Test (UT) and Visual Test (VT) had found penetrating cracks in five 900 MW vessel heads and four 1300 MW vessel heads (after 30000–40000 h of operation). Due to vessel head and RPV homogeneity, EDF (the operating utility) decided to replace all of the vessel heads with Alloy 600 penetrations (54 out of 58 vessel heads are Alloy 600, the remaining four are Alloy 690 penetrations). In France, pressure vessel heads at Bugey Unit 5 have been replaced since 1993; in Japan, replacement of the pressure vessel heads started at Takahama-1 in 1996; in Spain at Almaraz-1 in 1996; in Sweden at Ringhals-2 in 1996; in Belgium at Tihange-1 in 1999; in the United States at North Anna-2; and in China at Guangdong-2 in 2003. As of 2005, the pressure vessel heads of 93 power plants across the world had been replaced with Alloy 600 penetrations.

The importance of carbide was first established for steam generator tubes, but this has also been applied to penetration of the upper heads. Interpretation was based on carbon content, as well as forging, temperature of rolling and resistance strength upon hot forming. Three classes were determined in terms of grain boundary carbide decoration: Class 1 – well-developed intergranular carbide structure; Class 2 – mainly prior grain boundary carbides with recrystallized grain; Class 3 – mainly intragranular carbides with recrystallized grain. Modelling for crack initiation probability took into account the composition of materials considering the influence of other classes, angle of penetration and location (i.e. near or opposite to the weld zone) and the influence of cold working treatment.

In 2004, after over 100000 operating hours, leakage of the 47th CRDM head penetration was detected in Ohi Unit 3 in Japan through visual inspection. Ohi Unit 3 is a power plant where RV head reactor coolant system (RCS) temperature had been revised from 289°C to 310°C in 1997. In order to distinguish the leakage ratio, helium leakage testing, eddy-current testing, dye-penetration testing and ultrasonic testing were carried out on the

J-weld of the 47th CRDM head penetration. The leak in the J-weld was found through helium leakage testing; ECT found some indications of cracks on the J-weld. After grinding of the J-welds, dye-penetration testing was carried out. Cracks were observed on the portion of dye penetrant indication located along the grain boundary. It was ascertained by additional grinding that the long crack was connected with other cracks. The reactor vessel head (RVH) was replaced in 2007. The new head had a penetration nozzle and J-weld made from Alloy 690. There are 23 in-service PWR plants in Japan. At the present, 14 plants have replaced RVHs and seven additional plants will replace RVHs with Alloy 690 TT in the near future. One power plant has had CRDM nozzle penetrations which have been thermally treated with Alloy 690 since plant construction. Other power facilities in Japan have solved the issues of CRDM head penetrations by reducing the temperature at penetration.

In April 2003, a small amount of boric acid sediment was found in two BMI penetrations (No. 1 and No. 46 of a total of 58) in South Texas Project Unit 1 (STP Unit 1). This is the only evidence of leakage of BMI nozzle penetration reported in facilities in the United States up to now. The BMI penetrations of STP Unit 2 were built with drilled Alloy 600 bar, and connected to the lower head of the nuclear reactor vessel by welding Alloy 82/182 J-grooves.

In January 2003, a small cracking signal was detected on the internal surface area of the BMI penetration nozzle, and 50 BMI penetrations were found in Takahama Unit 1 in Japan through ECT. This indicator was within the permissible limit (≤3 mm depth) but it was concluded that the facility in this unit was likely to be at the beginning of PWSCC. The utility applied water jet peening on the surface in the BMI penetration nozzle after removing the crack indication. Laser and water jet peening are used for relief in other Japanese PWR plants. The peening method has been carried out in welding J-grooves in this location.

In September 2003, the thirteenth regular inspection of the PZR nozzles in Tsuruga Unit 2 in Japan was conducted. Cracks were found in the weld zone of the pressure relief line nozzle stub. This was the first case where sediments of boric acid were found. In ultrasonic testing on the relief line stub, two indicators were found located on the repair weld zone. In ultrasonic testing of other nozzle stubs, an indicator on the safety valve was found, but nothing was detected elsewhere. According to the observation, cracks remained in the weld zone only and developed in a circumferential direction of the pipe. The cause was analysed as SCC created in nickel-based alloy (600 type), also of the same type as the material welded in the PWR first coolant environment. The welding metal for the weld zone of the pipe nozzle on the pressure relief valve, the pipe nozzle stub on the pressure relief valve and the safe end was changed to nickel-based alloy (690 type) which has resistance to SCC. At the end of 2008, most of the pressurizer nozzles would be supplemented with an Alloy 52M weld cover. Minor pressurizers have been replaced with Alloy 690 material.

Many incidences of cracks in Alloys 182 and 82 have been found in in-service PWR power plants. In July 2000, cracks at the outlet nozzle to pipe safe end weld of Ringhals Unit 4 were found. Many small axial cracks were found and removed with a boat sample through electro discharge machining (EDM). The first cracking had in fact been discovered in Ringhals Unit 3 in June of that year, but the power plant was permitted to continue to operate without any repair because the crack seemed superficial and shallow in depth. In both cases, welding used Alloy 182 and cracks were axial. The cracks in Ringhals Unit 3 and Unit 4 were removed in 2003 and in 2004, respectively, and the cracks were repaired by welding inlay using Alloy 52 M.

The next largest incident happened in October 2000. Penetrating cracks and leakage were found in the V.C. Summer power plant in the same part as in Ringhals Unit 3. Initial UT was carried out on the internal surface of the pipe and as a result, an axial defect near the upper part of the pipe was discovered. The next test was conducted in spring 2001 and many defects were found. All of the defects were axial, and the largest defect was penetrated. The defects were removed, and a new spool piece was welded. The part was restored to its original condition. Alloy 52 was used for the V.C. Summer repair from the exit nozzle to the pipe weld zone; Alloy 82 was used in some parts for thickness and for the rest of the weld zone. The other V.C. Summer exit nozzle was repaired by using mechanical stress improvement process (MSIP).

7.5 References

ASTM 1988c, *Standard Guide for the In-Service Annealing of Light – Water Cooled Nuclear Reactor Vessels* Annual Book of ASTM Standard, ASTM E 509–86, Vol.12.02, American Society for Testing and Materials, Philadelphia.

Badanin V.I. (1989), 'Application of annealing for WWER vessels life extension', Transactions of the 10th International Conference on Structural Mechanics in Reactor Technology, August 1989, Atomic Energy Society of Japan, Tokyo, pp. 129–34.

Cole N.M. and T. Friderichs (1991), Report on Annealing of the Novovorenezh Unit-3 Reactor Vessel in the USSR, NUREG/CR-5760.

Electric Power Research Institute (2008), Materials Reliability Program: Pressurized Water Reactor Internals Inspection and Evaluation Guidelines (MRP-227-Rev. 0).

Hwang S.S. (2003), 'Degradation of alloy 600 steam generator tubes in operating pressurized water reactor nuclear power plants'. *Corrosion*, **59**, 9, 821.

Hwang S.S. *et al.* (2007), KAERI/RR-2903/2007 *Failure Analysis of Retired Steam Generator Tubings.* Daejeon, Korea, KAERI.

Hwang S.S. *et al.* (2010), *Guideline on Management of the Internals of the Nuclear Reactor of Korean Power Plant.* Daejeon, Korea, KAERI.

IAEA (2007), Tec Doc Series 1556, *Assessment and Management of Ageing of Major Nuclear Power Plant Components Important to Safety: PWR Pressure Vessels 2007 Update*. Vien, Austria, IAEA.

IAEA (2011), Technical Document, No. NP-T-3.13, *Stress Corrosion Cracking in Light Water Reactors: Good Practices and Lessons Learned*. Vien, Austria, IAEA.

Morgan W.C. and Livingston J.V. (1995), *A Review on Information for Managing Aging in Nuclear Power Plants, PNL-10717*. PNNL, USA. Batelle.

Moylan *et al.* (1987), *Reactor Vessel Life Extension, Pressure Vessel and Piping Conference*. San Diego, California, 28 June–2 July, 1987, ASME 87-PVP-15.

Robert Gérard *et al.* (2009) 'Situation of the baffle-former bolts in Belgian units'. Proceedings of the 17th International Conference on Nuclear Engineering, ICONE17, 12–16 July, 2009, Brussels, Belgium.

Shah V.N. (1992), 'Assessment of primary water stress corrosion cracking of PWR steam generator tubes,' *Nuclear Engineering and Design*, **134**, 2–3,199–215

Shah V.N. and MacDonald P.E. (1993), *Ageing and Life Extension of Major Light Water Reactor Components, Amsterdam, Elsevier. Science Technology.* Sections 3.6–3.7. Amsterdam. Elsevier Science Publishers.

Wood (1990), *PWR Primary Water Chemistry Guidelines: Revision 2, EPRI NP-7077.* EPRI Palo Alto, California, USA.

8
Materials management strategies for VVER reactors

T. J. KATONA, MVM Paks Nuclear Power Plant Ltd, Hungary

DOI: 10.1533/9780857097453.3.335

Abstract: The strategic goal of the VVER operator is to extend its operational lifetime beyond the design life. Here, technical and regulatory conditions and methods for ensuring long-term operation of the VVER plant are presented plus an overview of the basic technical design features of VVER relevant to long-term operation. Degradation mechanisms of structures and components which limit the operational lifetime of the plants are identified. The method for evaluating ageing of the plant, a review of existing plant activities for ensuring the required performance of safety-related systems, development of ageing management programmes and other related plant programmes are described. The integration of plant programmes into a system that ensures safe long-term operation is shown through examples. Trends and need for future research are presented.

Key words: VVER, ageing mechanism, ageing management, long-term operation, in-service inspection, maintenance, environmental qualification, time-limited ageing analyses.

8.1 Introduction

The VVER reactors (Vodo-Vodyanoi Energetichesky Reaktor, which translates as Water moderated Water Cooled Energetic Reactor or WWER) are light water moderated and cooled, that is, pressurized water reactors (PWRs). A summary of basic features of VVER reactors is given by Katona (2010, 2011). VVERs were developed in the 1960s. The first three were built in Russia and Eastern Germany in the period 1964–1970, and operated up to 1990. There are 52 Russian-designed, VVER-type, pressurized water nuclear power plants operating in the world today, out of a global total of 443 nuclear power plants (for the latest operational statistics on VVER plants, see IAEA PRIS database) (IAEA PRIS, 2011). The cumulative time of safe operation of VVER reactors currently exceeds 1200 reactor-years.

The first standard series of VVERs had a nominal electrical capacity of 440 MW (and are therefore referred to as 440 units, 440 reactors, 440 designs, etc.) and reactors in the second standard series have a capacity of 1000

MW (and are thus referred to as 1000 units, etc.). There are two basic types of VVER-440 reactors, which are based on different safety philosophies. The VVER-440/230 design comprise the Generation I reactors, while the VVER-440/213 represents the Generation II reactor design with reduced pressure containment. There are two specific VVER-440 designs currently in operation: the Finnish Loviisa NPP with reduced pressure western-type containment and the Armenian Medzamor NPP. In the VVER 1000 MW series, there was a gradual design development through the five oldest plants (small series), while the rest of the operating plants represent the standardized VVER-1000/320 model. More VVER-1000 units were commissioned recently and those currently under construction are improved versions of the VVER-1000/320. For example, the Tianwan (China) plant with AES-91 type units and the Kudankulam (India) plant with AES-92 type units. New VVER models, such as the AES-2006 design, are being considered for future bids; these new evolutionary models of large VVERs already exhibit Generation III features.

The design operational lifetime of the VVER plants is generally 30 years, with the exception of the new VVER-1000 type units which have 50 or 60 years of designed operational lifetime. A great majority of VVER plants are quite old, nearing the end of their design lifetime, except for some in Russia. The VVER operating countries are dependent on nuclear power production, for example the Paks Nuclear Power Plant in Hungary provided 40% of domestic production in 2010. The nuclear power capacities in these countries ensure the necessary diversity of power generation and contribute to the security of supply. Therefore, the VVER owners in Central and Eastern Europe are keeping their plants in operation via implementing plant lifetime management (PLiM) programmes, with the intention of ensuring a safe and financially viable operation in the long term. The PLiM practice of VVER plants is presented by Katona (2010) and Katona and Rátkai (2008, 2010).

The possibility of extending the operational lifetime of VVER-440/213 plants was recognized in 1992. It was based on an assessment of the robustness of the design, good technical condition of the plants and synergy between safety upgrading measures and overall condition of the plants (Katona and Bajsz, 1992). In all VVER operating countries, lifetime management had the explicit goal of ensuring the extension of operational lifetime (Rosenergoatom, 2003).

The operational licence of the four VVER-440/213 units at Paks NPP in Hungary, is nominally limited to the design lifetime of 30 years. Extension of the lifetime of this particular plant by an additional 20 years is feasible. The first formal step of licence renewal of the Paks NPP was made in 2008 and the relicensing process is still ongoing. In Ukraine, the nuclear share of domestic production of electricity is approximately 48%, while this nuclear power plant comprises 26.6% of total installed capacity. There is a keen interest in extending the operational lifetime of all Ukrainian NPPs. The operational licence of the VVER-440/213 type Units 1 and 2 at Rivne NPP in Ukraine has been renewed by an additional 20 years with the condition of performing a safety assessment after ten years of prolonged operation. The extension of operational lifetime is a generic strategy of operators of VVER-440/213 plants in the Czech Republic and Slovakia. The Loviisa NPP in Finland (a non-standard VVER-440 design) has been allowed to prolong operation up to its next Periodic Safety Review (10 years).

The operational lifetime of the VVER plants in Russia will be extended by 15–25 years. The four oldest VVER-440/230 units, Novovoronezh NPP Units 3 and 4 and Kola NPP Units 1 and 2, have already received a 15 year licence for extended operation. The VVER-440/213 type units (Kola NPP Units 3 and 4) are also prepared for 15 years extension to the operational licence. Among VVER-1000 plants, Novovoronesh Unit 5 is prepared for a 25-year extension of operation, after an extensive safety upgrading and modernization programme.

The VVER operators performed a comprehensive assessment of plant condition and safety, while making their decisions about the extension of operational lifetime. A decision on the preparation of feasibility studies for long-term operation (LTO), was based on the recognition of the following VVER features and experiences:

- robust design of VVER plants
- good plant condition due to well-developed maintenance, in-service inspections, careful operation and extensive modernization and reconstruction
- implementation of safety upgrading measures, resulting in an acceptable level of safety.

Safety of the plants and compliance with international standards has been considered as the decisive precondition for LTO. The comprehensive modernization and safety upgrading programmes (Vamos, 1999) implemented by the VVER operators during the last two decades, resulted in gradual decreases in the CDF of these plants. The level 1 probabilistic safety analysis (PSA) study establishes the resulting CDF for all VVER-440/213 units at Dukovany NPP of $1.47–1.67 \times 10^{-5}$/a, as stated in national reports compiled under the Safety Convention (Czech National Report, 2010). The same achievements are published for other VVER plants. Extensive modernization and safety upgrading programmes have been implemented in Ukraine (2011), Russia (Rosenergoatom, 2003) and Bulgaria (Popov, 2007). The safety deficiencies do not inhibit the LTO of the VVER plants; the VVER operators have a strong commitment to continuous improvement of safety and are ready to meet the future challenges in this respect.

One of the issues related to the current licensing basis at VVER plants outside of Russia was the inadequate knowledge of the design basis. The design of VVER-440/213 and the older VVER-1000 plants was generally based on the former USSR regulations of the early 1970s, the General Requirements on Safety of NPP Design, Construction and Operation (OPB-73) and the General Safety Rules for Atomic Power Plants (PBYa–74). OPB-73 marked the beginning of a transition to the generally accepted international practice in nuclear safety (e.g. defence in depth, single failure criterion). Knowledge of the design base is absolutely critical for the preparation of LTO and licence renewal, especially for the review of time-limited ageing analyses. Operators of VVER-440/213 units have to perform a specific project for design base reconstitution. In many countries, the design base has to be entirely recreated, taking into account all essential changes in the licensing requirements. For example, in the case of the Paks NPP, seismic loads had not been considered in its design. The current design/licensing base includes safe shutdown during an earthquake with 0.25 g horizontal acceleration. Availability of a state-of-the-art Final Safety Analysis Report (FSAR), and regular updating thereof is required for the control of compliance with the current licensing basis and configuration management.

The condition of the plant and appropriate plant programmes are also preconditions for LTO, especially surveillance of reactor pressure vessel (RPV) embrittlement and monitoring the condition of long-lived passive structures and components. The most important ageing management (AM) activities are performed at the VVER plants from the very beginning of their operation. The early AM activity was focused on known degradation of the main systems, structures and components (SSCs), like the RPV embrittlement, or on the early recognized issues, for example leaking of the confinement due to the liner degradation, outer surface corrosion of the steam generator heat-exchange tubes. Most of the early AM programmes were state-of-the-art, for example the RPV surveillance programme. In the course of the first periodic safety reviews, the definition of the most critical SSCs for operational lifetime and the dominating ageing mechanisms were explained. Adequate assessment of the aged condition and forecast of safe lifetime of structures and components (SCs) can only be performed if the ageing process is monitored properly from the very beginning of the operation. The operational history of SCs has to be documented in sufficient detail for the trends in ageing to be discovered.

There are several non-technical conditions which affected the strategy of VVER operators and can be considered as motivation for the decisions on LTO. The positive international tendencies, with regard to LTO of existing nuclear power generation capacities, stimulated the LTO of VVERs too. (This tendency might be changed by the Fukushima nuclear accident following the Great Tohuku earthquake in Japan March 2011.) Accumulation of the experiences and scientific evidence for justification for longer than designed operation of NPPs, provides a good basis for LTO of the VVER. Good market positions of NPPs overall in the VVER operating countries, with high levels of public acceptance and positive public attitudes, help in supporting the operation of NPPs in these countries.

Considerable progress has been achieved at VVER plants with respect to the improvement of the performance and plant reliability. The load factor of the majority of VVER plants is over 80%; in some places for example at Paks and at Dukovany NPP it is around 90%.

The national regulation for allowing the approval of an extension beyond designed operational lifetime is also a condition of the LTO. According to Šváb (2007) and IAEA (2006, 2007a), there are two principal regulatory approaches to LTO, depending on the legislation for the operational licence. The operational licence in VVER operating countries may be either limited or unlimited in time. In countries where the operational licence is not time limited, the basis of regulatory approval is the periodic safety review (PSR). In those countries where the operational licence has a limited validity in time, a formal renewal of the operational licence is needed.

The internationally accepted rules and requirements regarding PSR are documented in the IAEA Safety Guide NS-G-2.10 (IAEA, 2003). One of the objectives of the PSR is to review the condition of the SSCs, and whether it is adequate to meet their intended safety functions. This includes knowledge of any existing or anticipated ageing and obsolescence of plant systems and equipment. In particular, the objective of the review of PSR Safety Factor 4: 'Ageing,' is to determine whether the ageing of SSCs is being effectively managed. This means whether or not the required safety functions are maintained, and whether an effective ageing management programme is in place for future plant operation (NS-G-2.10 para 4.21 of IAEA, 2003). The design lifetime is a technical limit for the operation, which is based on assumptions by the designer regarding time limit of performance and functionality of systems, structures and components due to ageing. The PSR used for justification of extension of operational lifetime beyond the design lifetime has to demonstrate that the prolonged operation is safe, despite expiration of the design lifetime. It means the PSR has to review all the time limiting analyses made by the designer. When reviewing the ageing of the plant, both programmatic aspects and technical aspects of ageing management should be evaluated. Rules for developing and establishing and attributes for adequacy of ageing management programmes are given in the IAEA Safety Guide NS-G-2.12 (IAEA, 2009).

Examples of the licence renewal approach are the Russian and Hungarian cases. For licence renewal, the regulations require the performance of integrated plant assessment, focusing on the review of plant condition, effectiveness of ageing management programmes and validation of time-limited ageing analyses for the extended period of operation. In Hungary, the national rules for licence renewal have been developed on the basis of 10CFR54, the licence renewal rule of the U.S. Nuclear Regulatory Commission. In Russia, the rules are defined within the context of national regulation.

In this chapter – after an overview of the basic technical features of VVER plants – the basic issues and methods for ensuring LTO of VVER plants will be presented. The dominating degradation mechanisms of structures and components limiting the operational lifetime of the plants will be identified, on the basis of operational experience and research results. The method for evaluating the condition of the plant; review of existing plant activities for ensuring the required performance and functionality of safety-related systems, structures and components; and development of ageing management programmes and other related plant programmes are described. Integration of particular plant programmes into a system that ensures safe LTO is shown on the basis of particular examples. Trends and needs for future research are also presented.

The presentation of the ageing issues will focus on the older VVER-440/213 and VVER-1000 plants. The VVER-440/230 plants (Kozloduy NPP, Bulgaria and Bochunice V1 NPP, Slovakia) are already on permanent shutdown. In contrast to this, the Kola 1 and 2 and Novovoronesh 3 and 4 units in Russia have already received licences to operate for a further 15 years. This was after implementation of modernization and safety enhancement programmes (Rosenergoatom, 2003) to cope with the safety issues relevant to this design (IAEA, 1992). The LTO and plant lifetime management of VVER-440/230 is not a generic practice and will be discussed below, although only to a limited extent. The LTO of the VVER-440/213 plants requires specific engineering effort and will be discussed in detail. From the point of view of LTO, the newly designed and constructed VVER plants are also of less interest. Obviously, they have been designed and manufactured taking into account the ageing lessons learned from operational experience. The question about the need and possibility of longer than designed operation of these plants is not on the agenda today.

8.2 Description of operating VVER reactors

In the sections below, the basic design characteristics of VVER plants are presented. The design and manufacturing features which are relevant from the point of view of LTO are discussed.

8.2.1 Basic design features of the VVER-440

The V-179, V-230 and V-213 types of VVER plants are equipped with a six-loop VVER-440 reactor. In each loop, there are main isolating valves (MIV) on the cold and hot legs, one main circulation pump (MCP) per loop and horizontal steam generators (SG). The pressurizer, with safety valves, is connected to the primary loop. The two generations of the VVER-440 type of reactors have very similar layouts in their primary systems. Typical operating parameters are T_{hot}=297°C, T_{cold}=266°C, p=12.3 MPa. However, the design bases of the VVER-440/230 and the VVER-440/213 are essentially different, which manifests in the design of safety systems and confinement (IAEA, 1992; 1996a).

There are 16 nuclear power plant units of type VVER-440/213, namely, four in Hungary, four in the Czech Republic, four in Slovakia, two in Russia and two in Ukraine. The owners of these plants are preparing for the LTO of these units.

The design bases for the VVER-440/213 safety systems are similar to those used in Western PWRs, including the postulating of a double-end guillotine break of the main circulation line in the reactor coolant system. The safety systems exhibit triple redundancy and the reactors have bubbler condenser-type, pressure suppression containments capable of withstanding the imposed loads and maintaining containment functionality, even following large break LOCA events. The design of the VVER-440/213 plants considered internal and external hazards to some extent. Protection against single failures in the auxiliary and safety systems has generally been provided in the design. The safety concerns with VVER-440/213 plants are discussed in the IAEA report (1996a). The VVER-440/213 has essentially inherent safety characteristics, for example robustness of the design, low heat flux in the core, large water inventory in the primary system and a large containment volume, which compensates to a large extent for other deficiencies in the containment concept. At all of the plants, most of the safety deficiencies have been addressed by retro-fitting and plant modifications. Due to the robustness of the design, it was feasible to upgrade the safety of the original VVER-440/213 design to a level comparable with the PWR plants of the same age. The latest constructed units of VVER-440/213, such as Mochovce NPP Units 1 and 2, had several improvements and modifications made during the design and construction phase.

There are specific modifications of the VVER-440 design: the Finnish nuclear power plant at Loviisa, represents a combination of the VVER-440/230 basic design and nuclear island equipment with a Westinghouse-type, reduced pressure, ice-condenser containment and several other western-designed and manufactured systems, like the complete instrumentation and control (I&C) systems. These units have a very successful operational history and excellent safety features. A comprehensive lifetime management programme was launched in the very early stages of operation, and has allowed LTO of the Loviisa units. The Armenian reactor also represents a modification of VVER-440 with an enhanced seismic capacity. The shut down Units 3 and 4 at Kozloduy NPP, Bulgaria represent an intermediate type between 230 and 213 series.

It should be noted that, the VVER-440s have certain inherent safety characteristics that are superior to most modern PWR plants, for example robust design, large water inventory in the primary system relative to the reactor power and large volume of the confinement.

8.2.2 Basic design features of the VVER-1000

The VVER-1000 model exists in several versions. The 'small series' plants could be considered as pioneers of this model. The VVER-1000/320 is the large series version of the design. Developed after 1975, VVER-1000/320 type plants are operated in Bulgaria, the Czech Republic, Russia, Ukraine and China. Modernized versions of VVER-1000 plants are under construction in five countries (Bulgaria, China, India, Iran and Russia).

In regard to lifetime management, the VVER-1000/320 plants have greatest practical importance. The 'small series' plants show some specific design features, but the lifetime management practice of these plants does not differ essentially from the VVER-1000/320 version.

The VVER-1000 is a four loop PWR with horizontal steam generators. Each loop consists of a hot leg, a horizontal steam generator, a main circulating pump and a cold leg. Main isolating valves on the hot and cold legs of each loop equip the non-standard VVER-1000 primary loops. The standard V-320 design and the new clones of the VVER-1000 do not have isolating valves on the primary loop. A pressurizer is connected to the hot leg of one of the loops and the spray line to the cold leg. Operating conditions are T_{hot}=322°C, T_{cold}=290°C, p=15.7 MPa. The reactor, the primary and the safety systems are all placed within a full pressure, dry, pre-stressed concrete containment.

The design bases, and also the technical solutions applied, are very similar to the PWRs operated in Western countries. The safety concerns about the VVER-1000 plants are discussed in detail in IAEA reports (1996b; 2000). The main safety concern regarding the VVER-1000 plants lies in the quality and reliability of the individual equipment, especially the I&C equipment. The plant layout has weaknesses that make the redundant system parts vulnerable to hazardous systems interactions and common cause failures by fires, internal floods or external hazards. At all plants, many of these deficiencies have been addressed by plant modifications and an acceptable safety level has thus been achieved.

There are several advanced VVER-1000 plants presently under construction, more than 20 new projects of advanced VVER design are under preparation or consideration and several are in the bidding phase. The most advanced versions of VVER design, showing features of Generation III reactors, are being considered for future bids for large generating capacity reactors.

8.3 Ageing of the VVERs – plant operational experience

Operational experience provides the basis for preparing the strategy of ageing management. The experiences of the plants regarding degradation of the lifetime-limiting structures and components have primary importance. These are the non-replaceable long-lived structures and components. In the VVER-440 plant design, lifetime-limiting structures and components are the containment building, reactor pressure vessel and the steam generator (Katona *et al.*, 2005, 2009b; Katona and Rátkai, 2008). Unlike the VVER-1000 and PWRs, the steam generators are practically irreplaceable in the VVER-440/213 design. In the case of the VVER-1000, the most important lifetime-limiting structures and components are the containment and the reactor pressure vessel. The proven design solutions of the VVER-440 were incorporate in the VVER-1000 design: the horizontal steam generator and also materials selection.

Alongside this, ageing the mechanical commodities, structures other than containment and electrical equipment, are also important for the development of an ageing management strategy. With this in mind, the operational experience of the plants varies because of the design variation of these components and structures at different plants.

8.3.1 Method for the evaluation of actual plant condition

Evaluation of actual/aged condition in safety-critical SCs is the basic method for identifying the ageing mechanisms and their effects on the intended functions. Plant condition has to be reviewed for the feasibility study of LTO. Review and evaluation of plant condition is an obligatory part of both the periodic safety review (Safety Factor 2 in the PSR, see IAEA, 2003) and the justification of safe operation in the licence renewal process.

The scope of review and evaluation of actual plant condition covers the safety- and seismic-classified SSCs and non-safety SSCs, failure of which may in turn jeopardize the safety functions. The review of plant condition is based on the information related to the health of components from the following sources:

- Results of operational information, records of the operational events.
- Failure data, root-cause analysis, failure statistics.
- Outage and maintenance records.

The inspection programme for safety Class 1 SCs is the most rigorous. It includes the following:

- Data of the non-destructive testing of the SCs.
- Evaluation of the results of the in-service inspections.
- Evaluation of the results/findings of the maintenances.
- Evaluation of the results of the ageing management programmes.
- Evaluation of failure data and other lifetime information.
- Evaluation of operational information.

Non-destructive testing is a regular activity at nuclear power plants. However, in the context of the plant review for the justification of LTO, some additional tests might be necessary. Individual programmes can be useful and developed for the Class 1 SCs, the reactor, main isolation valves (if such exist), main pipelines of the primary loops, steam generators and pressurizer.

In the case of SSCs in safety Classes 2 and 3, the most practical review method is visual on-site inspection. Application of the graded approach is useful, so that, in the case of higher importance or safety relevance, the inspection has to be performed for each particular item, whereas the review can be limited to the inspection of a representative sample of the commodity. The selection of the representative sample has to be made taking into account its type, the material, dominating degradation mechanism, environmental stressors, etc.

There are minor aspects to be checked during the inspections, for example:

- symptoms of leakages
- condition of the insulation
- paint condition
- condition of unpainted surfaces
- condition of welding
- condition of components at junction points of different materials
- condition of bolted joints.

After performing all of the on-site inspections, the findings have to be evaluated and any corrective measures identified. The final result of their evaluation can result in:

- modification of the maintenance procedures
- modification of the periods of maintenance
- introducing new diagnostic/monitoring measures in order to determine the necessary additional actions
- performing additional evaluation of the situation
- modifications such as implementing new sealing
- replacement of the component for a different type.

The information obtained has to be taken into account while reviewing and developing the ageing management programmes. Review and revalidation of the time-limited ageing analyses can also be considered as part of the evaluation of the part conditions. Feedback from experience of other VVER plants and the research results provide some guidance and background information for the review and evaluation of plant condition.

8.3.2 Ageing of mechanical components

The VVER-440/213 reactor pressure vessel

The design of the VVER-440 RPV is rather specific: the relatively small RPV diameter has to allow its transportation on rails. As a consequence of its limited diameter, the water gap between the RPV and the core is small, so the fast neutron flux (E>0.5 MeV) on the RPV is rather high at 10^{15} $m^{-2}s^{-1}$ and the RPV base material should therefore be more resistant to irradiation embrittlement. The RPV is assembled from forged rings without longitudinal welds. The coolant from the low-pressure emergency core cooling systems and hydro-accumulators is directly injected into the RPV, and from the high-pressure system into the cold leg of the loops. The inlet and outlet nozzles of the loops are separated on different levels. The penetrations for the instrumentation for core control are on the RPV head.

The ferritic steel reactor pressure vessel is clad internally with austenitic stainless steel. The RPVs are made from low alloy steel (15Cr2MVA; at Loviisa NPP 12Cr2MFA) and the circumferential submerged arc welding was made using Sv-10CrMoVTi wire. The RPV was covered internally by a welded clad of two stainless steel layers. The inner layer is a non-stabilized stainless steel (Sv-07Cr25Ni13, similar to AISI 309) and that, when in contact with the coolant, is a niobium stabilized stainless steel (Sv-08Cr19Ni10Mn2Nb; Sv-07Cr19Ni10Nb at Loviisa; both equivalent to

AISI 347). Components of the primary circuit in contact with the primary coolant, other than the RPV, are also made of austenitic stainless steel, that is the piping of the primary loop, the main circulating pumps, gate valves and the emergency and auxiliary systems pipework.

From the point of view of longer-term operation, the main deficiency of VVER-440/230 was the high irradiation exposure of the reactor pressure vessel wall by fast neutrons, and the relatively quick embrittlement of the RPV material. The issue had been aggravated by the lack of a proper RPV surveillance programme at these plants. Several attempts have been made to assess the embrittlement of the base and weld material of those RPVs. For the first generation RPVs, essential data for RPV materials were absent, for example transition temperature, concentration of copper and phosphorus; the archive metal of the RPVs was not available. The phosphorus and copper contents in the welds of VVER-440/230 are in the range 0.030–0.048% and 0.10–0.18%, respectively. In the case of VVER-440/213, the same concentrations are in the range 0.010–0.028% for P and 0.03–0.18% for Cu (Brumovsky *et al.* 2005; Vasiliev and Kopiev, 2007).

Reactor pressure vessel surveillance programmes became obligatory in all VVER plants that had been commissioned after Units 1 and 2 at Loviisa. Proper RPV surveillance programmes have been implemented at VVER-440/213 plants outside of the former Soviet Union from the commencement of plant operation.

An 'Extended Surveillance Specimen Programme' was prepared with the objective of validating the results of the standard programme (Kupca, 2006). It aimed to increase the accuracy of the neutron fluence measurement, make a substantial improvement in the determination of the actual temperature of irradiation, fix the orientation of RPV samples to the centre of the reactor core, minimize the differences in neutron dose between the Charpy-V notch and crack-opening-displacement specimens and evaluate any dose-rate effects. For Units 1 and 2 of the Mochovce NPP, a completely new surveillance programme was prepared, based on the philosophy that the results of the programme must be available during the entire service life of the NPP. The new, advanced surveillance programme deals with the irradiation embrittlement of both the weld area heat affected zone and the austenitic stainless steel cladding of the RPV, which were not previously evaluated in surveillance programmes.

Several measures were implemented for the resolution of the RPV embrittlement issue:

- Reducing neutron flux on the RPV, low leakage core design, dummy shielding assemblies.
- Annealing, that is effecting a change of material properties.

- Decreasing the stressors, for example, heating up the water in the emergency core cooling system (ECCS) to lessen thermal shock in a pressurized thermal shock (PTS) situation; steam-line isolation; system solutions interlocks.
- Introduction of volumetric non-destructive testing for in-service inspection.

Annealing of RPV has been implemented at Loviisa NPP and Kola NPP (also at the shut down plant Bochunice V1). Annealing in the case of the VVER-440 reactor vessel weld was performed at a temperature of 475±15°C and the holding time was 150 hours. Assessment of annealing effectiveness (level of properties recovering after annealing), determination of re-irradiation re-embrittlement rates after annealing, and the behaviour of VVER-440 weld materials, showed the real possibility of recovering RPV toughness properties of irradiated VVER-440 RPV materials. Measures were also taken to improve the knowledge of the vessel material by vessel sampling. A more detailed description of the RPV neutron irradiation embrittlement issue is provided by Erak *et al.* (2007), for example. Based on the results of the VVER-440/213 plants, annealing of the RPV has been implemented at Rivne NPP.

In order to determine the time limit of operation of the RPV, it is necessary to consider and analyse the neutron irradiation damage, thermal ageing and low-cycle fatigue in decreasing the fracture toughness of the RPV materials.

Pressurized thermal shock (PTS) is the most critical lifetime limiting event for the RPV. Since the PTS screening requirement (pressure-temperature-loading limits) is the lifetime limiting process for the RPV of VVERs, the methodology of PTS evaluation has to be established in the national regulations. This will take into account the applicable best practices, features of the RPV and the thermal-hydraulic peculiarities of the VVERs. The assumptions of renewed PTS analyses have been confirmed with mixing tests. International research projects supported the effort of VVER plants in the evaluation of PTS for the RPV (IAEA, 2005).

The results of the PTS calculations, based on the analysis of postulated embedded flaws, endorse the possibility of 50 years of operation for all of the units, without annealing of the 5/6 welds. At the Paks NPP, the assumption of the embedded postulated crack (under-cladding semi elliptical type) was justified by the results of qualified in-service inspections, which followed the procedure of European Network for Inspection Qualification (ENIQ). Two types of inspection were applied to the full cladding area: (1) ultrasonic inspections from the inner surface and (2) Eddy current inspection, overlapping the first 5 mm thickness of the RPV inner-wall area. There is

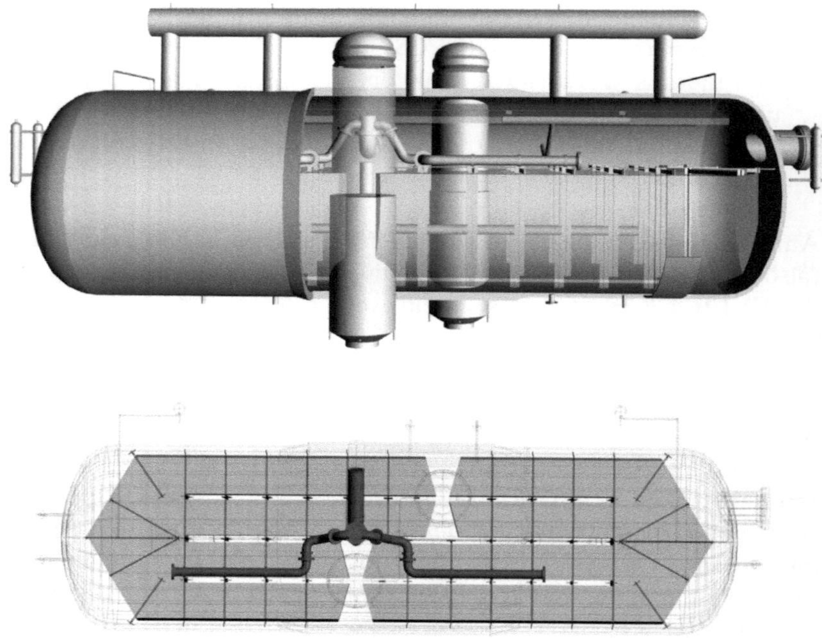

8.1 The steam generators in VVER-4403.

no generic need for the heating up of the emergency core cooling water. It was introduced as an example at Rivne NPP in Ukraine, however it seems unnecessary at Paks NPP in Hungary.

The critical locations when considering fatigue are the welds of the inner tubes of the control rod drive nozzles.

Steam generator of VVER-440/213 design

The steam generators in VVER-440 are horizontal (see Fig. 8.1). The advantages of the VVER horizontal steam generator design are the high reliability, absence of vibrations, no accumulation of sludge at the tube sheet and ease of access for maintenance. The SG design has a positive impact on safety as well, for example the design allows reliable natural circulation, effective gas removal, large water inventory and essential thickness of the heat-exchange tubes.

The heat-exchange tubes and the steam generator tube headers (collectors) are manufactured from austenitic stainless steel (18% Cr, 10% Ni stabilized with titanium) in VVERs, instead of the nickel-based alloys (Alloy 600 and 690) and higher chromium-containing alloys (Alloy 800) as used in PWRs. The material of the SG heat-exchange tubes in VVER-440

Table 8.1 A comparison of steam generator materials in VVER-440, VVER-1000 and typical PWR

	VVER-440	VVER-1000	PWR
Heat-exchange tubes	08H18N10T	08H18N10T	Alloy 600, 690 or 800
Tube sheet, collector	08H18N10T	10GN2MFA, 08H18N10T and cladding	Low alloy steel and cladding
SG vessel	22K	10GN2MFA	Low alloy steel
Tube-grid	08H18N10T	08H18N10T	Carbon or stainless steel

is equivalent to AISI 321. A comparison of the SG materials selected for VVER-440, VVER-1000 and a typical PWR is shown in Table 8.1.

The oldest VVER-440 type steam generators at Novovoronesh NPP Units 3 and 4 have been operating for 40 years. The condition of the oldest VVER-440 steam generators at the Kola and Novovoronesh plants has allowed a 15 year extension of operation for these plants. According to the operational history, the feed-water distributor inside the SG shows accelerated ageing due to erosion. These elements were replaced at almost all VVER-440 plants (see the dark coloured new distributor in Fig. 8.1). The experience regarding the ageing of VVER steam generators is summarized in TECDOC-1577 from the International Atomic Energy Agency (IAEA, 2007b).

At VVER-440 plants, the lifetime limiting ageing mechanism of the SGs is outer diameter stress corrosion cracking (ODSCC) of the austenitic stainless steel heat-exchanger tubes. The ODSCC indications appear typically (80%) at the grid structure supporting the tube bundle, where the secondary circuit corrosion products (with concentrated corrosive agents) are deposited. An eddy current inspection programme is implemented for monitoring the tubes. Samples have been removed from plugged tubes to facilitate investigations into the phenomenon. The rate of the ODSCC was essentially slowed down by a series of modifications and actions, implemented at different plants and to different extents. The measures implemented are as follows:

- Replacement of the condensers: the new condensers have austenitic stainless steel tubes.
- Removal of copper and copper-bearing alloys from the secondary circuit.
- Replacement of the feed-water distributor (the old one was manufactured from carbon steel).
- Cleaning the heat exchanging surface of the SGs.

- Introducing high pH secondary water chemistry.
- Replacement of the high-pressure pre-heaters (with erosion-corrosion resistant tubes).

All of these measures have been implemented at the Paks NPP, and have completely changed the conditions and rate of ODSCC in the SGs. Consequently, a better (i.e. decreasing) plugging trend is experienced, which can also be expected in the long term. The gaps between the tubes and support grid are still the critical sites, since any remaining corrosion products will accumulate there. It is therefore difficult to forecast the ODSCC rate in the gaps and the ageing process has to be closely monitored in the future. Under the new conditions, sludge may accumulate at the bottom area of the SG and an effective method for draining it must be found. The reserve in heat-exchanger surfaces of the SG is relatively large (more than 15%). Considering past experience and the recent plugging trend of the heat-exchange tubes, none of the SGs would exceed 10% of plugged tubes by the end of 50 years operation, due to measures implemented (Katona *et al.*, 2003; Trunov *et al.*, 2006b). The number of allowable plugged tubes became more important at the plants where the primary energy output is increased for the power up-rate. Therefore, establishing an adequate performance criterion for the steam generators is very important.

Ageing of mechanical components of VVER-1000

In the VVER-1000 models, all primary circuit surfaces are either made from, or are clad in, stainless steel. The 08X18H10T type stainless steel (08Cr18Ni10Ti, AISI 321) is used for the core structures, main circulating pumps and steam generator tubing, whilst the main loop pipework and steam generator collectors are manufactured from 10GN2MFA type carbon steel and the cladding is made from 08Cr18Ni10T stainless steel. The pressurizer is also made from 10GN2MFA carbon steel, covered by cladding, with an inner layer of Sv-07Cr25Ni13 (similar to AISI 309) stainless steel and two layers of Sv-08Cr19Ni10Mn2Nb niobium stabilized stainless steel (similar to AISI 347). The reactor pressure vessel and head is made from the low alloy steel 15Cr2MNFA. The cladding of the reactor head has an inner layer of Sv-07Cr25Ni13 stainless steel and two layers of the niobium stabilized stainless steel Sv-04Cr20Ni10Mn2Nb (again similar to AISI 347). The phosphorus and copper contents in the welds of VVER-1000 RPVs are 0.005–0.014% and 0.03–0.08%, respectively.

It has been recognized that the standard surveillance programmes for VVER-1000/320 reactor pressure vessels have some deficiencies related to the design of the surveillance assemblies, for example the non-uniformity of neutron field within individual specimen sets, large gradient in neutron flux between specimens and containers, lack of neutron monitors in most of containers and no suitable temperature monitors (Brumovsky and Zdarek, 2005). The location of surveillance specimens does not assure similar conditions as the beltline region of reactor pressure vessels. A modified surveillance programme for VVER-1000/V-320C type reactors was designed and implemented at the Temelin NPP in the Czech Republic. The technical features of the surveillance test assemblies provide opportunities for implementation of an integrated surveillance programme, using samples from several VVER-1000 units: Temelin 1 and 2 (Czech Republic); Belene (Bulgaria); Rivne 3 and 4, Khmelnitsky 2 and Zaporozhie 6 (Ukraine); and Kalinin 3 (Russia). Irradiation of these archive materials together with the a reference steel JRQ (of ASTM A 533-B type) and reference steel VVER-1000 allowed a comparison of the irradiation embrittlement of these materials, and an opportunity to obtain more reliable and objective results, as no reliable predictive formulae exist up to now because of a higher nickel content in the welds. Irradiation of specimens from the cladding region will help in the evaluation of resistance of pressure vessels against PTS regimes.

Several mitigation measures have been identified for the VVER-1000 RPVs. Based on the fracture mechanics analysis, heating up the hydro-accumulator water to 55°C was recommended to prevent injection of ECCS water with temperatures below 20°C for all the plants. The use of low neutron leakage core loading patterns in VVER-1000 reactors would reduce RPV wall fluences by approximately 30%. For reducing the neutron flux on the reactor vessel, low leakage core design was introduced at some plants (i.e. fuel assemblies with high burn-up to be placed at the core periphery). In addition, the quality of manufacturing and alloy composition ensure the possibility of LTO for VVER-1000 reactors (Vasiliev & Kopiev, 2007).

The steam generators for VVER-1000 have been designed on the same principles as the VVER-440 plants, however the SGs at VVER-1000 plants are replaceable. At some units, throughout the design service life of the SG, there were problems resulting in necessary SG replacement. At the same time, the SGs at some plants could be operated beyond design service life. As operating experience has shown, it is the water chemistry of the secondary circuit that is the main factor influencing operability of the SG tubing, as in the case of VVER-440 plants. Tube integrity is inspected by the eddy current method; the results of the testing can be used to determine the plugging criterion for defected tubes. Proper definition of the plugging criterion is an important challenge.

The ageing problems of the SGs at VVER-1000 plants are as follows (Trunov *et al.*, 2006a):

- cracking at headers of the cold collectors of the heat-exchange tubes
- degradation of the welded zone at hot collector headers

- corrosion of the heat-exchange tubes
- formation of deposit
- difficulties in measuring and regulating the SG water level.

A study performed by the International Atomic Energy Agency summarizes the status of knowledge on steam generator ageing: TECDOC-1577 (IAEA, 2007b).

8.3.3 Ageing of the structures

VVER-440/213 containments

The reduced pressure containment of VVER-440/213 is made of reinforced concrete and the steel liner ensures its leak tightness. Therefore, the basic concern is the effect of ageing on the containment leak-tightness. The leak rates of the VVER-440/213 containment, allowed by the design and justified by the regular integral tests, is equal to 14.7%/day at the post large-break LOCA, when the design internal containment pressure equals 2.4 MPa. It is clearly higher at some plants than what is allowable for Western NPP containments. Therefore, the goal of the VVER operators is to improve the leak tightness. (It should be noted that comparison with Western NPP containments is not straightforward. This is because, in connection with the design basis accidents, the pressure suppression system tends to cause pressures below atmospheric, rather than overpressure, at the time period when the atmosphere of the containment has its highest contents of radioactive aerosols, and when the potential for radioactive releases would thus be the highest.)

Containment leakage has a complex origin. Investigations carried out at the Paks and Bochunice NPPs, almost from the time of start-up tests, show that the poor sealing of doors and hatches mainly cause the containment leakage and thus the leakage is a maintenance problem rather than an ageing issue.

Some VVER plants are built on relatively soft soil. Geodetic control of the settlement of the main building of these plants was started during construction and it is periodically performed. The phenomenon might be a concern when there is uneven settlement, that is the differential movement causes unacceptable additional deformation of the structures. Experience shows that the differential movement may cause cracks in non-structural masonry walls. Another concern might be if the non-uniform settlement results in non-allowed tilting of the RPV vertical axis, which would cause problems for control rod drive mechanisms (CRDMs). The operating experience and analysis of settlement with extrapolation to extended operational lifetime is discussed for the Paks NPP (Katona *et al.*, 2009a).

Operational experience is that ageing of neither the reinforced concrete load bearing structure nor the liner would limit the LTO of the VVER-440/213 plants.

Ageing of the containment structures of VVER-1000

At the VVER-1000 plants, ageing may affect the pre-stressing of the containment. Important ageing mechanisms of the pre-stressed containment resulting in loss of pre-stress are the relaxation of tendons, shrinkage, creep of steel. Requirements for the testing of the containment pre-stressing system are defined both by the designer and regulation (Orgenergostroy, 1989a; 1989b). The scope of inspection should be extended if defects are observed, and/or if average loss of tension force is more than 15%. If further testing verifies the results obtained, it is necessary to test 100% of the tendons. Tendons with force losses of more than 15% should once again be controlled after straining. If a force loss at 24 hours is more than 10%, the tendon should be replaced.

In order to enable monitoring of the level of the containment pre-stressing, measurement systems are installed permanently on the structure and these systems measure structural deformations and pre-stressing force in the cables. At VVER-1000 plants, detailed field investigations and analyses have been carried out for the assessment and evaluation of the condition of pre-stressing tendons. There are design solutions for the replacement of tendons. Thus, all existing defects leading to a loss of stressing force and rupture of tendons have been avoided. At some plants, new pre-stressing systems and an additional system for automatic control of stressing forces is installed in the bundles.

8.3.4 Ageing of electrical systems and I&C

Electrical components and I&C are replaceable and the required performance of these commodities can be ensured via maintenance and scheduled replacement. The qualified condition of the electrical and I&C equipment has to be ensured.

Full scope ageing studies had been prepared for the Paks NPP for the following electrical and I&C items:

1 Equipment of electric power and transmission systems:
   - Bus cabinets
   - Overhead-line towers, medium- and high voltage insulators
   - LV and HV cables of power supply systems
   - Cables for containment electrical penetration
   - Cable joints and assemblies

- Enclosed electrical equipment
- Battery packs.

2 Equipment of the technological systems:
- Fixtures for transmitters
- Impulse pipes and assemblies
- Operation monitors
- Relay boards
- Cables for E, I&C equipment
- Cables of containment electrical penetration for E, I&C
- Cable joints and assemblies
- Terminal boxes.

The basic issue at all VVER plants regarding electrical and I&C equipment is the lack of or insufficient environmental qualification. Lack of initial qualification of the VVER equipment was recognized in the 1980s at all VVER-440/213 plants as well as at VVER-1000 "small series" and VVER-1000/320 models.

Establishing the initial qualification is understood as a current licensing basis requirement at all VVER plants. This consists of the following steps:

- Definition of environmental parameters characteristic of the installation site.
- In the case of safety equipment, definition of environmental parameters characteristic to the installation site under accidental (loss-of-coolant) conditions.
- Definition of accelerated thermal and radiation ageing test parameters.
- Performing laboratory tests with the above parameters (accelerated thermal and radiation ageing, radiation exposure with accident condition and simulation of loss-of-coolant conditions).
- Performance checks on tested samples to verify conformity with acceptance criteria.

The maintenance of qualified condition of the cables for harsh environmental conditions is a critical issue at VVER plants.

In regard to the cables, the technical task of qualification is rather difficult. For example at the Paks NPP there are 130000 cables and among them, several hundred types. The first necessary measure related to the cables was to develop a comprehensive database, instead of having the cable sheets on paper. The database identifies for each cable the safety classes, types and routes. The environmental conditions to which the particular cables are exposed are identified in the database. It also shows whether these safety-related cables are affected by the harsher conditions after accidents. Examples of ageing mechanisms of important cables are shown in Table 8.2.

Table 8.2 Examples of ageing mechanisms of NPP cables

Item	Site of degradation	Mechanism of degradation	Worst consequences
XLPE I&C cables in harsh environment	Cover and core isolation	Thermal ageing; change of the material properties due to heat or irradiation.	Crack/loss of function under loss-of-coolant condition
6 kV PVC power cables in channels/ humidity/	Metal structure of cables	Humidity penetration; corrosion of metal structure	Decreased isolation resistance/loss of function/
Cable connection in harsh environment	Corrosion of metal structure	Humidity/chemical corrosion of joints	Increased transit resistance of connectors

In its current state, the database at the Paks NPP covers approximately 19000 safety-related cables. The database allows the formation of commodity groups for cables. Currently there are 45 commodity groups related to safety cables. A similar approach is implemented as an example at VVER plants in Ukraine. For each group of cables, a particular sample cable is identified which is under worst-case condition. The condition of the sample cable is monitored during the operation.

The VVER plants replaced the frequently criticized, obsolete I&C systems. At the Paks NPP, nearly all safety-related I&C systems have been replaced: the reactor protection system, Engineered Safety Features Actuation Systems (ESFAS) protection system and load sequencer programme of diesel generators. The new system is a digital one (Siemens TELEPERM XS) with multiple redundancy and diverse software features, and physical separation of hardware of different trains. The reactor protection logic was also reviewed and modified to assure diverse physical signals for detecting each postulated initiating event and to eliminate unnecessary input and output signals. Similar reconstruction programmes have been implemented in Slovakia and at Russian plants entering into an extended period of operation.

8.4 Ensuring safety for a long-term operation

Considering the best international practices and also tendencies in the development of management of ageing in general (including obsolescence), one can conclude that the required condition and functioning of all SSCs relevant to safety should be ensured via

- analyses defining time limits of safe operation (time-limited ageing analyses (TLAAs)) and corresponding monitoring of time limiting assumptions (e.g. fatigue monitoring)
- ageing management programmes (AMPs)
- environmental qualification and programmes for maintaining the qualification
- maintenance and control of effectiveness of maintenance with respect to safety criteria
- scheduled replacements and reconstructions.

This system should be comprehensive in the sense that ageing of any item in the list of safety-related SSCs, should be covered by at least one of the methods. The required safety function and performance of any selected SSC has to be ensured by one of the approaches listed above or a combination of the methods/programmes (e.g. AMP and TLAA). The safety functions are properly ensured if the non-safety classified items, which may affect the safety functions, are also covered by one of the programmes.

The operator should pay specific attention to those structures and components, the function and performance of which directly limit plant lifetime. These are the non-replaceable or not-to-replace SCs to which either an effective ageing management programme should be applied, or the required functions should be demonstrated for the extended operational time by analysis (fatigue, embrittlement, etc.) or by qualification (e.g. in case of cables).

There are different approaches to how the operator defines which method/programme or combination thereof is applicable for particular SSCs; the optimization of plant efforts may have economical aspects too. The existing plant programmes might be credited as appropriate for ensuring the required plant condition in the long-term, if they are reviewed and found to be adequate. The concept outlined above is illustrated in Table 8.3.

8.4.1 Ageing management strategies and ageing management programmes

Scope of ageing management

The scope of the review covers the following SSCs:

- SCs relevant for safety – Classes 1, 2 and 3
- those non-safety SCs which can jeopardize the safety functions.

The non-safety-related SSCs which can jeopardize the environment (oil pipelines and tanks, containers for storing different chemical substances)

Table 8.3 Concept for ensuring long-term operation

Goals	Ensuring		
	Safety functions/ performance	Production/ economy	Functioning of the operating organization
How	Review, assessment and amendment of the plant programmes Reconstitution of the TLAAs		
Principles	– All systems, structures and components (SSCs) have to be covered by certain plant programme(s), for example preventive/corrective maintenance, ageing management, scheduled replacement – All ageing mechanisms have to be considered – All plant activities have to be considered, that is the routine activities should be integrated with those specific to LTO – Synergies have to be utilized		
Results	PLiM Programme for LTO		

should also be considered when identifying the scope of AMPs. In some VVER operating countries, the SSCs which are important for production are also within the scope of AMP (e.g. turbine, cooling water system).

Identification of ageing mechanisms

The identification of ageing mechanisms and their effect on the safety is based on the following information:

- Analysis of the operational experience
  - Experience at the individual plant: events-related ageing, for example load cycles.
  - Experience at plants of the same design.
- Generic industrial experience.
- Research results.
- Analysis of results of destructive and non-destructive tests.
- Review of the design assumptions regarding ageing.

After analysing these sources of information, the dominating ageing mechanisms, critical locations and measures for ensuring the required status of SCs can be identified. A list of important mechanical systems and components and the relevant ageing mechanisms are given in Table 8.4.

Examples for the identification of the ageing mechanisms of cables are given in Table 8.2 and more fully listed below in the table.

Table 8.4 Important mechanical systems and components and relevant ageing mechanisms

	Fatigue	Thermal fatigue	Radiation embrittlement	Wear	Stress corrosion	Corrosion in boric acid environment
RPV	x	x	x	x	x	x
In-vessel structure	x	x	x	x	x	
Reactor supports	x		x			
CRDM	x	x		x		
Pressurizer	x			x	x	x
Steam generator	x	x		x	x	
MGV and MCP	x	x		x	x	
RCS main circulating pipes	x	x			x	
Pipes connected to RCS	x	x			x	
Hydro-accumulators					x	
ECCS quick-closing valves	x			x		
ECCS pumps	x			x		x
Low-pressure ECCS pumps	x					
Sprinkler pumps	x					
High-pressure boron pumps	x			x		
Pumps of make-up system	x			x		
Essential service water tank	x					
Essential service water pumps	x			x		
Containment quick-closing valves	x			x		
Normal + emergency feed-water pipes	x	x				
Normal + emergency feed-water pumps	x					
Safety classified piping and piping elements	x			x		

Erosion	Crevice corrosion	General corrosion	Embrittle- ment	Loosening	Change of properties	Stratifi- cation
			x	X	x	
		x				
		x	x	x		
x						
x	x	x	x	x	x	
x				x	x	
x						x
x						x
		x				
x		x		x	x	
x					x	
x		x			x	
x		x			x	
x					x	
x					x	
		x				
x					x	
x						
x		x				x
x		x				
x		x				

- Temperature: in the case of organic materials, commonly used as insulation and/or sealing parts of components, high temperature is the main factor of ageing
- Radiation: inside the containment, γ-rays are mainly taken into account. The most sensitive material is PVC and the least sensitive is XLPE. Therefore, PVC insulated cables are not used for safety-related functions inside the containment. Neutron radiation is be considered only for copper parts located next to the reactor, where these parts may be activated
- Pressure changes: extreme pressure changes may occur in loss-of-coolant conditions and may endanger the proper operation of systems and components by affecting the sealing materials of some equipment
- Humidity: humidity in the containment may change for several reasons, for example leakage or pipe breakage, unintended operation of fire extinguishing appliances. Penetrating humidity may result in malfunction of electrical and I&C equipment.
- Steam: under LOCA conditions, steam may condense on the surface of equipment causing rapid temperature rise and it may also penetrate into the equipment.
- Chemicals: the applied chemicals (boric acid, hydrazine, etc.) may penetrate into seals of electrical equipment, reducing dielectric strength, and causing corrosion.
- Seismic events: seismic effects and vibration may degrade the functionality of certain electrical and I&C equipment (relays, transmitters, motors, etc.)

Identification of the ageing mechanisms for civil structures and structural components is discussed by Katona *et al.* (2009a). Examples are given in Table 8.5 on the basis of Hungarian regulatory guide No. 1.26.

Structuring of ageing management programmes

The VVER plants developed different types and systems of ageing management programmes:

- Overall plant AMP
- AMPs addressing a degradation mechanism
- Structure- or component-oriented AMP.

Overall plant AMP

An AMP for an overall plant can be developed and implemented for: definition of goals of the operating company, distribution of responsibilities in

Table 8.5 Identification of the ageing mechanisms for civil structures and structural components.

Component	Degradation location	Degradation process/ageing effect
Reinforced concrete in the hermetic compartments	Reinforced concrete	Corrosion/boric acid corrosion/ material loss
		Change of material properties due to heat/decrease of strength, modulus of elasticity
		Change of material properties due to irradiation
		Fatigue/crack initiation and propagation
		Settlement/increasing stress levels breaking, cracking
	Insertion elements	Corrosion/boric acid corrosion/ material loss
	Liner	Fatigue/crack initiation and propagation
		Local corrosion/material loss/crack initiation and propagation
	Biological protection	Change of material properties due to heat/decrease of strength, modulus of elasticity
		Change of material properties due to irradiation
		Fatigue/crack initiation and propagation
		Corrosion/material loss
		Settlement/increasing stress levels, breaking, cracking
	Decontaminable coatings	Change of material properties due to heat and/or irradiation
Other reinforced concrete structures	Reinforced concrete	Corrosion/boric acid corrosion/ material loss
		Change of material properties due to heat/decrease of strength, modulus of elasticity
		Change of material properties due to irradiation
		Fatigue/crack initiation and propagation
	Insertions	Corrosion/boric acid corrosion/ chemical corrosion/material loss

(*Continued*)

Table 8.5 (continued)

Component	Degradation location	Degradation process/ageing effect
	Liner	Fatigue/crack initiation and propagation
		Local corrosion/material loss/crack initiation and propagation
	Coatings	Change of material properties due to heat and/or irradiation
Service shafts	Carbon steel cladding of spent fuel and refuelling pool and shaft number 1	Local corrosion/material loss/crack initiation and propagation
		Boric acid corrosion/material loss
	Syphon of refuelling pool	Local corrosion/material loss/crack initiation and propagation
		Boric acid corrosion/material loss
	Stainless steel cladding of shafts	Local corrosion/material loss
		Wear, cracking/material loss
	Welds and heat affected zone of stainless steel claddings	Local corrosion/material loss/crack initiation and propagation
	Supports and insertion elements	Local corrosion/material loss/crack initiation and propagation
		Wear, cracking/material loss
	Welds between the shaft cladding and connecting pipelines	Local corrosion/material loss/crack initiation and propagation
		Wear, cracking/material loss

Notes: Examples are based on Hungarian regulatory guide No. 1.26.

the organization and policy level activities, and definition of the programme system structure for ensuring the required plant condition, that is the implementation of the concept described in the introduction of section 8.4.

Several operating VVERs have utility- or even industry-level or umbrella type ageing management programmes. For example, in Ukraine the plant level programme has to be deduced from the overall one and the unit level programme from the plant level one. The overall plant AMP also includes the categorization of the SCs in accordance with safety relevance, importance and complexity. In considering the structuring and organization of AMPs, a graded approach should be applied according to the safety relevance of the

Table 8.6 Degradation mechanisms which an AMP may address

Low-cycle fatigue	Thermal ageing
Irradiation damage	Stress corrosion
Boric acid corrosion	Wear
Local corrosion	General corrosion
Irradiation-assisted stress corrosion	Loosening
Swelling	High-cycle fatigue
Thermal stratification fatigue	Erosion
Erosion-corrosion	Microbiological corrosion
Water hammer	Groundwater corrosion
Deposition	

given structure or component and plant lifetime limiting character of the given ageing mechanisms.

AMPs addressing a degradation mechanism

Some AMPs are based around addressing a particular degradation mechanism, examples of which are shown in Table 8.6.

Structure- or component-oriented AMP

Applying the graded approach, the SCs can be separated into two categories:

1 Highly important from a safety point of view, items with complex features and ageing mechanisms.
2 Items which have the same type, safety class, identical design features, materials, operating circumstances and dominating ageing mechanism could be grouped into commodity groups and for each commodity group a designated AMP can be implemented, for example pipelines, pipe elements, valves, heat exchangers, etc.

The highly important SCs like the reactor pressure vessel together with internals or components of main circulating loop (SCs of Safety Class 1 and some SCs of Class 2) can have dedicated, individual AMPs, for example:

- Reactor pressure vessels
- Steam generators
- Reactor pressure vessel internals
- Pressurizers
- Main circulation pipeline
- Main coolant pumps
- Main gate valves.

Table 8.7 Attributes for the definition of commodity groups

Safety classification	Type of SSC	Medium	Material
Safety Class 1	Valve body	Borated water	Stainless steel
Safety Class 2	Pump body	Prepared water	Cast stainless steel
Safety Class 3	Pipe and pipe elements	River/sea water	Carbon steel
Non-safety class, failure of which may inhibit intended safety function	Heat exchanger Tank	Steam, gas-steam mixture Acid or alkali Oil, other	

The pipelines, pipe elements (elbows, T-pieces), valves and heat exchangers can be grouped into commodity groups according to type, material and working environment. The SCs within a group have the same degradation mechanism and approximately the same operational and maintenance history. It is very reasonable to develop specific ageing management programmes addressing the ageing of commodity groups. The definition of the commodity groups is decided by applying the attributes given in Table 8.7 in all reasonable combinations.

Ageing management peculiarities of the VVER-440/213 plants

A peculiarity of the VVER-440/213 design is the extremely large number (over one hundred thousand) of safety-classified SSCs because of the design features and methodology of safety classification.

After screening out the active and short-lived systems from the total safety-classified SSCs, approximately 38 000 mechanical, 6500 electrical and 2000 structural SCs have been identified to be in scope at the plant in Paks, Hungary.

Ageing management of mechanical commodities might be ensured approximately by nine vessel specific, nine pump-specific, 14 valve-specific, 22 heat-exchanger-specific, 15 piping specific, nine filter-specific programmes. There are also 15 special components requiring individual AMP. The number of structural commodities exceeds 25. The AMPs and their hierarchical structure is plant specific, demonstrating that Paks NNP practise an adaptation of best international practice to VVER-440/213 instead of a copy-paste approach. At the same time, the Paks NPP is utilizing the ageing experience of other plants and elements of an adequate ageing management programme are in line with international practice.

The specific approach practicable in the case of the VVER-440/213 plants can be shown in the example of ageing management of civil structures. The VVER-440/213 design differs very much from the usual architecture of PWRs. In the example of the Paks NPP, practically all buildings, earth structures, etc., at the plant are within the scope. Most of these building structures are complex, and heterogeneous from the point of view of structural design, layout, manufacturing and construction of members, material composition and contact with environment (Katona *et al.*, 2009a).

In the case of the Paks NPP, it would be difficult to adopt the AMPs described in the GALL Report (US NRC, 2010), where nine groups of building structures and seven groups of structural components are defined, and ten ageing management programmes cover the whole scope. At the Paks NPP the large number and variety of building structures and structural components requires establishment of a hierarchical structure of ageing management programmes.

Type A programmes have been developed for foundations, reactor support structures, building movement, reinforced concrete structural members, high temperature concrete, equipment foundations, steel and reinforced concrete water structures, liners (Carbon-steel), prefabricated panels, masonry walls, earth structures, doors and hatches, steel-structures, cable and pipe supports, paintings and coatings, SS-liners, cable and pipe penetrations, fire protection structures, main building settlement, support structures of cabinets, seals and isolation and corrosion in a boric acid environment. These programmes are related to specific structures, that is structural commodities or specific ageing mechanisms (e.g. building settlement due to soft soil conditions). An exceptional A-type programme is the control of leak tightness of the containment, which is related to the containment only.

The buildings having identified safety functions are composed from structural commodities. Using these type A programmes for specific structures (commodities), 30 type B programmes have been developed which cover all plant building structures. These AMPs contain the identification of ageing effects and mechanisms to be managed, the lists and details of the proper application of type A AMPs to be applied, while managing the ageing of the given building. The type B AMP also contains logistical type information since the accessibility of certain buildings is limited.

8.4.2 Steps for the development of AMP

The AMP can be developed in the following sequence:

1 Identification of degradation mechanisms and locations susceptible to ageing
2 Identification of the mitigation and preventive measures
3 Identification of the parameters to be controlled
4 Definition of the method for the detection of ageing effects
5 Definition of the monitoring, trending, condition evaluation

6 Definition of the acceptance criteria
7 Identification of the corrective actions
8 Organising the administrative control
9 Organising the operational experience feedback.

In reality, the development is some kind of iterative process and steps overlap, as will be shown below.

Identification of ageing mechanisms

The development of AMPs has to begin with the identification of the ageing mechanisms, critical locations and effect of ageing on the intended safety function. When an AMP is developed for a complex structure or component, like the reactor or steam generator, several mechanisms and critical locations can be identified. The material, conditions and stressors are considered at this step of the AMP development. Examples for the mechanisms are listed in the Table 8.4.

As a matter of fact, the structuring of the AMPs together with the identification of the commodities is not independent from the identification of ageing mechanisms. For example, a commodity group can be defined as follows, see Table 8.7: Safety Class 3 + Piping and pipe elements + working in prepared water (e.g. feed-water line) + carbon steel. From experience, the dominating ageing mechanism of this group is flow-accelerated corrosion (FAC), a degradation process resulting in wall thinning of piping, vessels, heat exchanger and other equipment made of carbon and low alloy steel. This degradation mechanism of the identified commodity group should be addressed by proper AMP, which can be developed for example via application of the COMSY system (Zander, Nopper, Roessner, 2007) used by several VVER operators.

Preventive measures

The second step of the development of the AMPs is the identification of the means of preventing or controlling ageing. For example, the corrosion phenomena on the internal surfaces can be slowed down via adequate water chemistry parameters. General corrosion and soil corrosion may be reduced by coatings and ensuring the undamaged state of the coatings. The most effective way of avoiding boric acid corrosion is the timely detection and effective termination of leakages onto carbon steel elements, which are the subject of walk-down inspections.

Parameters to be controlled

Identification of the parameters allowing the control of the degradation process is an essential part of the AMP development. Some parameters indicate the evolution of degradation directly, for example the wall thickness of piping. The water chemistry parameters can be used as indirect controlling parameters of all internal surface corrosion mechanisms.

Definition of the method for the detection of ageing effects

Most of the postulated ageing effects can be detected during the execution of the current programmes of the plant, as follows:

- Non-destructive testing performed in the context of in-service inspection programmes
- Visual inspections performed in the frame of maintenance programmes
- Visual structural inspections
- Walk-down inspections.

Monitoring, trending and condition evaluation

A definition of the methods for monitoring, trending and condition evaluation is the fifth step in the development of the AMPs. For example, the monitoring of the trend of fast neutron fluence absorption in the critical components of the reactor pressure vessel is one of the most important indirect ageing management elements. The monitoring of load cycles defined during design and of their parameters belongs to the ageing management of fatigue degradation mechanism. The monitoring of the number and growth of crack-indications found during material inspections and visual inspections in the frame of in-service inspection can be assigned to each local degradation phenomenon. The monitoring and trending of the value of wall thickness reduction could be taken into account in the case of degradation forms with general material loss. In the case of heat exchangers, the monitoring of the number of plugged tubes can also be considered as an element of the ageing management programme.

Acceptance criteria

The acceptance criteria are expressed as a limit value for the controlled parameter of the ageing. The limit value corresponds to the performance or functioning with required margin. Acceptance criteria have to be defined for each component, or commodity, for each degradation mechanism in relation to fulfilment of the intended safety function. The acceptance criteria can be derived from stress calculations in case of allowable wall thickness of piping, or fatigue calculation regarding allowable load cycles. The acceptance criteria for degradation phenomena entailing decrease of the brittle toughness are determined by the relevant TLAA analysis results.

The compliance criteria for water chemistry parameters are defined in the relevant chemistry instructions.

The steps described in the sections above can be illustrated by the examples for civil structures given in Table 8.8.

Corrective actions

Any damage not in compliance with the acceptance criterion should be repaired if possible. In the case of fatigue cumulative usage factor (CUF) > 1.0, appropriate fatigue monitoring and a focused in-service inspection programme can be implemented.

Administrative control

The administrative and organization arrangements have to be defined for the performance of ageing management programmes. Appropriate plant procedures have to ensure the planning, staffing, performing, documenting and management control of the AMPs. Proper systems for documentation and reporting have to be established. A proper quality assurance plan also has to be developed for AMPs.

Operational experience feedback

A system for the verification of the effectiveness of AMPs and feedback of experience has to be in place at plants. In the case of any damage discovered, the degradation mechanism should be identified followed by an evaluation of whether the given degradation mechanism is appropriately managed by the AMP(s).

8.4.3 Reviewing and qualifying the ageing management activity

Attributes of adequate ageing management programmes are defined by the regulation; see for example the NUREG-1801 (US NRC, 2010) adapted by several VVER operating countries. An adequate AMP has to have the following elements:

1 Definition of SSCs that are subject to ageing management
2 Actions to prevent or mitigate specific ageing processes
3 Surveillance, monitoring and testing of all parameters related to the degradation of the function or serviceability of the SSCs
4 Investigation of ageing factors that may cause degradation or loss of function of SSCs

Table 8.8 Examples for ageing monitoring programmes of structures and structural components

Building or part of the building	Structure	Aim of the monitoring	Measurements	Evaluation and criteria
Main building complex: all buildings, including reactor and auxiliary buildings, stacks, diesel-building and other structures	Reference points	Building movements; settlement; control of stability of cracks caused by movements	Fixed geodetical measuring points;	3D evaluation of building movements; correlation with groundwater table; allowable declination of vertical axis of reactor pressure vessel defined by functioning of CRDM
Reactor building: floor slabs and walls	Heavy reinforced concrete	Interaction with boric acid media	Investigation of samples; inspection of check-holes	Control of mechanical and chemical properties and comparison with reference values
Reactor building, turbine building, intermediate building and galleries: floor slabs and walls	Reinforced concrete	Control of possible leakages and consequent leaching	Investigation of samples; inspection of check-holes	Control of mechanical and chemical properties and comparison with reference values
Reactor building: floor slabs and walls	Carbon steel liner	Control of corrosion rate, identification of possible leakages	Ultrasonic control of liner wall thickness at the identified places	Control of corrosion rate and thickness; focused investigation if the overall leak tightness is less than the reference value for the given unit (comparison with allowable leak rate)
Reactor building and auxiliary building: floor slabs and walls	Decontaminable coating and painting	Control of condition of coating and painting	Walk-down and visual control according to checklist	Expert judgment
Main building complex: all building parts	Hatches, gates, penetrations, fire protection doors;	Control of condition of doors	Walk-down and visual control according to checklist, fluorescent test	Expert judgment

5 Trend analysis to predict degradation processes and to perform corrections in time
6 Acceptance criteria to assure that the functions of the SSCs are maintained
7 Correction measures to prevent or solve problems
8 Feedback process to ensure that preventive actions are effective and appropriate
9 Administrative control of the processes
10 Information retrieval from operational practice to ensure that ageing management is properly carried out.

The same attributes can be applied while reviewing the adequacy of existing plant programmes.

8.4.4 Review and validation of the time-limited ageing analyses

TLAAs and their role in the justification of LTO

Although the wording is sometimes different, the term 'time-limited ageing analyses' is understood by the VVER operators in a very similar way to its definition in 10CFR54.3 (Requirements for Renewal of Operating Licenses for Nuclear Power Plants). The role of the review and revalidation of the TLAAs in the justification of LTO is also the same as international practice.

Existing TLAAs should be reviewed and revalidated with an assumed extended time of plant operation. The evaluation of each identified TLAA should justify that the safety function of the SC will remain within design safety margins during the extended period of operation. The plants have to demonstrate either in the context of the PSR or in the licence renewal application that:

- the analysis remains valid for the period of LTO;
- the analysis has been projected to the end of the period of LTO via removing the conservatism used in the TLAA analysis by less conservative assumptions and methods for analysis; or
- the effects of ageing on the intended function(s) will be adequately managed for the period of LTO.

The scope of the required analyses

The identified TLAAs cover the usual areas: fatigue calculations, assessment of embrittlement, changes of material properties, etc. However, the scope of TLAAs for some VVERs differs from the usual one either because of the peculiarities of the design or because of national regulation. For example, in case of Paks NPP, the scope of fatigue calculations is extended to the Safety Class 1 and 2 piping and components and includes analysis of thermal stratification, too. In regard to its RPV, besides of PTS analysis, the limits and conditions of safe operation, that is the p-T curve has to be re-analysed in the frame of revalidation of TLAAs.

The issue of the TLAAs

Review and validation of TLAAs is a rather complex task for the majority of VVER plants. The issue is related to the availability of design base information and incompleteness of the delivered design documentation. Often only the final results of the analyses are known; in some cases, the analyses are presumably obsolete. For the majority of VVER plants outside Russia the TLAAs have to be performed anew using state-of-the-art methods in accordance with the recent requirements. In comparing the practice of different VVER operating countries, the most complex cases are probably the Eastern-European VVER-440/213 plants since these plants have to overcome this issue. For instance, in the case of the Rivne NPP in Ukraine, full scope stress calculation and fatigue analysis had to be performed for the VVER-440/213 type units.

The case of Paks NPP Hungary will be discussed below on the basis of Katona, Rátkai and Pammer (2007), Katona *et al.* (2010) and Katona, Rátkai and Pammer (2011). TLAAs have to be reviewed and verified for the most important structures and components (SCs). Developing a methodology for TLAA reconstitution and defining the method of adaptation of ASME BPVC for a Soviet designed plant has been reported by Katona, Rátkai and Pammer (2007) and Katona, Rátkai and Pammer (2011). Hungarian regulations require application of state-of-the-art methods and standards in the time-limiting ageing analyses. ASME Boiler & Pressure Vessel Code, Section III, edition 2001 (ASME BPVC) had been selected for the reconstitution of TLAAs and associated strength verification. The code selection requires understanding of both the Russian (Soviet) design standards and the ASME BPVC code. Different studies were performed for ensuring the adequacy of ASME BPVC implementation for VVER-440/213. Calculations were performed for a 50 year extended operational lifetime with an additional margin of 10 years. The use of ASME BPVC is not a generic approach used by VVER operators. In some VVER operating countries the conservative PNAE G-7–002–86 standard is used by the operators and accepted by the regulators.

Mechanical components

To justify the safety of LTO, the scope of TLAAs which must be reconstructed or newly performed covers Safety Class 1 and 2 mechanical components. Examples of the calculations/analyses follow.

For low-cycle fatigue analysis of Safety Class 1 and 2 piping and mechanical components, ASME BPVC was adapted for the calculations (Katona, Rátkai and Pammer, 2011). This task also includes identification of needs for fatigue monitoring. The most critical ones are the high stresses in the body and sealing block of the main circulating pumps. These, however, could be managed via focused non-destructive examination programmes.

Analysis of thermal ageing of Class 1 and 2 components focuses on components manufactured from 15Ch2MFA, 22K, 08Ch18N9TL cast stainless steel materials and also on welds (Sv04Ch19H11M3, EA400/10T, Sv10ChMFT, IONI 13/55) which are sensitive to thermal embrittlement. Significant changes of material properties due to thermal embrittlement are to be expected above 220°C operational temperature in case of ferrit-pearlit materials or cast stainless steel. Only a few components match these conditions at the Paks NPP. According to fatigue analyses performed, there are no cases where crack propagation due to fatigue might be expected. The analysis performed for the main gate valve cast stainless steel body shows that crack propagation should not be expected even if the J-R curve for C8 steel is changing due to embrittlement and a crack is postulated.

For analysis of thermal stratification for Class 1 and 2 pipelines, a measuring system was operated at the Paks NPP Unit 1 pressurizer surge line in 2000–2001. Assessment of the measured data shows significant thermal stratification (110°C), which moved periodically from the pressurizer to the hot leg. This temperature swing was maintained by the swing of water level control in the pressurizer during the heat-up and cool-down. During normal operation, the temperature differences were decreased to a negligible level. A similar temperature monitoring system has been operating on both legs of the surge line at Unit 3 since 2007. Evaluation of the measured data and the subsequent fatigue analysis justify LTO for the pressurizer surge lines. Other pipelines have also been identified where thermal stratification might occur. These are the pipelines connecting coolant cleaning system No 1 to the primary system; the pipeline of the passive emergency core cooling system and the feed-water system pipeline and also the auxiliary emergency feed-water pipelines. Experience gained at other VVER-440/213 plants (Mochovce and Dukovany NPP) has been taken into account in identifying the pipelines of interest. Implementation of monitoring programmes is ongoing for these pipelines with temperature and displacement measurements.

High-cycle fatigue analysis of flow-induced vibration of internal structures of the steam generator tubes shows that the flow-induced vibration of the heat-exchange tubes does not cause significant stresses compared to those from operational loads. Taking into account 60 years of operation and

108% of reactor thermal power, the CUF is equal to 0.027 due to vibration even if a pipe wall thinning of 50% is assumed.

Analysis of the corrosion of piping wall must question whether the erosion-corrosion allowance applied in the design provides sufficient margin for 50+10 years of operation. Only a few cases are expected where the existing corrosion-erosion monitoring programme using COMSY software will have to be extended.

In analysing for material property change of the steam generator tubes, the main finding of the study is that the thermal ageing of 08H18N10T material used for heat-exchange tubes is negligible at operating temperatures ~290°C. Results of laboratory tests show that there is no change in the fatigue crack propagation rate due to LTO at 288°C (NPO Hidropress, 2007). An operational time of 60 years is justified in this respect.

Reactor pressure vessel and internals

For the justification of operability of RPV and RPV internals for extended operational lifetime, the following analyses have to be performed.

PTS analyses for RPV test the structural integrity against brittle fracture (fast fracture) of the RPV; it is ensured if the factual ductile-brittle transition temperature (DBTT) of its critical components is less than the maximum allowable component-specific DBTT. The analysis is based on the comparison of the static fracture toughness of the material and stress intensity factor calculated from the given loading situation (Linear Elastic Fracture Mechanics (LEFM) concept). The steps in the analysis are presented by Katona, Rátkai and Pammer (2011). The final conclusion of the analyses is that the RPVs at Paks NPP can be safely operated for at least 60 years. For the sake of completeness of the studies, some additional analyses are still ongoing regarding PTS sequences initiated by internal fires, flooding and earthquakes under shutdown conditions. The neutron fluences also have to be modified taking into account the new fuel design introduced after power up-rate.

Analysis of fracture toughness of structures within the reactor pressure vessel was undertaken. According to the preliminary results the irradiation-assisted stress corrosion cracking and void swelling may be of interest. The stud joints fixing the polygon mantle to the core basket can be critical in both ageing mechanisms. Measures may be identified after visual inspection of the core basket and review of inspection procedure. The possibility of implementation of a non-destructive volumetric test method for the bolts is also a consideration. With respect to void swelling, the possibility of implementation of ultrasonic measurements as well as gamma heating and a replacement programme are being investigated.

Analyses related to operational limits and conditions

Reviews of the Final Safety Analyses Report and reconstruction of design bases, which have been performed at the Paks NPP, resulted in a recognition of the need for justification of operational limits and conditions related to certain ageing phenomena via adequate thermo-hydraulic, stress and fracture mechanics analyses. These analyses have been included in the scope of TLAAs required for the justification of LTO of the Paks NPP. The task also includes the justification for modification of the limits and conditions in accordance with operational needs allowing rapid temperature changes in certain cases. The temperature measurements and the temperature rate control methodology have also been reviewed and amended.

The calculation methodology was based on an adaptation of ASME BPVC. For the calculation of temperature transients in the primary system, the RELAP5/mod3.3 code was used. A thermo-hydraulic model was developed for accident simulation. This model consists of a detailed model of the primary system, the heat removal system and the automatic control system and it takes into account operator actions during the heat-up and cool-down processes. The thermo-hydraulic model and the calculation method have been verified via comparison of the calculated transient time histories with the measured ones.

Containment, civil structures and structural components

Taking into account the specific features of the VVER-440/213 design of civil structures and also the lack of/missing analyses performed by the designer, eight analysis tasks were identified as necessary for the justification of LTO of the Paks NPP. The need to perform stress calculations for the validation of the rather sparse information available for containment and other safety-classified structures was also recognized. Considering their content, these calculations are not typical TLAAs, however, without sufficient information on the design of civil structures the newly performed TLAAs would not have the design basis. The scope of TLAAs for Paks NPP includes the generic tasks, like:

- Analysis of buildings classified into safety category for the verification of the design.
- Fatigue analysis for the containment penetrations.
- Fatigue analysis for the hermetic liner of the containment (welding, transition welding, area of anchors).
- Fatigue analysis for the liner of the spent fuel pool (welding, transition welding, area of anchors).
- Stress and fatigue analysis for the safety-classified crane in the reactor hall with capacity of 250/32/2 tons.

There are also several design-specific TLAAs in the case of the main reactor building at Paks NPP, for example:

- Fatigue analysis of the containment for increased pressure level during integral leak-tightness tests.
- Analysis of main reactor building settlement.

The allowable leakage value of VVER-440/213 containment is 14.7% per day at the design pressure of 2.5 bars. Each of the containments was tested at this design pressure in the start-up phase. The pressure of the yearly leakage tests is 1.2 bar and tests at a pressure of 1.7 bars are also carried out during the outages. The leakage value for the nominal pressure of 2.5 bars is calculated via extrapolation from the leak rate results of tests. This practice has been criticized regarding correctness of the leak rate extrapolated from the measured ones, and investigations of enhancement of the test pressure level has been proposed. Nevertheless, the recent reduced pressure test procedure has obvious advantages compared to the tests at enhanced pressure level: the time needed for the low-pressure test is short and the load on containment structures is moderate. According to the results of leak tests the correct leakage values at the nominal pressure of 2.5 bars can be determined from the results of tests carried out at considerably lower pressure values. This statement is based on analyses of numerous tests, including the results of tests carried out at the design pressure of 2.5 bars at Paks NPP Unit 2.

Regarding Paks NPP, the analysis of settlement of the main building complex has been identified as a TLAA requirement since an excessive inclination of the main building complex due to differential settlement may result in non-allowed tilting of the RPV vertical axis, which may in turn cause problems with the control rods. Additionally, excessive inclinations can also cause extreme local loading resulting in degradations of the building. It has to be mentioned that the VVER-440/213 type units at the Paks NPP have twin-unit-design, that is two main reactor buildings separated by a dilatation gap are built upon a common base mat. Detailed settlement control was started during the construction period at Paks. The measured results are to be evaluated and reported annually. A consolidation process, prolonged in time, was observed in case of the main reactor buildings, the settlement of which is still continuing. The phenomenon is related to the seasonal variation of the water level of the river Danube, which may reach a value of 9 meters. This variation of river water level influences the groundwater level. According to the data measured in the wells at and around the plant site, the groundwater level follows the variation of the water level in the Danube with a certain time delay. The water-table fluctuations influence the stress-deformation conditions in the subsoil. This can explain the successive settlement of the raft foundation measured during past years. The settlement at Unit 4 is somewhat larger than at Units 1–3, which is due to the slight in homogeneity of the subsoil and the highest alteration of the level of the water-table, occurring in the vicinity of Unit 4.

Detailed analyses have been performed for the subsidence and differential settlements of the main reactor buildings for the end-of-life situation taking into account the static loading (immediate settlement), groundwater fluctuation, seismic settlement, dynamic settlement due to machinery and tectonic subsidence. The calculation model and procedure has been calibrated to the measured time-history of subsidence. An appropriate constitutive model has to be defined for the soil, which includes the development of a non-linear hardening model and proper definition of the decay curve for cyclic loading due to groundwater fluctuation based on soil tests results.

In regard to LTO, the analyses show that a value of differential settlement that may cause non-allowed tilting of the RPV axis due to the inclination of the building should not be expected. The structural integrity of the foundation and the containment part of the main building structures is not affected by the settlement and is not expected as a result of further subsidence.

Basic findings of the revalidation/reconstitution of the TLAAs

Dedicated ageing management programmes already control some of the processes addressed by the time-limited ageing analyses presented above, for example the process of settlement of the main building and erosion-corrosion of piping wall. The results of the above analyses show that only a few non-compliances or lifetime-limiting cases have been found and all of them can be managed by the extension/amendment of the existing ageing management programmes and/or other plant programmes. For example, in relation to RPV and internals the stud joints fixing the polygon mantle to the core basket are the critical structures from the point of view of irradiation-assisted stress corrosion cracking and void swelling. In order to manage these mechanisms, review and extension of the present programmes are ongoing. Regarding operational limits and conditions for injection into the pressurizer, the margin to allowable stresses is minimal and the number of allowable cycles is rather small; consequently, the number of cycles should be monitored. It was also found that during certain heat-up and cool-down processes the averaging intervals of the temperature measurements have to be modified at certain components. With respect to the containment civil structures the existing ageing management programme should be extended for managing the change of material properties of heavy concrete structures and for the corrosion of the steel liner on a heavy concrete surface.

8.5 Plant programmes credited for long-term operation

Review of the existing plant programmes can qualify these programmes as adequate for ageing management. For example, the following programmes can be classified as AMPs or part of AMP:

- Preventive and predictive maintenance programme can be considered to be a part of AMP because it is one of the solutions for ageing mitigation and because AM requires information on preventive maintenance of SCs that is carried out
- In-service inspection programme
- Functional Testing Programme – for active components if they are in the scope of AM.

8.5.1 Review and modification of the ISI programmes

The in-service inspection (ISI) programmes delivered partly by the supplier or developed by Hungarian institutes basically follow the ex-Soviet regulation.

Recent review and overall updating of the ISI programmes adopt state-of-the-art techniques and methodologies (e.g. ASME Section XI). Extensive studies are ongoing to provide a solid basis for changing the rules and techniques of ISI. One practical question is the periodicity of the ISI programmes, which is four years at Paks NPP, in accordance to the ex-Soviet regulation. For practical reasons the new ISI period should be eight years. At the same time, the scope and depth of ISI programmes also have to be upgraded. This type of modification is not unique; moreover there are similar examples among the countries operating VVER-440 type NPPs (e.g. Finland); however, the change cannot be performed routinely, it requires careful justification.

8.5.2 Maintenance programmes

Maintenance is the subject of Maintenance Effectiveness Monitoring (MEM), the purpose of which is to control the effectiveness of maintenance on SSCs ensuring they are capable of performing their intended functions. This means ensuring that safety-related SSCs are capable of performing their intended functions; that failures of certain non-safety-related SSCs that could affect safety-related functions will not occur; and failures that could result in scrams or unnecessary actuations of safety-related systems are minimized. The systems within the scope of MEM might be divided into high and low risk-significant categories. The risk significance has been defined quantitatively by PSA or qualitatively by expert judgement. The MEM is an adaptation of 10CFR50.65 for the VVER-440/213 design features, Hungarian regulatory environment and plant practice.

There are two basic methods applied in the MEM: deterministic method, that is control of maintenance via testing/measuring performance parameters of components, and probabilistic method, that is assessing the effectiveness of maintenance via comparison of reliability/availability parameters at the level of component/system or plant. Performance parameters are defined in accordance with safety class and risk significance. The deterministic method is based on ASME OM Code. For example in case of pumps the performance criteria to be checked are the head, flow-rate and vibration level. Plant level deterministic performance parameters include the capacity factor, thermal efficiency of the unit and leakage of the containment (%/day). Risk significance and the probabilistic performance criteria are set on the basis of PSA. High risk significant SSCs are those which are in 90% cut-set, have a high contribution to CDF or high Fussell-Vessely rank. Performance criteria for MEM are based on the reliability or unavailability data of performing safety function. System level performance parameters are, for example, failure rates per demand (failure/start) or run failure rate (failure/time) during operation. Plant level performance parameters are the CDF or some selected contributors to the CDF and other safety factors (unplanned reactor scrams or safety system actuations per year). The MEM is being implemented at Paks; for the implementation of ASME OM Code, the existing in-service and post-maintenance testing programmes of the Paks NPP have to be modified and amended. Probabilistic performance criteria are under development at present. It is expected that the MEM will improve the safety factors and capacity factors for the plant while the maintenance effort will be optimal. MEM is a prerequisite for license renewal in Hungary, since it provides assurance for the correct functioning of active components.

8.5.3 Maintenance of environmental qualification

Performance and functioning of active systems can be tested during operation and can be ensured via maintenance under maintenance rule (MR), that is evaluation and assessment of the effectiveness of the maintenance along safety criteria, and/or via implementation of the programme for maintaining the environmental qualification (EQ).

For I&C components operating under harsh conditions environmental qualification should be implemented. When the older VVER-440 and

VVER-1000 NPPs were built, a large part of the originally installed electrical and I&C equipment did not have initial qualification or the qualification was not certified properly. The issue was recognized in the first safety reviews (see IAEA, 1992; 1996a; 1996b; 2000). The issue can be resolved in two steps:

- Restoring the initial qualification.
- Maintaining the qualified state of equipment.

The maintenance of the qualification means:

- Control of the capability of equipment to fulfil its safety function through:
  - periodic testing of systems and components
  - testing of the equipment following maintenance
  - results of service routes by maintenance personnel
  - diagnostics measurements.
- Development and implementation of a scheduled replacement programme taking into account the requirements for environmental qualification in purchasing the new equipment.
- Preventive maintenance of the equipment.

The environmental qualification should be reviewed and validated for the extended operational lifetime. There are different possible outcomes of the review:

- The qualification remains valid for the period of LTO.
- The qualification has been projected to the end of the period of long-term operation.
- The effects of ageing on the intended function(s) have to be adequately managed for the period of LTO via introducing new ageing management programmes.
- Replacement of the equipment.

8.5.4 Synergies: safety upgrading, reconstructions and power up-rate

There is a synergy between the possibility of LTO and different plant actions and measures implemented for safety upgrading, power up-rate, improving reliability and plant programmes. Implementation of the safety upgrading programme for ensuring the compliance with national and international requirements is a precondition for LTO. At the same time, safety is the most important aspect of public acceptance. The operator commitment in relation to safety is and will be the decisive point of judgement by the public.

Most of the safety upgrading measures result in positive technical effects too. Due to these modifications, the safety systems or essential parts thereof had been practically renewed or reconstructed. Consequently, a large part of safety systems is un-aged. In some cases, the safety upgrading measures have a direct influence on the lifetime limiting processes. For example, the new relief valves installed on the pressurizer for the cold over-pressurization protection eliminate the danger of brittle fracture of the reactor vessel. Some of the VVER plants implemented an extensive seismic upgrading programme involving the addition of a large number of new seismic fixes and other strengthening measures (see papers in IAEA, 1993). Fixing the building structures, the anchorage equipment, cabinets, racks and also the structural support of cable trays can be considered as reconstruction of these SCs.

The most important economic condition for LTO is preserving the present cost advantage of nuclear electricity generation within the market conditions. By exploiting reserves and advantageous features of the VVER-440/213 reactors, the electrical output of the plants can be safely increased up to approximately 500 MWe by improving the efficiency of the secondary circuit/turbine and increasing reactor thermal power via implementation of modernized fuel assemblies. Obviously the power up-rate should not result in a decrease of the plant safety level and should not cause stressors of ageing which affect the lifetime extension perspectives and the plant availability.

The VVER plants replaced the frequently criticized, obsolete I&C systems. The new I&C systems have proper environmental qualification. Aside from their obsolescence, the lack of environmental qualification was the basic issue in the case of the old systems at practically all plants.

One of the major causes of corrosion in the steam generator heat-exchange tubes local is the high concentration level of corrosion activators (chloride ions, sulphates, copper oxides, etc.) in the secondary circuit and partially in the hidden surfaces of the SG secondary side locations. This can be critical in the case of VVER-440 plants where the steam generators are not practically replaceable. To limit local corrosion, the high levels of deposition on tube surfaces should be eliminated to reduce the concentration of the corrosion activators. The most important measure implemented was to replace the main turbine condenser, for example at Paks NPP (Katona *et al.*, 2003). Unlike the old condensers with a copper alloy tube bundle, the new condensers with stainless steel tubing are leak tight. They in turn allowed the introduction of the high pH water regime in the secondary circuit providing better operational conditions for components of the feed-water system and for the steam generators as well.

8.6 Conclusion

A complex picture of ensuring and justification of LTO of VVER plants has been given in this chapter. The VVER-440/213 model especially is discussed in detail.

In VVER operating countries, proper regulatory frameworks and comprehensive plant lifetime management systems have been developed to ensure the safety of LTO of VVER-440 and VVER-1000 type plants. Detailed studies and already approved cases of extension to plant operational lifetime demonstrate the feasibility of LTO of VVER plants.

Generally accepted principles for safety have been followed while developing plant systems, which ensure that any SSCs will be covered by some of the plant programmes, and within the framework of the LTO programme, all conditions of safe operation will be ensured. Structures, systems and components are identified which are significant for safe LTO. An appropriate level of understanding of the ageing phenomena has been reached and adequate ageing management programmes developed for ensuring the required status and intended function for the long term. Revalidation of time-limited ageing analyses also justify the safety of LTO, which is completed by the monitoring of maintenance on performance criteria combined with the maintenance of environmental qualification and replacement and reconstruction programmes.

Best international practice and state-of-the-art methodologies have been applied while performing the particular tasks for preparation and justification of LTO and licence renewal. However, as demonstrated, any good examples and experiences should be adapted in creative ways, taking into account the design features, national regulations and existing plant practice. In this way the strategy of VVER operators to operate safely for as long as possible with economic advantage and at higher power levels will be ensured.

8.7 References

Brumovsky M. and Zdarek J. (2005). Integrated surveillance specimen program for WWER-1000/V-320 reactor pressure vessels, Technical Meeting on Material Degradation and Related Managerial Issues at Nuclear Power Plants, Vienna, 15–18 February 2005, IAEA Proceeding Series, 2006, pp. 171–178.

Brumovsky, M. *et al.* (2007) Prediction of irradiation embrittlement in WWER-440 reactor pressure vessel materials, Problems of Atomic Science and Technology, Series: Physics of Radiation Effect and Radiation Materials Science No 6, pp. 72–77.

Erak, D.Yu. *et al.* (2007). Radiation embrittlement and neutron dosimetry aspects in WWER-440 reactor pressure vessels life time extension, Second IAEA

International Symposium on Nuclear Power Plant Life Management, 15–18 October 2007 Shanghai China.

IAEA (1992). Ranking of safety issues for WWER-440 model 230 nuclear power plants IAEA-TECDOC-640 IAEA Vienna.

IAEA (1993). Proceedings of the SMiRT-12 Conference Seminar No. 16 on Upgrading of Existing NPPs with 440 and 1000 MW VVER Type Pressurized Water Reactors for Severe External Loading Conditions. 23–25 August 1993 IAEA Vienna Austria.

IAEA (1996a). Safety issues and their ranking for WWER-440 model 213 nuclear power plants IAEA-EBP-WWER-03 IAEA Vienna.

IAEA (1996b). Safety issues and their ranking for WWER-1000 model 320 nuclear power plants IAEA-EBP-WWER-05 IAEA Vienna.

IAEA (2000). Safety issues and their ranking for WWER-1000 model "small series" nuclear power plants IAEA-EBP-WWER-14 IAEA Vienna.

IAEA (2003). Periodic Safety Review of Nuclear Power Plants, Safety Guide, Safety Standards Series No. NS-G-2.10 International Atomic Energy Agency Vienna 2003.

IAEA (2005). Guidelines for prediction of irradiation embrittlement of operating WWER-440 reactor pressure vessels IAEA-TECDOC-1442 IAEA Vienna 2005 ISBN 92–0–105605–2; ISSN 1011–4289.

IAEA (2006). Plant Life Management for Long Term Operation of Light Water Reactors: Principles and Guidelines. Technical Reports Series No. 448. IAEA Vienna.

IAEA (2007a). Safety aspects of long term operation of water moderated reactors IAEA-EBP-SALTO IAEA July 2007 Vienna.

IAEA (2007b). Strategy for assessment of WWER steam generator tube integrity IAEA-TECDOC-1577 IAEA Vienna 2007.

IAEA (2009). Ageing Management for Nuclear Power Plants IAEA Safety Standards Series No. Ns-G-2.12 International Atomic Energy Agency Vienna 2009 (ISBN:978-92-0-112408-1).

IAEA PRIS (2011). <www.iaea.org >

Katona, T. (2010). Plant life management practices for water-cooled water moderated reactors (VVER). In: Understanding and Mitigating Ageing in Nuclear Power Plants, Woodhead Publishing ISBN 978 1 84569 511 8.

Katona, T. (2011). Long-term operation of VVER power plants. In: Nuclear Power, Deployment, Operation and Sustainability, edited by P.V. Tsvetkov, Intechweb.org, ISBN 978–953–307–474–0.

Katona, T. *et al.* (2003). Main Features of Design Life Extension of VVER-440/213 Units NPP Paks Hungary, ICONE 11th International Conference on Nuclear Engineering. Tokyo, Japan, 20–23 April 2003.

Katona, T. *et al.* (2005). Key Elements of the Ageing Management of the VVER-440/213 Type Nuclear Power Plants. In: 18th International Conference on Structural Mechanics in Reactor Technology (SMiRT 18). Beijing, China, 2005.08.07–2005.08.12. Paper D02–4.

Katona, T. *et al.* (2009a). Assessment and Management of Ageing of Civil Structures of Paks NPP. In: Proceedings of the ASME Pressure Vessels and Piping Conference. ASME 2009: Sustainable Energy for the Third Millennium. Prague, Czech Republic, 2009.07.26–2009.07.30. American Society of Mechanical Engineers p. CD-ROM. Paper PVP2009–77513 (ISBN:9780791838549).

Katona, T. *et al.* (2009b). Extension of Operational Life-Time of VVER-440/213 Type Units at Paks Nuclear Power Plant. In: Proceedings of the ASME Pressure Vessels and Piping Conference. ASME 2009: Sustainable Energy for the Third Millennium, Prague, Czech Republic 2009.07.26–2009.07.30. American Society of Mechanical Engineers, p. CD-ROM. Paper PVP2009–77911 (ISBN:9780791838549).

Katona, T. *et al.* (2010). Time-limited Ageing Analyses for Justification of Long-Term Operation of Paks NPP. In: ASME International Mechanical Engineering Congress and Exposition. Vancouver Canada 2010.11.12–2010.11.18. Paper IMECE2010–40201 (ISBN:978–0–7918–3891–4).

Katona, T. and Bajsz, J. (1992). Plex at Paks – making a virtue out of necessity. *Nuclear Engineering International* **37**(455) pp. 27–31.

Katona, T. and Rátkai, S. (2008). Extension of operational life-time of VVER-440/213 type units at Paks nuclear power plant. Nuclear Engineering and Technology 40(4) pp. 269–76.

Katona, T. and Rátkai, S. (2010). Programme of Long-term Operation of Paks Nuclear Power Plant. In: Transactions of ENC 2010: European Nuclear Conference 2010, Barcelona, Spain 2010.05.30–2010.06.02. Paper A0114 (ISBN:978–92–95064–09–6).

Katona, T., Rátkai, S. and Pammer, Z. (2007). Reconstitution of Time-limited Ageing Analyses for Justification of Long-Term Operation of Paks NPP. In: 19th International Conference on Structural Mechanics in Reactor Technology (SMiRT 19). Toronto, Canada, 2007.08.12. Paper D02/2–1.

Katona, T., Rátkai, S. and Pammer, Z. (2011). Reconstitution of time-limited ageing analyses for justification of long-term operation of Paks NPP. Nuclear Engineering and Design Volume 241 Issue 3 (March 2011) pp. 638–43.

Katona, T. *et al.* (2005). Key Elements of the Ageing Management of the VVER-440/213 type Nuclear Power Plants. In: 18th International Conference on Structural Mechanics in Reactor Technology (SMiRT 18). Beijing, China 2005.08.07–2005.08.12. Paper D02–4.

Kupca, L. (2006). Irradiation Embrittlement Monitoring Programs of RPV's in the Slovak Republic NPP's. In: 14th International Conference on Nuclear Engineering (ICONE 14), Miami FL (United States) 17–20 July 2006.

NPO Hidropress (2007). Steam Generator PGV-213. Analysis of Material Properties of Steam Generator Tubes (In Russian). NPO Hidropress Report No. U213D3 2007.

Orgenergostroy (1989a). Instruction of technical servicing for standardized units VVER-1000 NPP type B-320 containment pre-stressed system, Moscow, 1989.

Orgenergostroy (1989b). Instruction of technical servicing for main series (non-standardized) units VVER-1000 NPP type 302338 and 187 containment pre-stressed system, Moscow, Orgenergostroy, 1989.

Popov, V. (2007). The large projects at Kozloduy NPP – with focus on long time operation and ageing management, Second IAEA International Symposium on Nuclear Power Plant Life Management, 15–18 October 2007 Shanghai China.

Rosenergoatom (2003). Safety enhancement and lifetime extension of the power unit 1 of Kola NPP, Summary Report, Moscow, 2003.

Slovak Report (2010). National Report of the Slovak Republic Compiled in Terms of the Convention on Nuclear Safety June 2010 http://www.ujd.gov.sk/files/dokumenty/NS_NS_2010.pdf.

Šváb, M. (2007). Regulatory approach to the long-term operation of Czech nuclear power plants, Second IAEA International Symposium on Nuclear Power Plant Life Management, 15–18 October 2007, Shanghai China.

Trunov, N. B. *et al*. (2006a). Consideration of field experience in developing new projects of steam generators for nuclear power stations equipped with VVER reactors. *Thermal Engineering* Vol. 53 No. 1, pp. 37–42.

Trunov, N. B. *et al*. (2006b). WWER Steam Generators Tubing Performance and Aging Management. In: 14th International Conference on Nuclear Engineering (ICONE 14) Miami FL (United States) 17–20 July 2006.

Ukraine (2011). Состояние работ по реализации «Комплексной программы работ по продлению срока эксплуатации действующих энергоблоков атомных станций». [Complex programme of works for extension of operation of units at operating nuclear power plants.] http://www.energoatom.kiev.ua/ru/Length_Extension?_m=pubs&_t=rec&id=25436.

U.S. NRC (2010). Generic Aging Lessons Learned (GALL) Report — Final Report (NUREG-1801 Revision 2).

Vamos G. (1999). Safety Improvement of Paks Nuclear Power Plant. In: IAEAInternational Conference on the Strengthening of Nuclear Safety in Eastern Europe, 14–18 June 1999Vienna, Austria.

Vasiliev, V. G. and Kopiev, Yu.V. (2007). WWER pressure vessel life and ageing management for NPP long term operation in Russia, Second IAEA International Symposium on Nuclear Power Plant Life Management, 15–18 October 2007 Shanghai China.

Zander, A., Nopper, H. and Roessner, R. (2007). COMSY – A Software Tool for PLIM + PLEX with Integrated Risk-Informed Approaches, Transactions SMiRT 19 Toronto, August 2007 Paper # D02/4.

9
Materials-related problems faced by light water reactor (LWR) operators and corresponding research needs

S. RAY and E. LAHODA, Westinghouse Electric Company LLC, USA

DOI: 10.1533/9780857097453.3.385

Abstract: This chapter provides background on current materials-related problems faced by the nuclear industry. These issues have become more important as the current fleet of nuclear plants ages and as life extensions of 20 years each are added onto the current 40 year life. Materials issues requiring research and development are presented in terms of the fuel, the primary boundary, the containment and other general issues.

Key words: nuclear, power plant, industry, materials, issues, fuel, rods, cladding, containment, primary, secondary, corrosion, cracking, buried, pipe, wiring, concrete, steel.

9.1 Introduction

As the light water reactor (LWR) nuclear fleet reaches and surpasses the original 40 year lifespan that it was licensed for and embarks on its next 20 years with visions of yet another 20 years beyond that, the need increases for a scientific underpinning of the understanding of the degradation of materials in a nuclear environment. The need to generate this scientific underpinning becomes more compelling when one considers that the original 40 year life had no scientific basis (INL, 2009) and that the materials designs were not based on irradiated materials in real life chemistry conditions (Majumdar, 2011). There are indeed many existing models and correlations for determining what may happen as materials in a nuclear environment age, but most are based purely on empirical data. The ability to extrapolate these models is under question by the industry and the NRC, which will not grant licenses based on extrapolated models. Far too often, researchers have discovered unexpected effects, both good and bad, which should not have occurred based on extrapolation of models.

This chapter focuses on the materials in a nuclear system from the inside out of a nuclear plant. First the initial fission boundary of the fuel (Fig. 9.1),

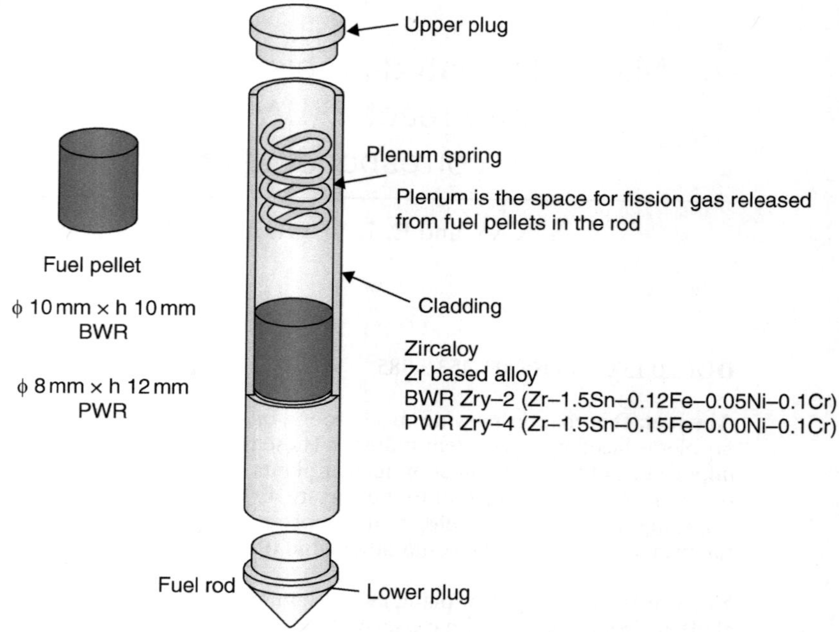

9.1 Nuclear fuel pellets and rods (Kazuya Idemitsu, Genshiryoku Zumenn Syuu, JAERO, p. 4, used by permission of the author).

both the fuel itself which contains most of the fission products, and then the fuel cladding which normally contains the rest of the fission products. Occasionally, the cladding leaks and the second boundary comes into play – the primary system of the pressurized water reactor (PWR) (Fig. 9.2) and the steam system of the boiling water reactor (BWR) (Fig. 9.3). After that, the boundary is the secondary system and the containment of the PWR or BWR. During postulated accident conditions, the innermost system (fuel and cladding) fails and then all that remains to avoid exposure to the public is the concrete and steel containment structure. Thus, a good understanding of all the containment systems is also needed.

9.2 Fuel and cladding materials – the first fission barrier

The first boundary for fission products is the fuel and the cladding (Fig. 9.1). The fuel normally holds up about 90% of the fission products during the normal operating cycle of an LWR. The main exceptions are elements that are gaseous at normal fuel operating conditions (~400°C to ~1200°C) such as I_2, Kr, and Xe which are held up by the cladding. Even under normal

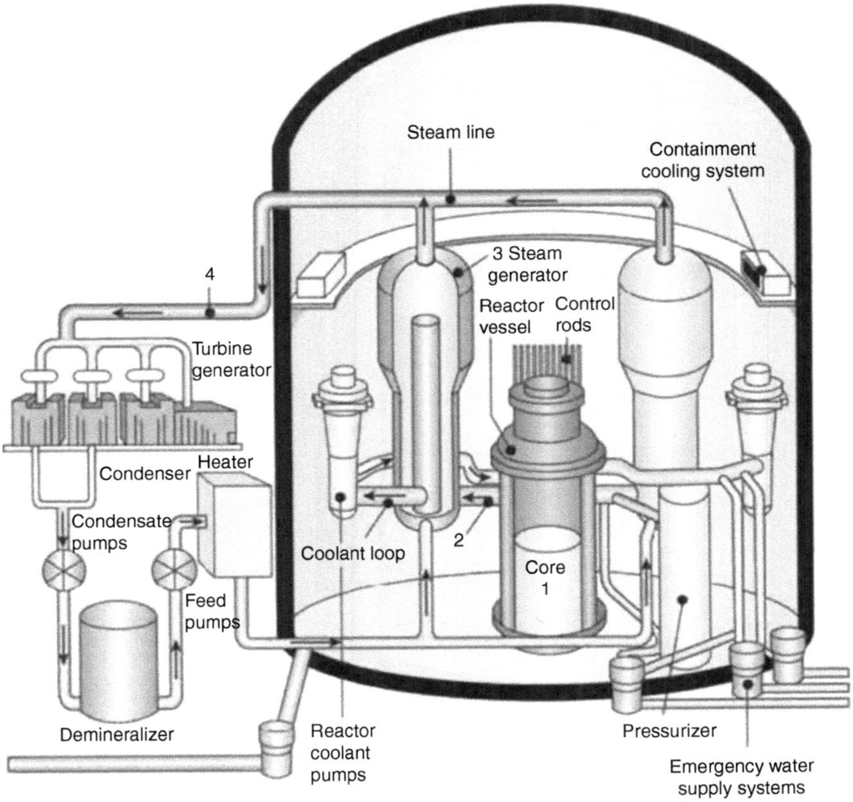

9.2 Graphic of the typical pressurized water reactor. (Published in NUREG -1350, Volume 23, August 2011.)

operation the cladding and other fuel assembly components operate under very extreme conditions. These components are constantly bombarded by neutrons (~10^{12} neutrons/cm^2/s), at moderate temperatures (~250–340°C), pressures (~15.5 MPa), mechanical conditions (boiling surfaces, high velocity two-phase flow with large amounts of vibration), and chemical conditions (up to ~2000 ppm of boron and up to 10 ppm Li for PWRs).

9.2.1 Understanding the fuel

A stable UO_2 fuel structure holds up most of the fission products and even a fair amount of the fission gases (up to ~90%) in the pores of the UO_2 pellet. Even though UO_2 pellets undergo a significant amount of cracking due to

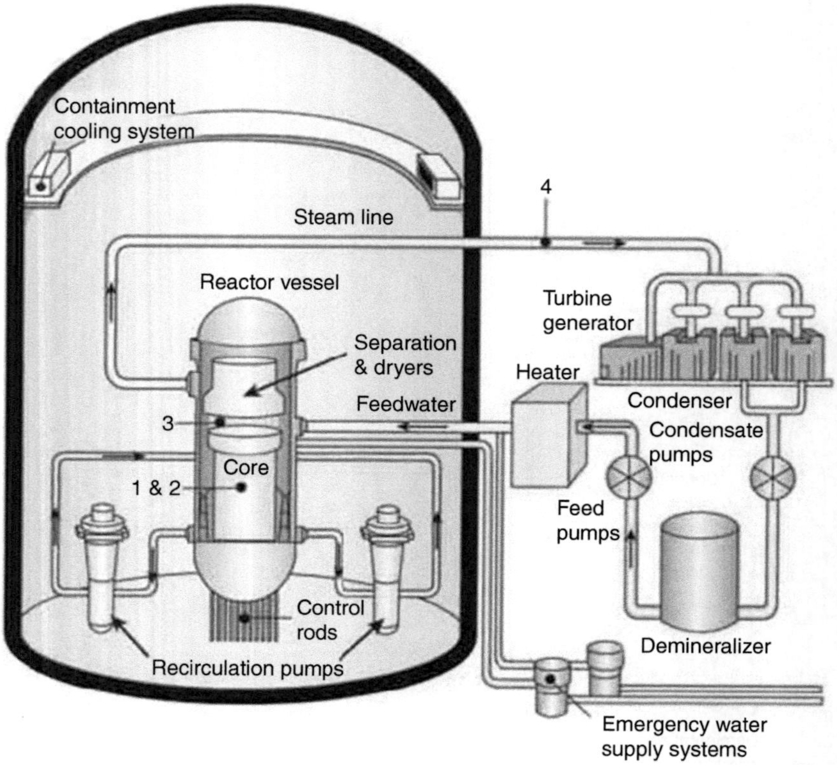

9.3 Graphic of the typical boiling water reactor. (Published in NUREG -1350, Volume 23, August 2011.)

the low thermal conductivity and the resulting stresses from the very steep temperature gradients experienced under normal operating conditions, the resulting pieces are relatively large and most of the pores of the UO_2 fuel act as reservoirs for fission gas products and the fission products themselves. Under circumstances where the cladding leaks and coolant enters the rod, the UO_2 fuel can further disintegrate releasing much of the soluble fission product (primarily Cs and Sr) and most of the gaseous fission products. This effect becomes more pronounced as the burnup of the fuel increases significantly above the current levels of about 50 MWtd/kgU. Therefore basic research work on fuel that is exposed to burnups >~60 MWtd/kgU is needed and includes:

- Structural changes in the UO_2 due to collection of solid fission products and the effect these structural changes have on the mechanical stability of the fuel during normal and abnormal operation.

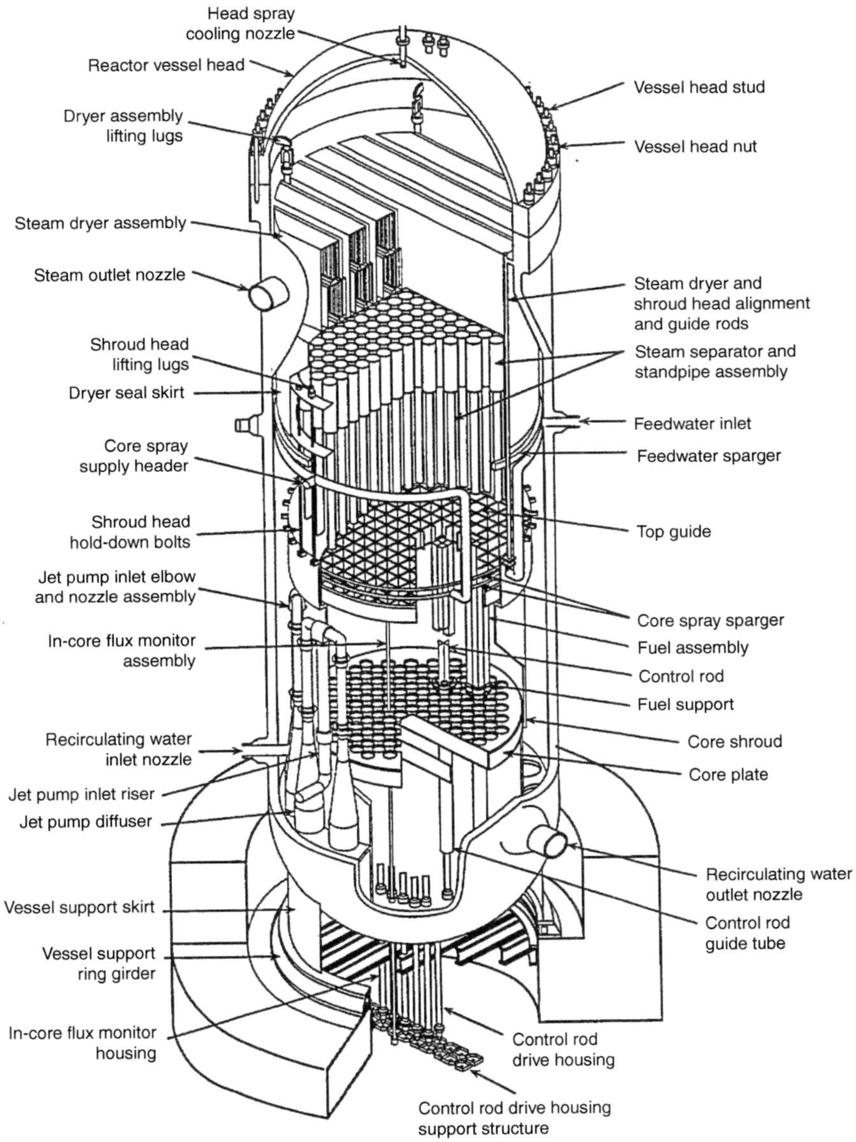

9.4 BWR/3 or BWR/4 reactor vessel (G.E. Technology Advanced Manual Differences/Introduction, USNRC Technical Training Center Rev 1195).

- How fission product gases are held up within the pores of the pellet and how they are released during upset events.
- Effect of coolant on fuel structure and stability under operating conditions when fuel cladding failure occurs.

- Extension of the understanding of these effects to new fuel pellet materials (e.g. uranium nitride (UN)).
- Effects of long-term wet and dry storage, as well as environmental conditions in any potential disposal site, on the integrity of the fuel in terms of its holdup of long-lived radioactive components (mainly U, Pu, Am, Cm, and Np).
- Interactive effects of non-homogeneous portions of the fuel (such as the rim after high burnups) on the performance of the fuel and its interaction with the cladding during upset events such as reactivity insertion accidents (RIAs).

Understanding of these effects in a mechanistic way is at the frontier of nuclear fuel research. Since the current level of burnup between 50 and 60 MWd/kgU is just about the practical and economic upper boundary of 5% enriched U-235 fuel today, this is the limit of our empirical knowledge. There is a need to extend this boundary if enrichments above 5% become accepted or if higher density fuels (such as UN) come into use. Phenomenological models, not correlation of empirical data, will be needed to allow predictions to be made without the huge cost associated with totally empirical approaches.

9.2.2 Getting longer life from zirconium alloys

Under current operating exposure times imposed by the five weight percent U-235 enrichment limit, the behavior of zirconium based fuel cladding alloys is reasonably understood on an empirical, and even somewhat phenomenological, basis. This understanding is based on predictions of the oxide layer and hydride content of the cladding, both of which affect its ability to withstand stresses due to normal and accident conditions. This does not mean that there are not unknown issues with the current zirconium alloy systems. Prediction of the cladding resistance to RIAs is not well understood; here large amounts of heat are almost instantaneously imposed on the fuel which causes rapid expansion of the pellets into the cladding resulting in cladding failure. As the exposure of the fuel increases and these rapid expansion effects become more pronounced and less well understood, the nuclear fuel manufacturers are left with the choice of understanding the mechanical behavior of highly exposed (and oxidized and hydrided) cladding to rapid stresses imposed by the fuel either on an empirical or phenomenological basis. The empirical basis used so far relies on the collection of immense amounts of data covering almost any potential event during fuel operation, a very time consuming and expensive approach. The phenomenological approach would be much better, but there is practically no such basis for understanding except at the very rudimentary level. Previous attempts to model and predict these effects have been unsuccessful due to the complexity of the interactions between the models of the various effects which lead to severe non-convergence problems. Major programs are currently underway funded by the U.S. Department of Energy (DOE) and industry to try to understand and model fuel behavior on a phenomenological basis using improved convergence algorithms (for instance the Consortium for Advanced Simulation of Light Water Reactors (CASL) and the Nuclear Energy Advanced Modeling and Simulation (NEAMS) Program).

Besides operating under extended burnup, the known operating limits of nuclear fuel cladding are being exceeded by high thermal duty conditions. This arises from the desire of nuclear utilities to get the most electrical generating capacity out of their current plants while still using a minimal amount of fuel. These higher thermal operating limits (in kilowatts per meter) increase the temperature of the cladding which, in turn, increases the build-up of crud on the surfaces, further increasing the cladding temperature and its oxidation rate and hydride content. Higher thermal rates also increase fretting issues between the fuel rods and their support structures causing unintended wear and fuel cladding failures. The added vibration is due not only to increased stress due to flow variations, but also to relaxation of the grid structures that support the fuel rods. In addition, there is a trend to move to slightly higher pH values by adding more lithium to the primary coolant. These higher lithium levels affect the corrosion rate of the fuel cladding though they also seem to inhibit stress corrosion cracking (SCC) of steam generator alloys in the primary circuit (EPRI, 2012).

Besides the more recent concern with RIA-type accidents and the ability to store used nuclear fuel for long times under both wet (spent fuel storage pool) and dry (spent fuel storage casks and perhaps internment in a final repository) conditions, there is the continuing concern of how fuel which has experienced high burnups will respond to such classic accident scenarios as large and small break loss of coolant accidents (LOCAs) due to loss in ductility and strength. Current knowledge of both oxidation layer thickness and hydride content is gained entirely through empirical testing. Development of new alloys is based entirely on an empirical approach with very little theoretical guidance.

Areas of research for zirconium (and any other metal) alloy claddings include the phenomenological understanding of:

- The oxidation behavior of zirconium alloys at all temperature conditions (including high temperature accidents).
- The hydriding of zirconium alloys at all temperature conditions.

- The effect of oxide level and hydride content on the mechanical behavior of zirconium alloys.
- The effect of rapid stress levels on the mechanical behavior of zirconium alloys.
- Effects of long-term wet and dry storage as well as environmental conditions in any potential disposal site on the integrity of the zirconium alloy cladding.
- Effect of mechanical stresses induced by flow vibration on the fuel rod cladding as well as the grid support structures.
- Effect of flow rate on corrosion and erosion of both the fuel and the fuel structure.
- Modeling of these effects to allow performance prediction in extended burnup and in transient operating conditions.

9.2.3 Development of advanced claddings and accident tolerant fuel

Recent (Fukushima) and not so recent events (Three Mile Island 2), have accelerated the quest for new cladding materials that will be much more resistant, if not totally tolerant, to current LOCA conditions (~1200°C for 400 s) as well as more extreme beyond-design basis accident conditions such as a long-term full station blackout where there is little or no supply of coolant to the fuel. The most notable of these materials is the use of SiC composites. As little as is known about the behavior of zirconium alloys on a phenomenological basis under normal or accident conditions, much less is known about SiC composites or any other ceramic materials that could potentially replace metal alloys as fuel cladding materials. SiC cladding along with higher density or higher enrichment fuel could provide:

- Resistance to accidents and departure from nucleate boiling (DNB) or dry-out incidents because of higher operating temperature (~2000°C) capability.
- Minimal hydrogen production due to a much lower rate of reaction with water.
- Ability to operate in much longer cycles due to the very low corrosion rate of SiC in water.
- Enrichment savings due to ~75% lower thermal neutron absorption.
- Uprate capability of ~30% due to the ability to operate at DNB or dry-out conditions.
- Immunity to debris or fretting failures.

These SiC characteristics allow simplification of safety systems, higher energy density (to reduce capital costs) and longer operating cycles with higher density or enrichment advanced fuels to reduce fuel and reactor operating costs. However, they do come at a cost – the need to develop whole new areas of understanding in behavior and manufacturing including:

- Modeling and design of a SiC composite that will be acceptable for use in reactors, that will not shatter, and can withstand normal handling and operations as well as transients and accidents. However, many difficulties remain before these composite structures are understood well enough to model, including:
  - The behavior of ceramic composites in a radiation field is not well known.
  - The interaction between the monolithic SiC tube and the composite layer is especially difficult to model.
- The interaction between pellet and ceramic cladding will require extensive tests and modeling.
- The impact of manufacturing variations on the ultimate performance of the ceramic structure under operating and accident conditions will need to be understood.
- Joining an end plug to the tube to form a hermetic seal in a cost effective way is an extremely challenging task.
- Methods to produce about 15 million feet of reactor cladding per year to very exacting specifications at acceptable costs will require significant manufacturing development.

Ultimately, the understanding of all these effects for this relatively new material must be integrated into a licensable fuel performance code. The data that is needed can be gained empirically and fitted into phenomenological models with empirical verification. At a minimum, the following is needed under irradiation and coolant conditions (Lahoda, 2011):

- Property standards for SiC/SiC-composite matrix ceramic (CMC) materials as applied to LWR's.
- Mechanical properties as a function of time, temperature and irradiation and use.
- Corrosion properties at high temperatures in oxidizing (steam/air) atmospheres.
- Thermohydraulic response under design basis (LOCA, RIA) and severe accident scenarios.
- Core melt progression and relocation during beyond-design basis accidents.

- The impact of core melt on reactor internals and reactor vessel integrity.
- The effects of long-term wet and dry storage as well as environmental conditions in any long-term disposal site on the integrity of the SiC cladding.
- Phenomenological models for the behavior of SiC-composite structures as a function of temperature, stress, and radiation.

9.2.4 Increasing the life and accident tolerance of control rods, blades and other fuel assembly structures

There are other components in the core which, while they do not directly contain fission products, do affect the ability of the cladding to do so and have a major effect on the ability of the nuclear plant operators to control the reactor as the plant ages. These components include the structural materials in the fuel assembly such as the top and bottom nozzles, springs, fasteners, grid structures, skeleton, and the control members (rods in a PWR and blades in a BWR). In a BWR there are also water channels that channel the coolant flow. Many of these structures are currently made of various zirconium alloys and some are of stainless or high alloy steels. The inside components of the PWR control rods are Ag-In-Cd alloy while BWR control blades contain B_4C. The research areas of interest are:

- Understanding and predicting the corrosion of the current alloys in their respective BWR and PWR environments.
- Development of new materials such as SiC composites for PWR control rod cladding and BWR water channels.
- Development of new control rod materials including gray rods (control rods with much lower thermal neutron absorption materials, such as tungsten, than control rods but higher than steel or zirconium) for PWRs.

9.3 The primary system – the second fission barrier

The second line of defense for nuclear plants is the primary system boundary in any LWR. The primary system for a PWR consists of the reactor vessel, the primary side piping, steam generators, pressurizer and coolant pumps. For a BWR, the primary system consists of the reactor vessel, piping, steam turbine, condenser and coolant feed pumps. In both cases, maintaining the integrity of this boundary is crucial to maintaining the cooling functions for the reactor and therefore its long-term ability to prevent the release of fission products into the environment. Loss of integrity of this boundary leads to the classic LOCA (or small break LOCA) and the possibility of eventual destruction of the core due to the inability of the reactor to provide continued cooling to remove the still considerable heat from the decay of residual fission products (about 7 MWt after 2 min for a 1000 MWe reactor). Once breached and after the ability to provide cooling to the core is lost, the loss of integrity of the primary system leads to the release of fission products from the core to the reactor containment building. Maintaining the integrity of the primary system is therefore considered a key safety (and regulatory) issue.

Of equal importance is the maintenance of clean heat transfer surfaces. Heat transfer is the reason why nuclear plants exist – ultimately to produce electric power for sale. Heat transfer issues, usually due to deposit formation, lead to a temperature rise on the surface of the fuel and on the secondary side of the PWR steam generator. Issues other than heat transfer are the build-up of boron in the fuel deposits in a PWR and of radioactive materials on both BWR and PWR fuels. The general response by PWRs has been cleaner chemistry (less Al, Si, and Ca in the primary water) and tighter pH specifications. Lately both BWRs and PWRs have explored the addition of materials such as Pt and Zn to maintain the integrity of system components.

The factors involved in maintaining this boundary intact have evolved over the last 50 years of reactor operation. Initially, the discipline of maintaining nuclear plants was viewed by both the utilities and the vendors as being similar to that of coal fired boilers. Due to the build-up of solids on fuel rods, primary side water specifications for PWRs and BWRs saw a drop in allowed levels of dissolved solids in the order of a factor of 1000 to low ppb levels. In the 1960s and early 1970s secondary side cooling water chemistry for PWRs was similar to that of any coal fired boiler which used phosphate chemistry. As deposit build-ups on the secondary side of steam generators occurred in PWRs (multiple tons in the early 1970s), problems began to emerge with the cracking of tubes at the support plates and U-bends of the steam generator tubes which necessitated the mass plugging of tubes in highly radioactive environments and the derating of some PWRs. It was then realized that different standards of water cleanliness would be required and all volatile chemistry was introduced, air leaks in condensers were repaired (and brass condensers were replaced with stainless steel ones), and feedwater specifications increased dramatically. Meanwhile cracking in coolant piping and in the welds of steam dryers (Fig. 19.4), bowing of water channels in BWRs (which would hinder the insertion of the control blades) also began to reinforce the notion that the care and feeding of nuclear plants required a very different approach than the care of fossil fired plants where boiler tubes might be routinely replaced every ten years with no worries about radioactivity.

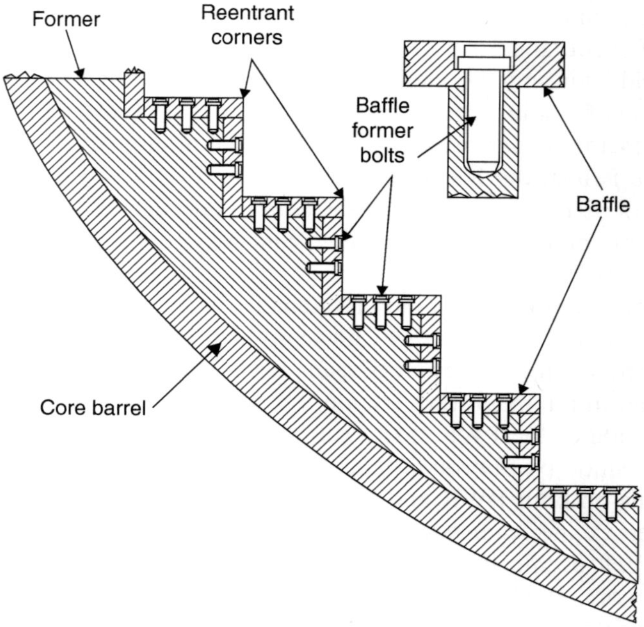

9.5 PWR baffle bolts (NUREG/CR–6897/ANL–04/28, Assessment of Void Swelling in Austenitic Stainless Steel Core Internals, H. M. Chung).

Recently, concerns have increased for the integrity of the primary vessels and baffle bolts for PWRs due to radiation hardening and loss of tensile strength (these bolts hold the flow baffles together on the outside of the core, see Fig. 9.5). Cracked welds around the penetration tubes through PWR reactor vessel heads have allowed internal corrosion of the alloy steel below the weld overlay material resulting in a potential small break LOCA situation (NRC, 2008a). In BWRs, breaking of welds due to excessive vibration from the large steam flow as well as boiling within the core and cracking of the coolant piping continue to be an issue.

A key integrity issue for PWRs is the reactor coolant pump seal. This is a component that must have a minimal flow since it is a leak path for primary coolant from the primary coolant system. This has been a long-standing issue that comes and goes for various plants; for some plants is the highest potential issue for a small break LOCA.

Over the past 40 years of LWR operations, these issues have led to a new emphasis on the importance of effective chemistry control; research into new additives such as zinc for nuclear service; the need for better design methods for components specifically to take into account vibration issues in both the core area and the primary side piping; and the introduction of new alloys that can withstand the rigors of very clean water. Examples of the alloy changes are the use of the 690 and 800 alloys to replace the 600 series alloys first used in steam generator tubes and the use of enhanced alloy steels for steam generator support tubes to replace the low alloy carbon steels that were previously used.

However, replacement of failed or deteriorating components is not always a feasible option and replacement is rarely cheap. For instance, a broken component such as a coolant injection nozzle deep inside a BWR reactor, or cracked baffle bolts inside the PWR reactor, cannot require the replacement of the vessel. Since these issues are occurring in highly radioactive and sometimes hard to access components, remote methods of repair (such as underwater laser welding) as well as methods for access (tethered robotic welders) have been developed. Replacement of steam generators costs $40 million to $50 million each (and there are two to four in every PWR). Replacement of reactor pressure vessel heads cost about $20 million each (NEI, 2010).

Finally, there is the design of the components. As those who take Six-Sigma® courses always learn, the root of most problems is in the initial engineering. For example BWRs have, for many years, had an issue with cracking of welds in the steam separator (see Fig. 9.4). While not directly a safety issue, this problem has led to replacement of the steam dryers, and continued cracking in the welds has been observed. Recently, modeling tools that provide a more accurate prediction of the vibration stresses affecting these welds have been developed and deployed on much more powerful computers. This capability has allowed better designs for the steam dryers to be made and installed which, so far at least, have led to elimination of this issue. Of course, better modeling tools alone cannot accomplish better designs. Better understanding of materials in the relevant chemistry and radiation environment of the LWR is needed. This is especially true for the understanding of environmentally assisted crack growth which presents the potential for catastrophic failure of key boundary components (NRC, 2008b). This understanding can then be translated into more phenomenological models for use in modeling components and perhaps even successful prediction of the response of these components to conditions outside those which were used to originally develop the models.

Another broader issue is the standards to which pressure vessels are designed. The current fleet of reactors was designed to the ASME Boiler and Pressure Vessel Code, Section III, in effect in the 1970s. This code was based on laboratory environment testing and not on tests using actual operating conditions (including radiation) (Majumdar, 2011). In the past 40 years, there has been enough data collected in LWR conditions that illustrates a need for changes in these codes, but more importantly, how actual LWR operating conditions have affected the current fleet. It is important to know what changes need to be made in current plants to keep them operating safely for the next 20, 40, or more years. Since the replacement of these plants would require an investment of over $500 billion, there is a huge economic driver for this work.

Much work remains to be done in this area including:

- Understanding of the effects of chemistry on the development of corrosion and especially SCC in piping, equipment, fasteners (bolts) and welds both inside and outside the core area. The likely result of better understanding of the chemistry will be the development of new additives for use in the primary and secondary systems of LWRs to prevent corrosion and cracking, the build-up of radioactive materials, and the maintenance of clean heat transfer surfaces.
- Long-term effects of radiation, temperature and pressure cycling, and chemistry on the integrity of the reactor vessel and its component materials. This also applies to the other components in the primary system, though radiation effects will be much less important.
- In-place maintenance methods (for instance, in-place annealing) that can be used to reverse the effect of irradiation on materials
- Maintenance methods for remote repair of components such as underwater welding and robotics
- Development of new alloys (for instance alloy 690 for steam generator tubes) or new materials (such as SiC composites) that can be used to eliminate potential issues due to radiation and chemistry
- Development of new materials for pump seals capable of withstanding high pressure drops, high wear, and primary system water chemistry
- Development of modeling tools to more accurately predict the stresses that components undergo during operation, and to design components that perform as well or better than the current components. Part of this effort should be on developing the phenomenological models required to predict the performance of current materials after having been in service for 40 or more years. This capability will be required for license renewals of current plants.
- Development of monitoring tools for in-use components that will provide sufficient warning to plant operators of impending maintenance and repair issues so that appropriate steps can be taken during scheduled outages, minimizing the potential for failure generated accidents and disruptions of the electrical supply.

9.4 The containment structure – the final fission barrier

The containment structure of an LWR acts as both a barrier to the spread of fission products from the reactor into the environment and as a shield to protect the nuclear components within it from missiles such as from aircraft and errant turbine blades. The issues with the containment include attack of the concrete by the surrounding environment (for instance, acid rain), attack of the rebar within the containment causing the concrete to spall off the outside (or inside), and corrosive attack of the steel liner. Besides the containment structure, which is the biggest and most obvious component, there are other components that must also work as designed to control the spread of fission products outside of the containment. Among these are:

- Steam isolation valves that block the main steam lines from the reactor inside the containment to the turbines outside of the containment.
- Various heat exchangers that provide cooling of systems inside the containment using cooling water from outside of the containment. One example of these is the aluminum air coolers that remove heat after a LOCA from the steam/air mixture inside the containment.

Maintaining the integrity of the containment structure and the other components that separate the reactor from the environment requires research and development in the following areas:

- Effect on the containment concrete of the external environment (for instance acid rain, bird feces, thermal cycling, etc.) as well as radiation over very long periods of time.
- Effect of high moisture levels on the containment steel liner, galvanized steels, and aluminum components found in the containment.
- Effect of age, temperature, erosive flow on large valves and piping with the associated welds throughout the system for both steam and water.
- Corrosion of stainless steel and aluminum heat exchangers both under normal operating conditions of high humidity, moderate temperatures and moderate radiation levels.
- Non-destructive examination of large structures.
- Methods to repair large concrete and steel structures.

9.5 Other nuclear reactor systems

The remainder of the nuclear plant is similar to any other power plant whether the steam is generated using coal, oil, natural gas or nuclear fuel. There will be issues with cooling tower decay, turbine corrosion, generator moisture, condenser corrosion, buried piping, etc. There are some issues that will be unique to nuclear power generation, however. For instance, decay of cable insulation in cable trays due to irradiation and high temperature is an important issue. Replacement of cabling in a nuclear plant is very expensive and takes a very long time and therefore should be avoided if at all feasible. The effects of soil and groundwater on buried cabling and the miles of buried piping on any nuclear plant site are also at issue (INL, 2009).

Another issue that arises comes from the use of secondary sources of water. For instance, so-called gray water (sewage that has been through a waste treatment plant) has been used and is being looked at as cooling water to reduce overall water use by large power plants. This water has the potential for higher and different salt concentrations than drinking water sources and can therefore cause unexpected corrosion or SCC issues (EPRI, 2007). Corrosion and decay of materials in the spent fuel pools, such as the boron containing structures that allow tighter packing of the spent fuel but which suffered unexpected degradation, also needs to be considered.

Outside of the nuclear plant itself but within the nuclear cycle, materials issues arise in such varied areas as zirconium metal production (the graphite receptors, ceramic gas injection nozzles, and ceramic linings in the chlorination of either ZrO_2 or zircon sand); the manufacture of nuclear fuel due to the common handling of mixtures of HF and nitric acid for UO_2 dissolution; solvent extraction systems; and incineration of radioactive waste materials. For the most part, material issues have been solved in other portions of the cycle, but the ones listed above continue to be very resistant to reasonable materials solutions.

In summary, the issues outside of the nuclear island that require research and development are:

- The effects on cable insulation decay of temperature and radiation inside the containment and due to groundwater or soil for buried cable outside the containment.
- Methods to monitor and repair buried piping and cabling.
- Understanding of the interaction between older materials used in the initial construction of the plant and newer materials that may be used to repair or upgrade current plant systems including spent fuel pools.
- Components in the zirconium conversion from oxide or silicates to chlorides.
- Systems that handle mixed HF and HNO_3 in the UO_2 fuel manufacture area.
- Development of phenomenological understanding of the behavior of these materials in their respective environments that can be used to predict their behavior over time.

9.6 References

Boiling Water Reactor (BWR) Systems (2012), Rep. U.S. NRC. Web. 17 May 2012. <http://www.nrc.gov/reading-rm/basic-ref/teachers/03.pdf>.

EPRI (2007), *EPRI Journal*, Summer 2007.
EPRI (2012), Fuel Reliability Program Update, 1 February 2012.
'Fuel Rod' (2012), European Nuclear Society. Web. 17 May 2012. <http://www.euro-nuclear.org/info/encyclopedia/f/fuel-rod.htm>.
INL (2009), Light Water Reactor Sustainability Research and Development Program Plan Fiscal Year 2009–2013, INL/MIS-08–14918 Revision 2, 2009.
'Industry's Efforts toward Technology Development Related to Aging Management of PWR Plants'. (2012) E-Journal of Advanced Maintenance (EJAM). Web. 17 May 2012. <http://www.jsm.or.jp/ejam/Vol.1.No.4/GA/10/article.html>
Lahoda, E., Johnson, S., and Ray, S. (2011), Challenges in the Development of Silicon Carbide Advanced Fuel Cladding for Light Water Reactor Application, 2011 Water Reactor Fuel Performance Meeting, paper #T5–010, Chengdu, China, 11–14 September 2011.
Majumdar, S. and Natesan, K. (2011), Report on Assessment of Environmentally Assisted Fatigue for LWR Extended Service Conditions, ANL-LWRS-47, September 2011.
NEI (2010), Status and Outlook for Nuclear Energy in the United States, July 2010.
NRC (2008a), Davis-Besse Reactor Pressure Vessel Head Degradation: Overview, Lessons Learned, and NRC Actions Based on Lessons Learned, NUREG/BR-0353, Revision 1, August 2008.
NRC (2008b), NRC/DOE, Life Beyond 60 Workshop Summary Report, 19–21 February 2008, Bethesda, Md.
'Pressurized Water Reactors' (2012), U.S.NRC: United States Nuclear Regulatory Commission. 29 March 2012. Web. 17 May 2012. <http://www.nrc.gov/reactors/pwrs.html>.

Index